AF595037

Exclusive Feature Papers in Colorants

Exclusive Feature Papers in Colorants

Editors

Jorge Bañuelos Prieto
Ugo Caruso

MDPI • Basel • Beijing • Wuhan • Barcelona • Belgrade • Manchester • Tokyo • Cluj • Tianjin

Editors

Jorge Bañuelos Prieto
Physical Chemistry
Basque Country University
Bilbao
Spain

Ugo Caruso
Department of Chemical Sciences
University of Naples "Federico II"
Napoli
Italy

Editorial Office
MDPI
St. Alban-Anlage 66
4052 Basel, Switzerland

This is a reprint of articles from the Special Issue published online in the open access journal *Molecules* (ISSN 1420-3049) (available at: www.mdpi.com/journal/molecules/special_issues/Exclusive_Feature_Papers).

For citation purposes, cite each article independently as indicated on the article page online and as indicated below:

LastName, A.A.; LastName, B.B.; LastName, C.C. Article Title. *Journal Name* **Year**, *Volume Number*, Page Range.

ISBN 978-3-0365-2723-9 (Hbk)
ISBN 978-3-0365-2722-2 (PDF)

Contents

About the Editors

Jorge Bañuelos Prieto

Dr. Jorge Bañuelos Prieto is an associate professor at the Physical Chemistry Department of the Basque Country University (UPV/EHU). He was born in 1976 and received his PhD in 2004 from the same university under the guidance of Profs. Iñigo and Fernando López Arbeloa. From then to 2006, he was an assistant lecturer, and after that lecturer until 2011, when he got his actual position in UPV/EHU. In 2006, he did a postdoctoral stay with Prof. Gion Calzaferri at Bern University (Switzerland). He belongs to the Molecular Spectroscopic Group at UPV/EHU and his main research topic is the computationally assisted design and photophysical characterization of organic fluorophores in solution and solid hosts for photonic and biophotonic applications. He has published more than a hundred research articles in reputed and indexed scientific journals, around five book chapters and supervised three doctoral thesis.

Ugo Caruso

Dr. Ugo Caruso graduated in Industrial Chemistry at the University of Naples "Federico II". He received his PhD degree in Chemical Science University of Naples "Federico II". Currently, he is an Associate Professor of Chemistry and Chemistry of Materials at the Department of Chemical Science of the University of Naples. His research activities concern the design, synthesis and characterization of novel conjugated materials for applications in different fields of organic electronics and photonics. He is author of more than 70 articles published on scientific journal of high impact factor. He is also author of about 100 Conference Proceedings.

Preface to "Exclusive Feature Papers in Colorants"

Natural colorants have been used from ancient times to colour objects owing to their ability to absorb light mainly within the visible window of the electromagnetic spectrum. Such organic and inorganic dyes and pigments were extracted from flora and fauna resources. In the 19th century, the first synthetic colorants were reported, being the beginning of a new industry related with the development of commercial colorants in the 20th century. The recent advances in strategic knowledge areas such as dyes and materials chemistry, computational resources and technological advances have greatly contributed to the design not only of modern colorants with improved photonic properties, but also to understand the involved photophysical processes via spectroscopic and microscopic techniques. Actually, photoactivatable colorants able to display fluorescence are involved in many light-driven devices and are the cornerstone of many ongoing research projects. These multifunctional colorants are applied in many fields related with biophotonics (diagnosis and therapy), optolectronics (lasers and sensing), photovoltaics (photosensitizers and solar cells) and nanotechnology overall.

This Special Issue project entitled "Exclusive feature papers in colorants" was motivated by such popularity and vast applications of colorants. Along with the following articles and reviews, we intend to highlight some cutting-edge developments in this field through selected works. In particular, the herein included works focus on the design and characterization of new nanomaterials and dyes as pigments and fluorophores, sensors of surrounding physicochemical properties and biomarkers for imaging. Finally, an overview of the state of the art of zinc cation-based complexes as fluorophores is also included.

We hope that the launch of this book could be an inspiration for the readers and a forum to share new ideas and finding around this exciting research topic. We would like to acknowledge Ms. Rachael Zhang for her kind assistance and support during the Special Issue project.

Jorge Bañuelos Prieto, Ugo Caruso

Editors

Article

Ready Access to Molecular Rotors Based on Boron Dipyrromethene Dyes-Coumarin Dyads Featuring Broadband Absorption

Ernesto Enríquez-Palacios [1], Teresa Arbeloa [2], Jorge Bañuelos [2,*], Claudia I. Bautista-Hernández [1], José G. Becerra-González [1], Iñigo López-Arbeloa [2] and Eduardo Peña-Cabrera [1,*]

1 Departamento de Química, Universidad de Guanajuato. Noria Alta S/N. Guanajuato, Gto. 36050, Mexico; e.enriquezpalacios@ugto.mx (E.E.-P.); cla_isb_6@hotmail.com (C.I.B.-H.); jg.becerragonzalez@ugto.mx (J.G.B.-G.)

2 Departamento de Química Física. Universidad del País Vasco-EHU, Apartado 644, 48080 Bilbao, Spain; teresa.arbeloa@ehu.es (T.A.); inigo.lopezarbeloa@ehu.es (I.L.-A.)

* Correspondence: jorge.banuelos@ehu.es (J.B.); eduardop@ugto.mx (E.P.-C.)

Academic Editor: Goze Christine

Received: 15 January 2020; Accepted: 8 February 2020; Published: 12 February 2020

Abstract: Herein we report on a straightforward access method for boron dipyrromethene dyes (BODIPYs)-coumarin hybrids linked through their respective 8- and 6- positions, with wide functionalization of the coumarin fragment, using salicylaldehyde as a versatile building block. The computationally-assisted photophysical study unveils broadband absorption upon proper functionalization of the coumarin, as well as the key role of the conformational freedom of the coumarin appended at the *meso* position of the BODIPY. Such free motion almost suppresses the fluorescence signal, but enables us to apply these dyads as molecular rotors to monitor the surrounding microviscosity.

Keywords: molecular rotors; BODIPY; viscosity sensors; dye chemistry; energy-electron transfer

1. Introduction

The design of molecules featuring two or more chromophoric units is of great interest [1–4]. One can envisage new applications of such systems by taking advantage of the interactions that may develop between the chromophoric units, i.e., energy transfer [5,6], or electron transfer pathways [7], enabling the design of light harvesters with broadband absorption and high *pseudo*-Stokes shifts [8–11], or photosensitizers for photovoltaic devices [12,13] mimicking the natural photosynthesis [14,15]. It is crucial to the design of multichromophoric systems to have flexible functionalization methods at one's disposal that are able to tailor the targeted analogues. In this regard, both boron dipyrromethene dyes (BODIPYs) [16] and coumarins [17] are in themselves two of the more widely used fluorophores (Figure 1), and hence have optimal candidates to design multichromophore ongoing energy transfer processes. Indeed, the former luminophores stand out due to their stability, chemical versatility and tunable photophysical signatures [18,19], whereas the latter fluorophores can be also deeply functionalized and display spectral bands at higher energies than the BODIPY core. Thus, both chromophores are complementary from a spectral point of view, and hence are suitable building blocks to be combined in a single molecular structure towards the promotion of intramolecular energy transfer hops.

Despite the fact that coumarin-BODIPY hybrids are known, and some examples of these dyads can be found in the literature applied as chemosensors [20,21], energy transfer cassettes [22,23] or fluorescent probes for bioimaging [24,25], the photonic performance, and hence the practical applicability of these multifunctional molecular assemblies, can be still improved. Indeed, one of the main drawbacks of these multichromophores is the low absorbance of the coumarin core [26], which falls in the ultraviolet (UV) region and it is usually masked (or at least overlapped) under the more energetic transitions of the BODIPY unit. This feature hampers the application of these dyads, for instance as broadband energy transfer cassettes since, while the energy transfer from the coumarin to the BODIPY is effective, the light harvesting is not greatly improved by the presence of coumarin.

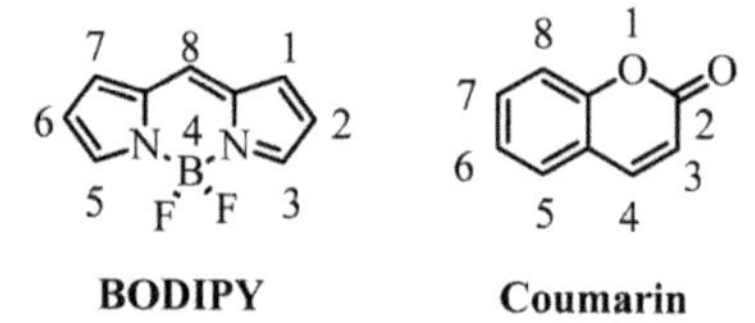

BODIPY Coumarin

Figure 1. Structure of boron dipyrromethene dyes (BODIPYs) and coumarins.

From the synthetic point of view, the utilization of salicylaldehyde as a versatile building block is well-documented [27]. Herein, we show that such an attractive starting material can be rendered fluorescent by attaching it to a BODIPY fragment, thereby opening up new possibilities for the synthesis of more complex products. In this first example, we used the BODIPY-containing salicylaldehyde to prepare BODIPY-coumarin hybrids. Thanks to this methodology, we have been able to decorate the chromene π-system of the coumarin with a battery of aromatic moieties (from functionalized aryl groups with electron donors, i.e., methoxy, or acceptors, i.e., cyano, groups, to pyridine, naphthalene, modified stilbenes, triphenylamine, or benzothiophene). The computationally-aided spectroscopic analysis of this set of dyads allows the selection of the best structural modification at the coumarin subunit to enhance the light harvesting ability along the UV-yellow spectral region towards applying these dyads as energy transfer cassettes.

Furthermore, in view of the conformational flexibility of these dyads around the linking bond between the 8-position of the BODIPY and the coumarin, we anticipated that they could behave as molecular rotors [28–30]. Thus, we have conducted additional measurements at different temperatures and controlled viscosities (increasing the amount of ethylene glycol in the medium) to test the viability of these hybrids as fluorescent sensors to monitor the viscosity of the surrounding environment.

2. Results and Discussion

2.1. Synthesis

Salicyladehyde was attached to BODIPY via the Liebeskind-Srogl cross-coupling (LSCC) reaction [31] between commercially available 8-methylthioBODIPY **1** and boronic acid **2** (Scheme 1).

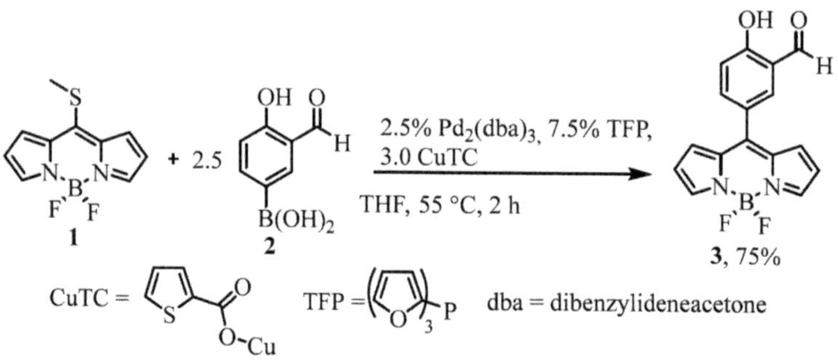

+ 2.5 — 2.5% $Pd_2(dba)_3$, 7.5% TFP, 3.0 CuTC — THF, 55 °C, 2 h → **3**, 75%

CuTC = TFP = dba = dibenzylideneacetone

Scheme 1. LSCC reaction conditions.

Key boronic acid **2** was prepared by treating commercially available bromide **4** with diboronic acid in the presence of Pd (Scheme 2) [32].

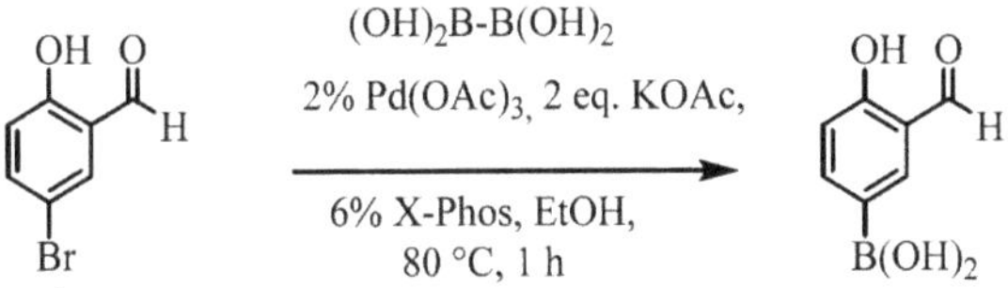

Scheme 2. Synthetic access to the boronic acid precursor.

Once we prepared **3**, we turned our attention to the method reported by Phakhodee et al. for the synthesis of coumarins starting from substituted salicylaldehydes (Scheme 3) [33].

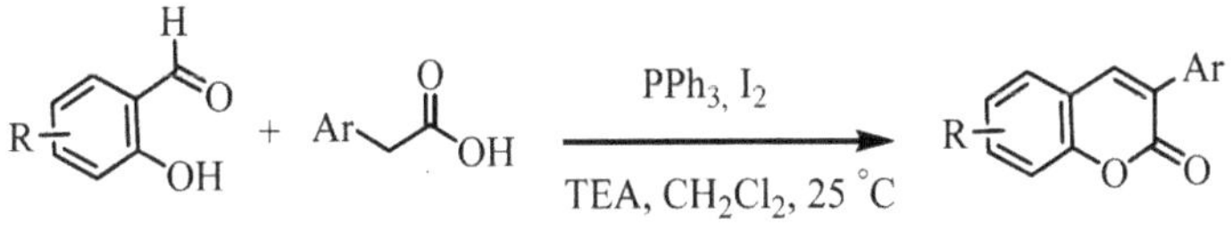

Scheme 3. Synthetic route to access aryl-functionalized coumarins.

In this fashion, BODIPY **3** was reacted with 3-bromophenylacetic acid **5** (Scheme 4). The reaction was operationally very simple and was performed between 0 °C and room temperature under air. BODIPY-coumarin **6** precipitates and after filtration and crystallization, it was obtained with a 59% yield.

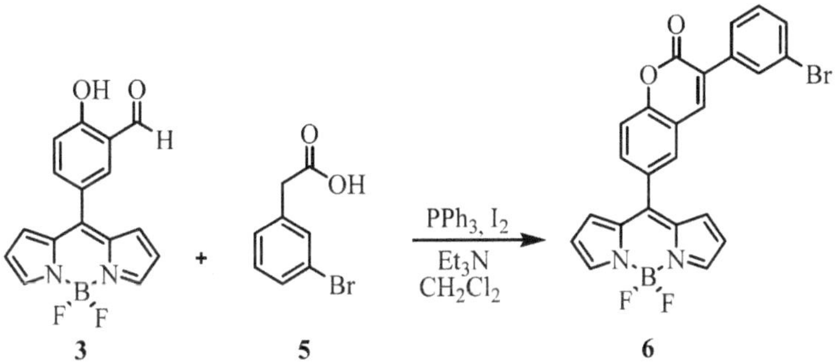

Scheme 4. Reaction between BODIPY and 3-bromophenylacetic acid.

Next, we studied the scope and limitations of the functionalization of **6**, under the Suzuki cross-coupling standard conditions (Table 1).

Table 1. Scope and limitations of the Suzuki cross-coupling reaction of BODIPI-coumarin **6**.[a]

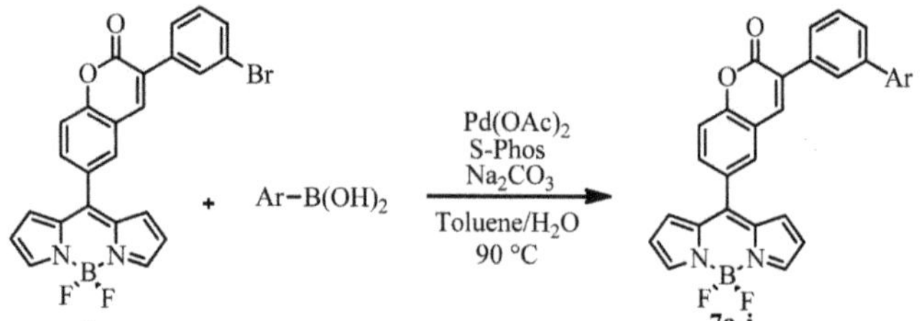

6 + Ar−B(OH)$_2$ → (Pd(OAc)$_2$, S-Phos, Na$_2$CO$_3$, Toluene/H$_2$O, 90 °C) → 7a-j

Entry	Ar-	Reaction time (h)	% yield [b]	cpd
1	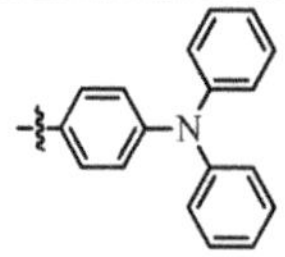 4-(diphenylamino)phenyl	21	75	**7a**
2	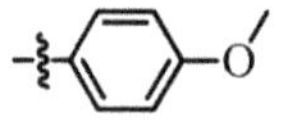 4-methoxyphenyl	16	85	**7b**
3	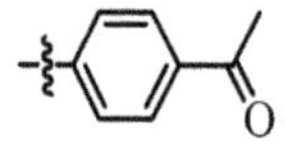 4-acetylphenyl	19	85	**7c**
4	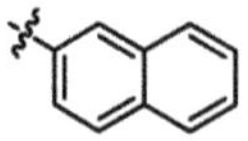 naphthyl	20	68	**7d**
5	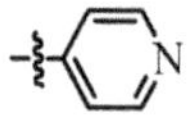 pyridyl	20	76	**7e**
6	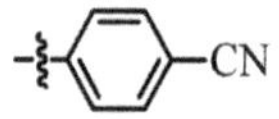 4-cyanophenyl (–CN)	17	62	**7f**
7	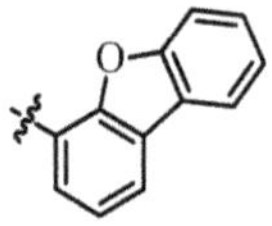 dibenzofuranyl	18	68	**7g**
8	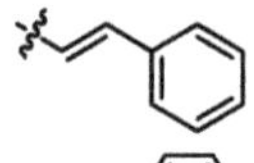 styryl	16	93	**7h**
9	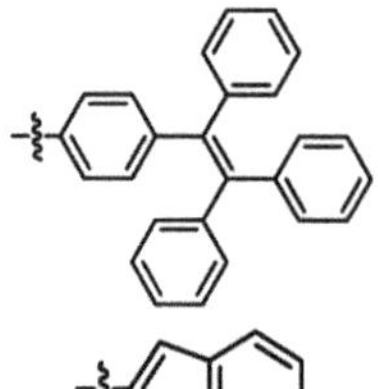 4-(triphenylvinyl)phenyl	6.5	82	**7i**
10	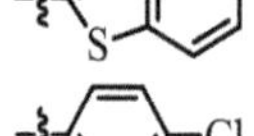 benzothiophen-2-yl	7	77	**7j**
11	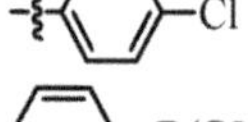 4-chlorophenyl (–Cl)	-	- [c]	**7k**
12	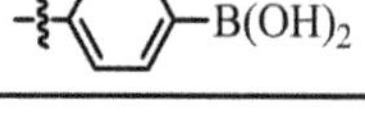 4-(B(OH)$_2$)phenyl	-	- [d]	**7l**

[a] Conditions: **6** (1 equiv), boronic acid (2.0 equiv), Pd(OAc)$_2$ (5 mol%), S-Phos (15 mol%), Na$_2$CO$_3$ (2.0 equiv.) /H$_2$O (4:1) at 90 °C. [b] Isolated yield. [c] Inseparable complex mixture. [d] The expected product was not detected by NMR.

The yields of the reactions ranged from good to excellent. The Suzuki reaction works efficiently regardless of whether electron-donating or electron-withdrawing boronic acids are used. Heteroarylboronic acids cross-coupled efficiently (entries 5, 7, and 10). Reaction with *p*-chlorophenylboronic acid (entry 11) gave a complex mixture without a major compound. Presumably, once the initial product is formed, the chlorine atom reacts further under the reaction conditions. An attempt to prepare a dimeric analogue using *p*-phenyldiboronic acid failed as well. No evidence for the formation of the desired product was observed by NMR of the crude material. Tetraphenylethene derivative **7i** was prepared in the hope that it would display aggregation-induced emission (AIE) [34–36], however, disappointingly, it did not.

2.2. Photophysical Properties

The spectroscopic properties of the BODIPY-coumarin hybrids in the visible spectral region are ruled by the spectral bands owned to the BODIPY subunit. Indeed, sharp absorption and fluorescence bands were registered at around 500 nm and 520 nm, respectively, regardless of the kind of coumarin appended at the *meso* position (Figure 2). Therefore, the coumarin subunit is electronically decoupled with the dipyrromethene backbone and the profile of the visible spectral bands of the dyads fully remained to those of the BODIPY alone. However, the presence of such moieties at the sterically unconstrained *meso* position has a deep impact on the fluorescence response. In fact, the fluorescence efficiency and lifetime harshly decreased due to the presence of the coumarin at an 8-position, yielding values lower than 5% and faster than 500 ps for all the tested compounds (Table S1 in Supplementary Materials). Such sudden enhancement of the non-radiative rate constants was attributed at first sight to the deactivation channels afforded by the free rotation of the 8-coumarin fragment directly linked to the BODIPY. Indeed, low fluorescence efficiencies have been reported for BODIPYs bearing unconstrained aryls at the said key *meso* position [37–39], as supported by the theoretically conducted potential energy curves, which highlight the key role on the photophysics of the conformational freedom around the linkage bond between the BODIPY and the 8-aryls [40–43].

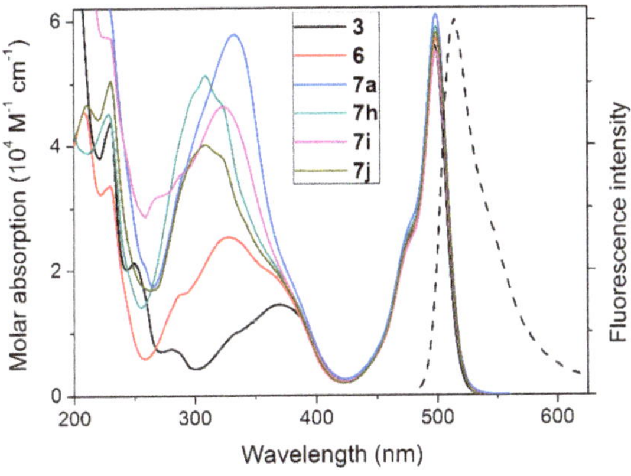

Figure 2. UV-Vis absorption and fluorescence (dashed line) of representative BODIPY-coumarin hybrids in diluted solution of cyclohexane. The reference BODIPY not bearing 8-coumarin **3** is also included for comparison. Note that the shape and position of the fluorescence spectra does not change nor with the kind of tethered coumarin neither the excitation wavelength (UV or Vis) owing to the ongoing intra-EET. The absorption spectra of the rest of the dyads are collected in Figure S1 in Supplementary Materials.

Whereas no change was detected in the visible absorption and fluorescence, the UV absorption remarkably changed depending on the kind of coumarin placed at the *meso* position, in particular, on the aromatic functionalization added to the chromene core at 3-position (Figure 2 and Figure S1 in Supplementary Materials). The coumarin absorption band was detected at 325 nn, with its long-wavelength tail overlapped with the more energetic transitions of the BODIPY ($S_0 \rightarrow S_2$ and $S_0 \rightarrow S_3$, energetically close, and giving a broad and weak band placed at around 375 nm, see **3** in Figure 2). Nevertheless, in the dyads where the coumarin is functionalized with electron-rich groups, like triphenylamine (**7a**), stilbene

(**7h** and **7i**) and benzothiophene (**7j**), a marked increase in the absorption within 275–375 nm was clearly recorded (up to 2-fold regarding to **6,** bearing the simplest coumarin unit, and up to 5 fold with respect to the BODIPY in **3**). Indeed, in these last dyads, the molar absorption of the band attributed to the modified coumarin became almost equal to the Vis band of the BODIPY. It is likely that these functionalizations of the chromene core with aromatic groups promote a more π-extended system of the coumarin. Strikingly, such spanning of the conjugation was not reflected in the ensuing pronounced bathochromic shift, but led to a marked enhancement of the absorption probability of the coumarin. It is noteworthy that the spectral profiles of all the compounds remained the same after prolonged UV irradiation or long aging times, evidencing their chemical stability and photostability.

The computationally predicted absorption profiles matched the experimentally recorded ones and support the aforementioned assignment of the spectral bands and their trends with the functionalization of the coumarin (Figure 3 and Figure S2 in Supplementary Materials). The theoretical simulation of the energy gap for the band placed at the UV region was much accurate than for that located in the visible region. This is a typical drawback of the Time Dependent (TD-DFT) method; as the spectral band is shifted to lower energies, the method overestimates the energy gap [44,45]. The predicted visible absorption owned exclusively to the molecular orbitals of the BODIPY (HOMO→LUMO), whose position remained invariant with the type of appended 8-coumarin. Additionally, a UV band was predicted in the spectral region where the highest electronic transition of the BODIPY were placed (Figure 3), but with growing intensity when the coumarin is decorated with electron-rich groups (Figure 3 and Figure S2 in Supplementary Materials). The analysis of the molecular orbitals involved in such transition revealed that it was the consequence of many configurations. For instance, in dye **6,** the occupied orbitals were located preferentially at the coumarin moiety (HOMO-1 and -2), but with the electronic density shifted to the pendant phenyl (Figure 3), and eventually reaching the appended aromatic functionalization of such a ring in more complex coumarins. The overlapping with the highest transitions of the BODIPY can be clearly visualized in HOMO-3 (Figure 3), which is delocalized through the whole molecule, although the dipyrrin and the coumarin are electronically decoupled in the ground state. It is noticeable that the virtual orbitals involved in this UV transition are preferentially placed at the BODIPY (LUMO).

Strikingly, just in those dyads bearing electron-rich groups (like the aforementioned triphenylamine, stilbene and benzothiophene), the predicted energetic ordering of the molecular orbitals suggests that they are able to switch on a reductive photoinduced electron transfer (PET) [46,47] pathway. In these dyes, the HOMO is located at the coumarin (**7a**, bearing a strong electron donor like triphenyalmine) rather than in the dipyrrin as expected. In other words, the energy of the highest occupied orbital of the coumarin falls between the energy gap of the frontier orbitals of the BODIPY. In the rest of dyads with less electron rich coumarins (like **6**), the frontier orbitals of the coumarin are placed up and down the orbitals responsible of the visible absorption of BODIPY, hence not interfering with them (Figure 4). As a matter of fact, the presence of electron donor triplenylamine **7a** at the coumarin raises the energy of the highest occupied orbital placed at the coumarin around 1.6 eV, being the dyad where the PET is more feasible (Figure 4). Thus, in dyads like **7a, 7i** or **7j**, after selective excitation of the BODIPY (HOMO-1→LUMO in this case), the electron-rich moieties can inject an electron into the BODIPY (a thermodynamically feasible hop), hampering the fluorescence deactivation. Such PET can be also anticipated from the analysis of the molecular orbitals involved in the UV absorption, since excitation implies transfer of electronic density from the coumarin to the dipyrrin core, supporting the electron donor nature of the 8-appended coumarins (Figure 3). This quenching pathway adds another non-radiative channel to the aforementioned internal conversion enhancement prompted by the 8-aryl free motion, explaining the recorded extremely low fluorescence efficiencies for these BODIPY-coumarin hybrids (Table S1 in Supplementary Materials).

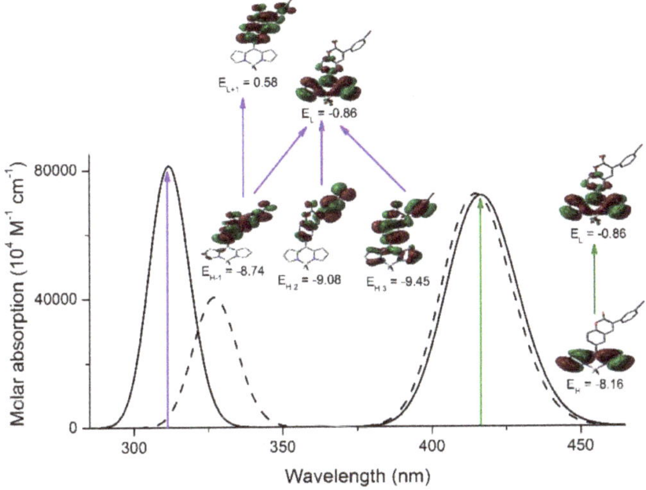

Figure 3. Predicted absorption spectrum (td wb97xd/6-311+g*) and main molecular orbitals and energies (in eV) involved in the electronic transitions for dyad **6**. The corresponding spectrum for reference compound **3** (dashed line) is added for comparison. The predicted absorption spectra and molecular orbitals for other representative dyads with π-extended coumarins are collected in Figure S2 in Supplementary Materials.

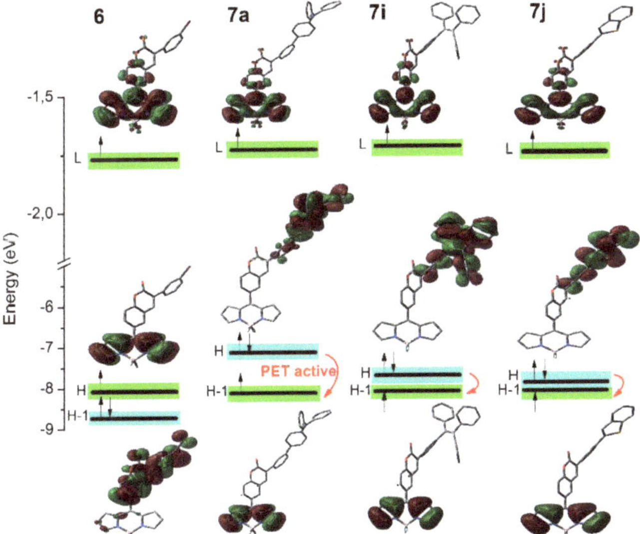

Figure 4. Main calculated molecular orbitals from the optimized geometries (wb97xd/6-311+g*) of the dyads with π-extended coumarins **7a, 7i** and **7j**, compared with those computed for the dyad bearing the simplest coumarin **6,** to illustrate the viability of the electron transfer upon selective excitation of the BODIPY in the former dyads.

Therefore, these dyads bearing π-extended coumarins (mainly **7a, 7h, 7i** and **7j**) improve the light harvesting efficiency of the BODIPY-coumarin hybrids, guaranteeing a better and broader collection of the incoming light to promote the ulterior energy transfer. Indeed, in all the dyads regardless of the excitation wavelength and the selectively excited subunit, just the visible emission from the BODIPY was recorded, without any sign of the emission from the coumarin (Figure 2). Thus, the intramolecular excitation energy transfer (intra-EET) from the donor coumarin to the acceptor BODIPY is highly efficient, although the fluorescence output is low owing to the said non-radiative pathways (conformational free rotation around the 8-position of the BODIPY and eventually PET).

Non-fluorescent dyes owing to the said conformational freedom (including dyads, such as coumarin-rhodamine) [48,49], are being currently applied as molecular rotors to monitor the microviscosity of the surrounding environment, even in the cellular media [50,51]. Accordingly, we hypothesized that, in the herein reported BODIPY-coumarin dyads, as the viscosity of the media increases, the free rotation of the 8-aryl should be hampered, and consequently the fluorescence quantum yield should increase and the lifetime lengthen, with this last property being very sensitive to such environmental property in view of the reported results in the bibliography. Therefore, we have tested the performance of dyad **6**, as a representative compound of the herein reported BODIPY-coumarin dyads not undergoing PET, as a molecular rotor to monitor the viscosity of the surrounding media. To this aim, firstly we have measured its photophysics in a viscous solvent like ethylene glycol. Successfully, the fluorescence quantum yield and lifetime in this viscous solvent were clearly higher and longer, respectively, (up to around 0.15 and 1 ns, in comparison with the rest of solvents in Table S1 in Supplementary Materials, with values lower than 0.05 and 450 ps, respectively). Such improvement of the fluorescence emission is nicely supported by the recorded fluorescence spectra and decay curves in media with controlled viscosity by means of ethanol/ethylene glycol mixtures (Figure 5). As the viscosity of the media progressively increases the emission spectra becomes more intense and the decay curves slower, suggesting that the free motion of the 8-aryl group is hampered and consequently the associated non-radiative relaxation is also hindered.

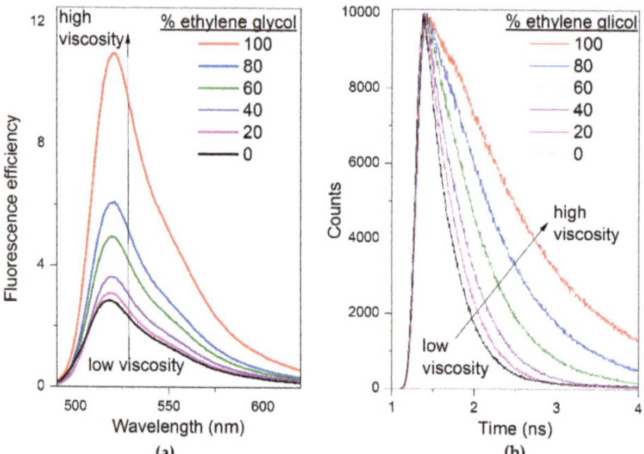

Figure 5. Evolution of the (**a**) fluorescence spectra (scaled by the fluorescence quantum yield) and (**b**) decay curves of dyad **6** with the content of ethylene glycol in diluted ethanolic solutions.

Such key influence of the viscosity in the fluorescence response can be also visualized following the evolution of the emission intensity with the temperature in ethylene glycol. Indeed, as the temperature increases, there is more energy available to rotate the 8-aryl and to overcome the impediment afforded by the environmental viscosity. Thus, a heating of the solution progressively decreased the emission efficiency owing to the discussed enhancement of the internal conversion processes (Figure 6). In fact, an activation energy of around 5.3 kcal/mol has been calculated from the evolution of the non-radiative rate, constant with the temperature in ethylene glycol (Figure 6).

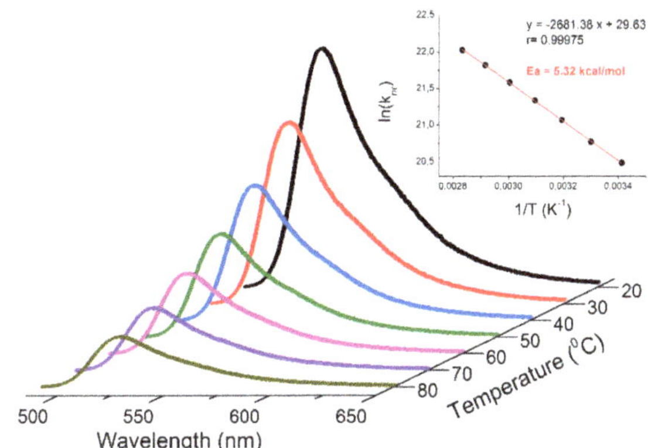

Figure 6. Evolution of the fluorescence spectra of dyad **6** with the temperature in ethylene glycol. The corresponding Arrhenius plot for the non-radiative rate (k_{nr}) constant is also enclosed. The k_{nr} data at each temperature was calculated after checking that the absorption spectra are the same regardless of the temperature, and assuming that the radiative rate constant (k_{fl}) does not change with the temperature. In other words, the loss of fluorescence signal upon heating is due solely to an increase of the internal conversion related to the 8-aryl motion.

3. Materials and Methods

3.1. Materials

Starting 8-methylthioBODIPY, CuTC, tri(2-furyl)phosphine, and boronic acids are commercially available. Solvents were dried and distilled before use.

3.2. General Procedure for the Suzuki Reaction

In a reaction tube under N_2, we dissolved **6** (1.0 equiv), the corresponding boronic acid (2.0 equiv), $Pd(OAc)_2$ (5 mol%), S-Phos (15 mol%), Na_2CO_3 (2.0 equiv) in a mixture toluene/H_2O (4:1, 2.5 mL). The reaction was heated at 90 °C until the starting material was consumed as indicated by thin-layer chromatography (TLC), cooled to room temperature, and then water was added. Then it was extracted with ethyl acetate (3 × 10 mL), washed with brine, dried over $MgSO_4$, filtered and evaporated to dryness. The crude was filtered through a short silica gel column, and eluted with dichloromethane (DCM). The product was crystalized using DCM/petroleum ether.

3.3. Synthesis and Characterization

$^{1.}$H and ^{13}C Nuclear magnetic Resonance (NMR) spectra (collected in the Supplementary Materials) were recorded on a Bruker (Billerica, MA, US) Avance III HD 400 (^{1}H, 400MHz; ^{13}C 100 MHz) or Bruker Ultrashield 500 (^{1}H, 500 Mhz; ^{13}C 125 MHZ) in deuteriochloroform ($CDCl_3$), with either tetramethylsilane (TMS) (0.00 ppm ^{1}H, 0.00 ppm ^{13}C), chloroform (7.26 ppm ^{1}H, 77.00 ppm ^{13}C). Data are reported in the following order: chemical shift in ppm, multiplicities (br (broadened), s (singlet), d (doublet), t (triplet), q (quartet), m (multiplet), exch (exchangeable), app (apparent)), coupling constants, *J* (Hz) and integration. Infrared spectra were recorded on a Perkin Elmer (Waltham, MA, US) Spectrum 100 Fourier-transform infrared (FTIR) spectrophotometer. Peaks are reported (cm^{-1}) with the following relative intensities: s (strong, 67–100%), m (medium, 40–67%), and w (weak, 20–40%). Melting points are not corrected. TLC was conducted in Silica gel on TLC Al foils. Detection was done by UV light (254 or 365 nm). High-resolution mass spectrometry (HRMS) samples were determined on a MaXis Impact ESI-QTOF-MS (Bruker Daltonics) by electrospray ionization in positive mode (ESI+) and recorded via the time of fly (TOF) method.

The corresponding reaction conditions for each compound as well as their characterization data are detailed in the Supplementary Materials.

3.4. Spectroscopic Measurements

The photophysical properties were registered using quartz cuvettes with optical pathways of 1 cm in diluted solutions (around 2×10^{-6} M), prepared by adding the corresponding solvent to the residue from the adequate amount of a concentrated stock solution in acetone, after vacuum evaporation of this solvent. Ultraviolet-visible (UV-vis) absorption and fluorescence spectra were recorded on a Varian model CARY 4E spectrophotometer (Agilent Technologies, Santa Clara, CA, US), and an Edinburgh Instruments (Livingston, England) spectrofluorimeter (model FLSP920), respectively. Fluorescence quantum yields (ϕ) were obtained using PM546 as a reference (Exciton, $\phi^r = 0.85$ in ethanol). Radiative decay curves were registered with the time correlated single-photon counting technique, as implemented in the aforementioned spectrofluorimeter. Fluorescence emission was monitored at the maximum emission wavelength by using a microchannel plate detector (Hamamatsu C4878) of picosecond time-resolution (20 ps), after excitation with a Fianium pulsed laser (time resolution of around 150 picoseconds). The fluorescence lifetime (τ) was obtained after the deconvolution of the instrumental response signal from the recorded decay curves by means of an iterative method. The goodness of the exponential fit was controlled by statistical parameters (chi-square) and the analysis of the residuals. Radiative (k_{fl}) and non-radiative (k_{nr}) rate constants were calculated as follows: $k_{fl} = \phi/\tau$; $k_{nr} = (1 - \phi)/\tau$.

3.5. Computational Simulations

Ground state energy minimizations were performed using a functional range-separated hybrid wb97xd within the Density Functional Theory (DFT), using the triple valence basis set with a polarization and a diffuse function (6-311+g*). The optimized geometries were taken as a true energy minimum using frequency calculations (no negative frequencies). The conformational search around the linkage bond between the coumarin fragment and the BODIPY at 8-positions suggests that the aforementioned geometry corresponds to the most stable conformer. The absorption profile was simulated with the Time Dependent (TD-DFT) method using the same calculation level and basis set. The Polarizable Continuum Model (PCM) was considered to have a solvent effect (cyclohexane) in all the calculations. All the calculations were performed in Gaussian 16, using the "arina" computational resources provided by the UPV-EHU.

4. Conclusions

Salicylaldehyde was efficiently functionalized with a BODIPY unit via a Liebeskind-Srogl cross-coupling reaction. In a first example of the application of **3** as a building block, a *meta*-bromophenyl BODIPY-coumarin was prepared from which 10 novel analogues were attained using the Suzuki reaction. The addition of coumarins, functionalized with aromatic moieties, to the *meso* position of BODIPYs is a suitable synthetically accessible strategy to ameliorate the light harvesting ability of BODIPY-coumarin hybrids. The proper selection of the functional aromatic groups to extend the π-system of the chromene core, enables the enhancement of the absorption probability at the UV-blue region, providing a more efficient and broader light collection for the subsequent excitation energy transfer to the BODIPY. The low fluorescence response of the herein reported dyads is attributed to the conformational freedom of the coumarin around the key *meso* position and the activation of electron transfer processes when electron-rich groups are tethered at the coumarin subunit.

However, the detrimental impact of the conformational flexibility on the fluorescence response of the BODIPY-coumarin hybrids paves the way to apply them as molecular rotors for the monitorization of the environmental microviscosity. In fact, in those dyads not undergoing PET, the fluorescence efficiency and lifetime progressively increases and lengthens, respectively, with the viscosity of the media. Therefore, these BODIPY-coumarin dyads behave as versatile molecular rotors to quantify the viscosity following the changes of the fluorescence signatures upon excitation almost along the whole UV-yellow spectral region, owing to their broadband light harvesting and ensuing energy transfer.

Supplementary Materials: The following are available online; synthetic details and characterization data (IR, NMR, HRMS) of each compound, ^{1}H and ^{13}C-NMR spectra, photophysical data (Table S1), absorption spectra (Figure S1) and computed absorption spectra (Figure S2).

Author Contributions: E.E.-P., C.I.B.-H. and J.G.B.-G. synthesized the compounds and carried out their characterization (IR, NMR, HRMS). T.A. conducted the photophysical measurements. I.L.-A. supervised the spectroscopy study. J.B. carried out the theoretical simulation and wrote the original draft. E.P.-C. designed the compounds, supervised the organic synthesis and wrote the manuscript. All authors have read and agreed to the published version of the manuscript.

Funding: This research was funded by Spanish Ministerio de Economia y Competitividad (project MAT2017-83856-C3-3-P), Gobierno Vasco (project IT912-16), CONACyT (grants 253623, 123732) and Dirección de Apoyo a la Investigación (DAIP-UG CIIC318/2019).

Acknowledgments: The authors thank SGIker of UPV/EHU for technical support with the computational calculations, which were carried out in the "arina" informatic cluster. E.E.-P. and J.G.B.-G. thank CONACyT (Mexico) for graduate fellowship.

Conflicts of Interest: The authors declare no conflict of interest. The funders had no role in the design of the study; in the collection, analyses, or interpretation of data; in the writing of the manuscript, or in the decision to publish the results.

References

1. Terenziani, F.; Parthasarathy, V.; Pla-Quintana, A.; Maishal, T. Cooperative Two-Photon Absorption Enhancement by Through-Space Interactions in Multichromophoric Compounds. *Angew. Chem. Int. Ed.* **2009**, *48*, 8691–8694. [CrossRef] [PubMed]
2. Schwartz, E.; Le Gac, S.; Cornelissen, J.J.L.M.; Nolte, R.J.M.; Alan, E.; Rowan, A.E. Macromolecular Multi-chromophoric scaffolding. *Chem. Soc. Rev.* **2010**, *39*, 1576–1599. [CrossRef] [PubMed]
3. Teo, Y.N.; Kool, E.T. DNA-Multichromophore Systems. *Chem. Rev.* **2012**, *112*, 4221–4245. [CrossRef]
4. Mirkovic, T.; Ostroumov, E.E.; Anna, J.M.; Van Grondelle, R.; Govindjee; Scholes, G.D. Light absorption and energy transfer in the antenna complexes of photosynthetic organisms. *Chem. Rev.* **2017**, *117*, 249–293. [CrossRef] [PubMed]
5. Fan, J.; Hu, M.; Zhang, P.; Peng, X. Energy transfer cassettes based on organic fluorophores: Construction and applications in ratiometric sensing. *Chem. Soc. Rev.* **2013**, *42*, 29–43. [CrossRef] [PubMed]
6. Sen, E.; Meral, K.; Atılgan, S. From Dark to Light to Fluorescence Resonance Energy Transfer (FRET): Polarity-Sensitive Aggregation-Induced Emission (AIE)-Active Tetraphenylethene-Fused BODIPY Dyes with a Very Large Pseudo-Stokes Shift. *Chem. Eur. J.* **2016**, *22*, 736–745. [CrossRef] [PubMed]
7. D'Souza, F.; Smith, P.M.; Zandler, M.E.; McCarty, A.L.; Itou, M.; Araki, Y.; Ito, O. Energy transfer followed by electron transfer in a supramolecular triad composed of boron dipyrrin, zinc porphyrin, and fullerene: A model for the photosynthetic antenna-reaction center complex. *J. Am. Chem. Soc.* **2004**, *126*, 7898–7907. [CrossRef] [PubMed]
8. Iehl, J.; Nierengarten, J.-F.; Harriman, A.; Bura, T.; Ziessel, R. Artificial light-harvesting arrays: Electronic energy migration and trapping on a sphere and between spheres. *J. Am. Chem. Soc.* **2012**, *134*, 988–998. [CrossRef]
9. Qu, X.; Liu, Q.; Ji, X.; Chen, H.; Zhou, Z.; Shen, Z. Enhancing the Stokes' shift of BODIPY dyes via through-bond energy transfer and its application for Fe^{3+}-detection in live cell imaging. *Chem. Commun.* **2012**, *48*, 4600–4602. [CrossRef]
10. Han, J.; Engler, A.; Qi, J.; Tung, C.-H. Ultra pseudo-Stokes shift near infrared dyes based on energy transfer. *Tetrahedron Lett.* **2013**, *54*, 502–505. [CrossRef]
11. He, L.; Zhu, S.; Liu, Y.; Xie, Y.; Xu, Q.; Wei, H.; Lin, W. Broadband light-harvesting molecular triads with high FRET efficiency on the coumarin-rhodamine-BODIPY platform. *Chem. Eur. J.* **2015**, *21*, 12181–12187. [CrossRef] [PubMed]
12. Bessette, A.; Hanan, G.S. Design, synthesis and photophysical studies of dipyrromethene-based materials: Insights into their applicattions in organic photovoltaic devices. *Chem. Soc. Rev.* **2014**, *43*, 3342–3405. [CrossRef] [PubMed]

13. El-Khouly, M.E.; El-Mohsnawy, E.; Fukuzumi, S. Solar energy conversion: From natural to artificial photosynthesis. *J. Photochem. Photobiol. C* **2017**, *31*, 36–83. [CrossRef]
14. Alstrum-Acevedo, J.H.; Brennaman, M.K.; Meyer, T.J. Chemical approaches to artificial photosynthesis. 2. *Inorg. Chem.* **2005**, *44*, 6802–6827. [CrossRef]
15. Benniston, A.C.; Harriman, A. Artificial photosynthesis. *Mater. Today* **2008**, *11*, 26–34. [CrossRef]
16. Bañuelos, J. BODIPY Dye, the Most Versatile Fluorophore Ever? *Chem. Rec.* **2016**, *16*, 335–348. [CrossRef]
17. Patil, P.O.; Bari, S.B.; Firke, S.D.; Deshmukh, P.K.; Donda, S.T.; Patil, D.A. A comprehensive review on synthesis and designing aspects of coumarin derivatives as monoamine oxidase inhibitors for depression and Alzheimer's disease. *Bioorg. Med. Chem.* **2013**, *21*, 2434–2450. [CrossRef]
18. Loudet, A.; Burgess, K. BODIPY dyes and their derivatives: Syntheses and spectroscopic properties. *Chem. Rev.* **2007**, *107*, 4891–4932. [CrossRef]
19. Boens, N.; Verbelen, B.; Ortiz, M.J.; Jiao, L.; Dehaen, W. Synthesis of BODIPY dyes through postfunctionalization of the boron dipyrromethene core. *Coord. Chem. Rev.* **2019**, *399*, 213024. [CrossRef]
20. Cao, X.; Lin, W.; Yu, Q.; Wang, J. Ratiometric Sensing of Fluoride Anions Based on a BODIPY-Coumarin Platform. *Org. Lett.* **2011**, *13*, 6098–6101. [CrossRef]
21. Bai, Y.; Shi, X.; Chen, Y.; Zhu, C.; Jiao, Y.; Han, Z.; He, W.; Guo, Z. Coumarin/BODIPY hybridization for ratiometric sensing of intracellular polarity oscillation. *Chem. Eur. J.* **2018**, *24*, 7513–7524. [CrossRef] [PubMed]
22. Zhao, Y.; Zhang, Y.; Lv, X.; Liu, Y.; Chen, M.; Wang, P.; Liu, J.; Guo, W. Through-bond energy transfer cassettes based on coumarin–Bodipy/distyryl Bodipy dyads with efficient energy efficiences and large pseudo-Stokes' shifts. *J. Mater. Chem.* **2011**, *21*, 13168–13171. [CrossRef]
23. Esnal, I.; Duran-Sampedro, G.; Agarrabeitia, A.R.; Bañuelos, J.; García-Moreno, I.; Macias, M.A.; Peña-Cabrera, E.; López-Arbeloa, I.; De la Moya, S.; Ortiz, M.J. Coumarin-BODIPY hybrids by heteroatom linkage: Versatile, tunable and photostable dye lasers for UV irradiation. *Phys. Chem. Chem. Phys.* **2015**, *17*, 8239–8247. [CrossRef] [PubMed]
24. Lee, H.; Yang, Z.; Wi, Y.; Kim, T.W.; Verwilst, P.; Lee, Y.H.; Han, G.-i.; Kang, C.; Kim, J.S. BODIPY-coumarin conjugate as an endoplamic reticulum membrane fluidity sensor and its application to ER stress models. *Bionconjug. Chem.* **2015**, *26*, 2474–2480. [CrossRef] [PubMed]
25. Zhang, Y.; Song, N.; Li, Y.; Yang, Z.; Chen, L.; Sun, T.; Xie, Z. Comparative study of two near-infrared coumarin-BODIPY dyes for bioimaging and phototermal therapy of cancer. *J. Mater. Chem. C* **2019**, *7*, 4717–4724.
26. Wang, P.; Guo, S.; Wang, H.-J.; Chen, K.-K.; Zhang, N.; Zhang, Z.-M.; Lu, T.-B. A broadband and strong visible-light-absorbing photosensitizer boosts hydrogen evolution. *Nat. Comm.* **2019**, *10*, 3155. [CrossRef]
27. Momahed Heravi, M.; Zadsirjan, V.; Mollaiye, M.; Heydari, M.; Taheri Kal Koshvandi, A. Salicylaldehydes as privileged synthons in multicomponent reactions. *Russ. Chem. Rev.* **2018**, *87*, 553–585. [CrossRef]
28. Alamiry, M.A.H.; Benniston, A.C.; Copley, G.; Elliott, K.J.; Harriman, A.; Stewart, B.; Zhi, Y.-Z. A molecular rotor based on an unhindered boron dipyrromethene (Bodipy) dye. *Chem. Mater.* **2008**, *20*, 4024–4032. [CrossRef]
29. Aswathy, P.R.; Sharma, S.; Tripathi, N.P.; Sengupta, S. Regioisomeric BODIPY benzodithiophene dyads and triads with tunable red emission as ratiometric temperature and viscosity sensor. *Chem. Eur. J.* **2019**, *25*, 14870–14880. [CrossRef]
30. Miao, W.; Yu, C.; Hao, E.; Jiao, L. Functionalized BODIPYs as fluorescent molecular rotors for viscosity detection. *Front. Chem.* **2019**, *26*, 825. [CrossRef]
31. Cheng, H.-G.; Chen, H.; Liu, Y.; Zhou, Q. The Liebeskind-Srogl Cross-Coupling Reaction and its Synthetic Applications. *Asian J. Org. Chem.* **2018**, *7*, 490–508. [CrossRef]
32. Gary, A.; Molander, G.A.; Trice, S.L.J.; Kennedy, S.M. Scope of the Two-Step, One-Pot Palladium-Catalyzed Borylation/Suzuki Cross-Coupling Reaction Utilizing Bis-Boronic Acid. *J. Org. Chem.* **2012**, *77*, 8678–8688.
33. Phakhodee, W.; Duangkamol, C.; Yamano, D.; Pattarawarapan, M. Ph_3P/I_2-Mediated Synthesis of 3-Aryl-Substituted and 3,4-Disubstituted Coumarins. *Synlett* **2017**, *28*, 825–830. [CrossRef]
34. Hong, Y.; Lam, J.W.Y.; Tang, B.Z. Aggregation-induced emission: Phenomenon, mechanism and applications. *Chem. Commun.* **2009**, 4332–4353. [CrossRef] [PubMed]
35. Mei, J.; Leung, N.L.C.; Kwok, R.T.K.; Lam, J.W.Y.; Tang, B.Z. Aggregation-induced emission: Together we shine, united we soar. *Chem. Rev.* **2015**, *115*, 11718–11940. [CrossRef] [PubMed]

36. Liu, Z.; Jiang, Z.; Yan, M.; Wang, X. Recent progress of BODIPY dyes with aggregation-induced emission. *Front. Chem.* **2019**, *7*, 712. [CrossRef]
37. Kee, H.L.; Kirmaier, C.; Yu, L.; Thamyongkit, P.; Youngblood, W.J.; Calder, M.E.; Ramos, L.; Noll, B.C.; Bocian, D.F.; Scheidt, W.R.; et al. Structural control of the photodynamics of boron-dipyrrin complexes. *J. Phys. Chem. B* **2005**, *109*, 20433–20443. [CrossRef]
38. Zheng, Q.; Xu, G.; Prasad, P.N. Conformationally restricted dipyrromethene boron difluoride (BODIPY) dyes: Highly fluorescent, multicolored probes for cellular imaging. *Chem. Eur. J.* **2008**, *14*, 5812–5819. [CrossRef]
39. Reddy, G.; Duvva, N.; Seetharaman, S.; D'Souza, F.; Giribabu, L. Photoinduced energy transfer in carbazole-BODIPY based dyads. *Phys. Chem. Chem. Phys.* **2018**, *20*, 27418–27428. [CrossRef]
40. Li, F.; Yang, S.I.; Ciringh, Y.; Seth, J.; Martin III, C.H.; Singh, D.L.; Kim, D.; Birge, R.R.; Bocian, D.F.; Holten, D.; et al. Design, synthesis, and photodynamics of light-harvesting arrays comprised of a porphyrin and one, two or eight boron-dipyrrin accessory pigments. *J. Am. Chem. Soc.* **1998**, *120*, 10001–10017. [CrossRef]
41. Ramírez-Ornelas, D.E.; Alvarado-Martínez, E.; Bañuelos, J.; López-Arbeloa, I.; Arbeloa, T.; Mora-Montes, H.M.; Pérez-García, L.A.; Peña-Cabrera, E. FormylBODIPYs: Privileged building blocks for multicomponent reactions. The case of Passerini reaction. *J. Org. Chem.* **2016**, *81*, 2888–2898. [CrossRef] [PubMed]
42. Prlj, A.; Vannay, L.; Corminboeuf, C. Fluorescence quenching in BODIPY dyes: The role of intramolecular interactions and charge transfer. *Helv. Chim. Acta* **2017**, *100*, e1700093. [CrossRef]
43. Ramírez-Ornelas, D.E.; Sola-Llano, R.; Bañuelos, J.; López Arbeloa, I.; Martínez-Álvarez, J.A.; Mora-Montes, H.M.; Franco, B.; Peña-Cabrera, E. Synthesis, photophysical study, and biological application analysis of complex borondipyrromethene dyes. *ACS Omega* **2018**, *3*, 7783–7797. [CrossRef] [PubMed]
44. Laurent, A.D.; Adamo, C.; Jacquemin, D. Dye chemistry with time-dependent density functional theory. *Phys. Chem. Chem. Phys.* **2014**, *16*, 14334–14356. [CrossRef]
45. Momeni, M.R.; Brown, A. Why do TD-DFT excitation energies of BODIPY/aza-BODIPY families largely deviate from experiment? Answers from electron correlated and multireference methods. *J. Chem. Theory Comput.* **2015**, *11*, 2619–2632. [CrossRef]
46. Mao, M.; Song, Q.-H. The structure-property relationships of D-π-A dyes for dye-sensitized solar cells. *Chem. Rec.* **2016**, *16*, 719–733. [CrossRef]
47. Greene, L.E.; Lincoln, R.; Cosa, G. Tuning photoinduced electron transfer efficiency of fluorogenic BODIPY-α-tocopherol analogues. *Photochem. Photobiol.* **2019**, *95*, 192–201. [CrossRef]
48. Scholz, N.; Jadhav, A.; Screykar, M.; Behnke, T.; Nirmalananthan, N.; Resch-Genger, U.; Sekar, N. Coumarin-rhodamine hybrids–Novel probes for the optical measurement of viscosity and polarity. *J. Fluoresc.* **2017**, *7*, 1949–1956. [CrossRef]
49. Dwivedi, B.K.; Singh, V.D.; Kumar, Y.; Pandey, D.S. Photophysical properties of some novel tetraphenylimidazole derived BODIPY based fluorescent molecular rotors. *Dalton Trans.* **2020**, *49*, 438–452. [CrossRef]
50. Kuimova, M.K.; Yahioglu, G.; Levitt, J.A.; Suhling, K. Molecular rotor measures viscosity of live cells via fluorescence lifetime imaging. *J. Am. Chem. Soc.* **2008**, *130*, 6672–6673. [CrossRef]
51. Kuimova, M.K.; Botchway, S.W.; Parker, A.W.; Balaz, M.; Collins, H.A.; Anderson, H.L.; Suhling, K.; Ogilby, P.R. Imaging intracellular viscosity of a single cell during photoinduced cell death. *Nat. Chem.* **2009**, *1*, 69–73. [CrossRef] [PubMed]

Sample Availability: Samples of the compounds are available from the authors.

Article

Spectroscopic Behaviour of Two Novel Azobenzene Fluorescent Dyes and Their Polymeric Blends

Rosita Diana [1], Ugo Caruso [2,*], Stefano Piotto [3], Simona Concilio [4], Rafi Shikler [5] and Barbara Panunzi [1]

1 Department of Agriculture, University of Napoli Federico II, NA 80055 Portici, Italy; rosita.diana@unina.it (R.D.); barbara.panunzi@unina.it (B.P.)
2 Department of Chemical Sciences, University of Napoli Federico II, 80126 Napoli, Italy
3 Department of Pharmacy, University of Salerno, SA 84084 Fisciano, Italy; piotto@unisa.it
4 Department of Industrial Engineering, University of Salerno, SA 84084 Fisciano, Italy; sconcilio@unisa.it
5 Department of Electrical and Computer Engineering, Ben-Gurion University of the Negev, POB 653 Beer-Sheva 84105, Israel; rshikler@ee.bgu.ac.il
* Correspondence: ugo.caruso@unina.it; Tel.: +39-081-674366

Academic Editor: Radosław Podsiadły
Received: 14 February 2020; Accepted: 16 March 2020; Published: 17 March 2020

Abstract: Two novel symmetrical bis-azobenzene red dyes ending with electron-withdrawing or donor groups were synthesized. Both chromophores display good solubility, excellent chemical, and thermal stability. The two dyes are fluorescent in solution and in the solid-state. The spectroscopic properties of the neat crystalline solids were compared with those of doped blends of different amorphous matrixes. Blends of non-conductive and of emissive and conductive host polymers were formed to evaluate the potential of the azo dyes as pigments and as fluorophores. Both in absorbance and emission, the doped thin layers have CIE coordinates in the spectral region from yellow to red. The fluorescence quantum yield measured for the brightest emissive blend reaches 57%, a remarkable performance for a steadily fluorescent azo dye. A DFT approach was employed to examine the frontier orbitals of the two dyes.

Keywords: azobenzene; dye; fluorophore; colorant; polymeric blend

1. Introduction

Azobenzene derivatives are π-conjugated molecules that have aroused enormous research interest owing to their fascinating characteristics. First, the unique broad UV/Vis absorbance spectra related to the considerable number of vibronic states in each energy level. The color is due to the presence of the N=N chromophore chemical group absorbing light in the visible spectrum. Suitable substituents can achieve the tuning of the color to cover the entire visible spectrum.

For this reason, from the past until today, azo compounds have been widely used as dyes and pigments. A significant amount of azo dyes is used in manufacturing processes of textile, paper, packaging pharmaceutical, and even food industries [1]. Excellent compatibility with the matrix and chemical and thermal stability are the main requirements. Despite this, photo-catalyzed degradation through illumination by solar light was found in most of the azo pigments, with the resulting discoloration of the dyed substrate.

On the other hand, the ability of azobenzene compounds to reversibly change from the stable *trans*-isomer to *cis* form upon photoirradiation (photoisomerization) causes nonradiative deactivation and very small/negligible fluorescence. The rapid photoisomerization of the azo bridge finds applications in on-off photoswitching techniques, such as in optical data storage, dye-sensitized solar cells, pharmaceuticals, and non-linear optics [2–8]. Azo-containing molecules can play the role of energy dissipator (quencher) associated with other emitting chromophores, through rapid nonradiative pathways, while the donor is excited for the occurrence of a Förster resonance energy transfer (FRET) [9]. Unlike many chromophores commonly utilized as fluorescent materials, there are a limited number of examples of steadily fluorescent azobenzene compounds.

The structural versatility and the tunable spectroscopic properties of the azobenzene derivatives could be exploited if they are themselves fluorescent. Attempts to increase and to modulate the fluorescence performance of azobenzene compounds are quite recent [10–15] and based on preventing fast photoisomerization. This is possible with bulky substituents close to the N=N group [16–21] or intramolecular hydrogen bonding preventing the rotation around the nitrogen–carbon bond [19]. The influence of hydrogen-bond interactions on the excited-state dynamics of azo dyes has been examined [22,23]. It appeared that by restricting rotation and isomerization, the excited-state intramolecular proton-transfer (ESIPT) [21,24] lead to emissive azo dyes in solution [10,16,24–29]. Conversely, examples of solid-state azo fluorophores are rare in the scientific literature. Those few are typically red emitters [30–33], particularly advantageous in the context of biological and medical measurements [34–37].

In previous contributions [10,13], we studied azobenzene scaffolds for their unique structural pattern related to their photophysical properties. In this work, two novel azobenzene dyes, A1 and A2 in Scheme 1, were explored. To fulfill the criteria as fluorescent dyes, we have sought stable, processable, and sterically encumbered structures. Based on the same symmetrical bis-azobenzene skeleton (AB in Scheme 1), the two chromophores differ for the terminal substituents, both electron-donor (NEt_2) or electron-acceptor (NO_2) groups. DFT computational study was employed to get information on HOMO-LUMO localization at the different conjugated patterns. To impede photoisomerization, we added methoxy substituents on the AB moiety and two terminal Schiff bases ESIPT undergoing sites to restrict molecular rotations.

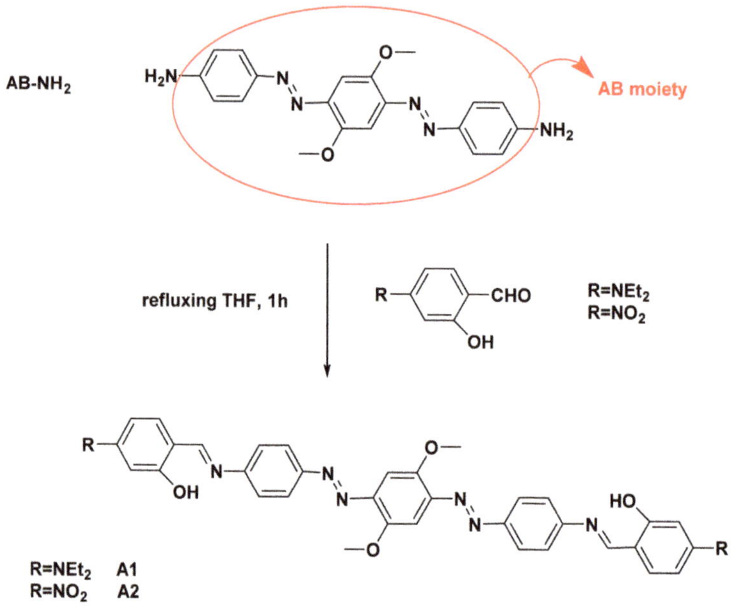

Scheme 1. Synthesis and structure of the dyes A1 and A2.

Under sunlight, the dyes are stable over three months both in solution and in the solid-state, retaining their orange-red color (see CIE: coordinates, International Commission on Illumination, in Tables 1 and 2). As fluorophores, a significant emission in solution and the crystalline phase was recorded. Different host matrixes were employed to produce polymeric blends from A1 and A2. In addition to classical non-conductive hosts such as polyvinyl chloride (PVC) and poly (styrene) (PS), we also checked conductive, poly (vinylcarbazole) (PVK), and even emissive, poly (9,9-dioctylfluorene) (PFO), polymers. Their coloring ability was examined dissolved in PVC, a white matrix used in many industrial processes. The emission in the solid-state was evaluated in PS, PVK, and PF blends. The last one recently emerged as a useful polymeric matrix for optoelectronic devices [38]. In PFO, the photoluminescence quantum yield (PLQY) of the A1 orange-red blend increases up to an outstanding 57% value.

Table 1. Optical data for A1 and A2 in solution and for polyvinyl chloride (PVC) blends.

Sample	λ_{abs} (nm) [a]	ε (10^5 $cm^{-1}M^{-1}$) [b]	λ_{em} (nm) [c]	PLQY% [d]	λ_{abs} (nm) [e]	CIE [f]
A1	448	6.29	511 (542)	3.0 ± 0.1	444	0.57; 0.39
A2	(348) 453	9.57	619	1.2 ± 0.1	435	0.52; 0.40
A1-PVC(10%)	-	-	-	-	455	0.51; 0.44
A2-PVC(10%)	-	-	-	-	455	0.60; 0.37
A1-PVC(30%)	-	-	-	-	426	0.51; 0.44
A2-PVC(30%)	-	-	-	-	425	0.60; 0.37

[a] Wavelength of UV-Visible absorbance maxima in THF solution. [b] Molar absorption coefficient. [c] Wavelength of emission maxima in THF solution. [d] PLQYs in THF solution. [e] Wavelength of UV-Visible absorbance maxima measured on the spin-coated film. [f] Absorption CIE coordinates on the spin-coated film.

Table 2. Optical data for A1 and A2 as neat solid samples and in 10 wt%. poly (styrene) (PS), poly (vinylcarbazole) (PVK), and poly (9,9-dioctylfluorene) (PFO) film blends.

Blend	λ_{abs} (nm) [a]	λ_{em} (nm) [b]	PLQY% [c]	CIE [d]
A1	444	585	3.4 ± 0.1	0.48; 0.50
A2	435	578	0.7 ± 0.1	0.50; 0.49
A1-PS	439	523–551	22.0 ± 0.1	0.35; 0.60
A2-PS	415	578	11.0 ± 0.1	0.47; 0.49
A1-PVK	440	536	40.0 ± 0.4	0.41; 0.56
A2-PVK	444	605	10.0 ± 0.1	0.55; 0.44
A1-PFO	390	551	57.0 ± 0.5	0.43; 0.50
A2-PFO	390	434, 452, (571)	63.0 ± 0.7	0.23; 0.19
PS [e]	290	-	-	-
PVK [e]	295, 340	388	7.6 ± 0.5	0.20; 0.16
PFO [e]	365	459	68.0 ± 0.4	0.20; 0.24

[a] Wavelength of UV-Visible absorbance maxima; [b] Wavelength of emission maxima; [c] Photoluminescent quantum yield; [d] Emission CIE coordinates; [e] Optical data for films of PS, PVK, and PFO, obtained in the same conditions as the blends.

2. Results and Discussion

2.1. *Synthesis and Optical Behavior of the Dyes and Their PVC Blends*

As summarized in Scheme 1, the dyes A1 and A2 were obtained by condensation of the diamino derivative AB-NH_2 [39] with 4-(diethylamino)-2-hydroxybenzaldehyde and 2-hydroxy-4-nitrobenzaldehyde, respectively. Though structurally similar, a different conjugation pattern is recognizable, D-π-D-π-D respectively for A1 and A-π-D-π-A for A2 (where D = electron donor moiety, A = electron acceptor moiety, and π = conjugated system).

Two ESIPT undergoing sites are generated by the condensation of the amino-terminal groups of the precursor AB-NH_2 with the salicylic aldehydes, guaranteeing emission in solution. According to a recent approach [26,39,40], it was found that the fast proton transfer in ESIPT sterically encumbered probes are impeded due to restriction of intramolecular rotation (RIR effect), also providing solid-state emission. The central methoxy groups cause steric hindrance on the conjugated skeleton without too much lowering solubility.

Identification and purity degree evaluation were assessed by mass spectrometry and ^{1}H NMR. Phase behavior was examined by optical observation and DSC/TGA analysis. All materials are thermally stable up to 330 °C under nitrogen flow with high melting points. The two dyes are soluble in most organic solvents, such as acetone, tetrahydrofuran (THF), methylene chloride, tetrachloroethane (TCE), dimetilformammide (DMF), 1-metil-2-pirrolidone (NMP), and dimethyl sulfoxide (DMSO). Their absorption and emission maxima, in solution and the solid films, are reported in Table 1.

In Figure 1, absorption and emission curves of A1 and A2 in THF solution and a picture of the related samples (inset) in natural light and under UV lamp at 365 nm are shown. The yellow THF solutions emit in the lime-yellow and the red region, respectively, with a negligible solvatochromic effect depending on the solvent polarity. PLQYs (see Table 1) have been measured in THF solution by relative methods using as standard quinine sulfate for A1 [41] and zinc phthalocyanine for A2 [42]. The solutions are stable and retain their optical characteristics up to three months under natural light at room temperature.

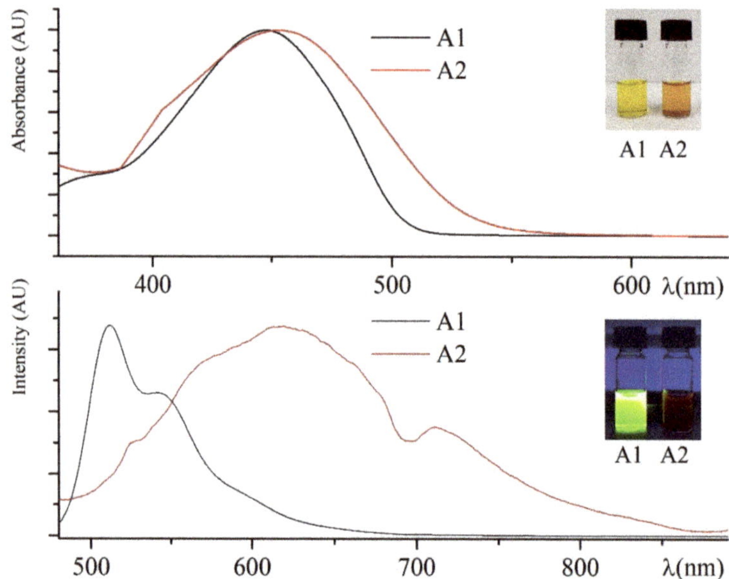

Figure 1. Absorption (above) and emission (below) curves of A1 (black curves) and A2 (red curves) in THF solution. The same samples in natural light (above) and under UV lamp at 365 nm (below) in the insets.

The spin-coated thin films (obtained as described in the Materials and Methods section) of crystalline A1 and A2 have a red and yellow-orange color, respectively (see CIE in Table 1). Images of the spin-coated crystalline films of neat A1 and A2 under polarized microscope are reported in the Supplementary Materials, in Figure S3. The absorption spectra are reported in Figure 2, compared with the spectra of PVC blends at different dopant percentage. PVC as an economical and versatile thermoplastic colorable polymer is an excellent candidate to test the two compounds as dyes. It is the world's third largest thermoplastic material widely used in construction, packaging, devices, and the textile industry. The demand for stable dyes for this white material is still high. The doped PVC homogeneous amorphous blends are an example of stable dyed blends with pigments at 10% and 30% by weight. In both cases, the dyes are soluble up to 30%. The diethylamino terminal groups make compound A1 more soluble than the dinitro derivative; hence A1 has a higher solubility limit (40%) in PVC. The PVC films have CIE coordinates in the orange-red region (see Figure 2), more red-shifted for A2-PVC, the same behavior recorded for THF solutions. The PVC films of blended A1 and A2 kept over three months in the air under natural light at room temperature perfectly retain transparency and optical characteristics. No swelling nor release of the same films were detected on samples kept in distilled water for 30 days at room temperature.

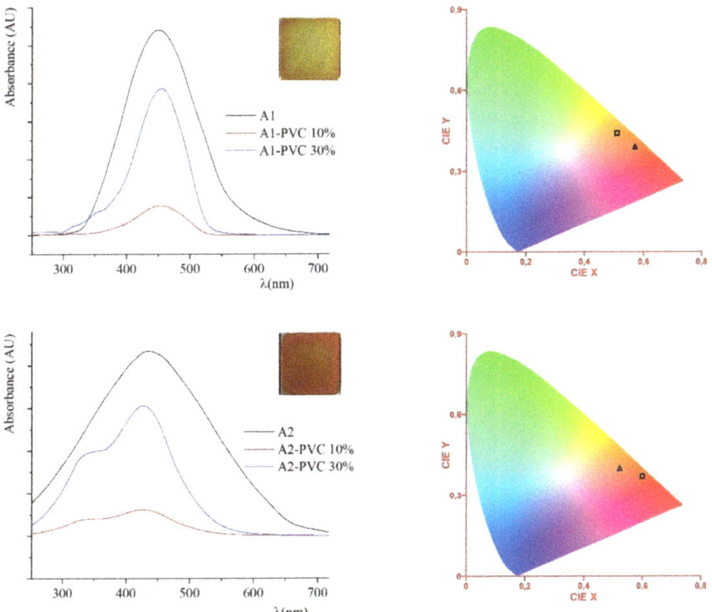

Figure 2. On the left: absorption curves of A1 (black curves), A1-PVC 10% (red curve) and A1-PVC 40% films (blue curve) above; absorption curves of A2 (black curves), A2-PVC 10% (red curve) and A2-PVC 30% films (blue curve) below. In the insets: 30% A1-PVC (above) and 30% A2-PVC (below) thin films. On the right: CIE diagram of A1 (triangle) and both its PVC blends (square) above; CIE diagram of A2 (triangle) and both its PVC blends (square) below.

In Figure 3, an SEM image of 30% A1-PVC film deposed onto quartz slide is reported. The SEM analysis confirms that there is no visible structuring in the spin-coated film, also after three months under natural light at room temperature. All PVC blended samples of A1 and A2 exhibit similar morphological characteristics.

As for PL performance of neat A1 and A2, the crystalline dyes emit in the orange-red region with very similar maxima, see Table 2. The Stokes Shifts are about 140 nm. Appreciable PLQYs (see Table 2) measured on the crystalline samples spin-coated on quartz slides are a good result for azobenzene dyes. The PL response for A1 is higher than the nitro derivative A2 as a result of the different conjugation patterns, as discussed in the DFT analysis section. In Figure 4a, the emission spectra of the crystalline dyes are reported, and their PL behavior will be discussed in the next section.

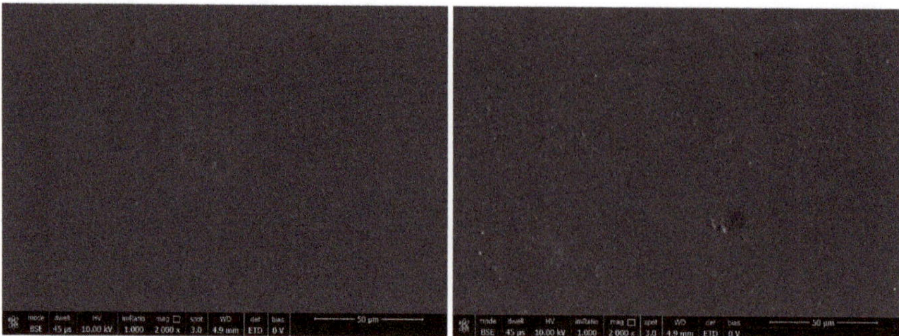

Figure 3. SEM image (2000x magnitude) of 30% A1-PVC film deposed onto quartz slide and metalized by Au-Pd sputtering (about 5 nm) (on the left). The same sample scanned after three months, on the right.

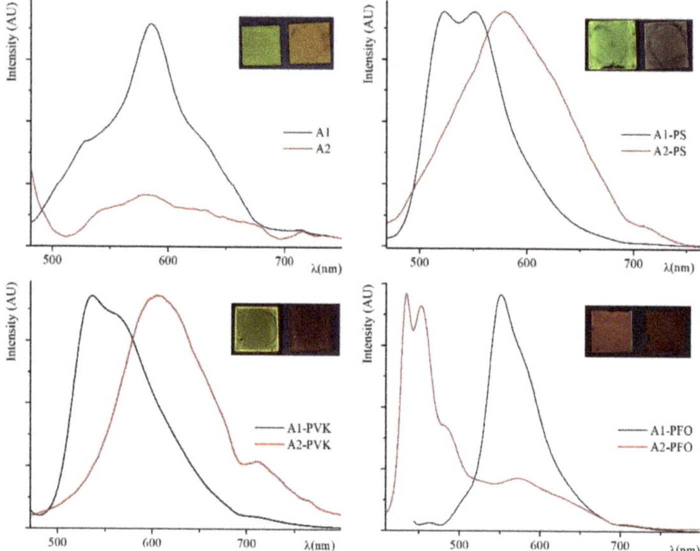

Figure 4. Emission spectra of the neat A1 (black curves) and A2 (red curves) samples (**a**), of their PS blends (**b**), PVK blends (**c**), and PFO blends (**d**). In the inset, the same spin-coated samples used for the measurements (A1 on the left and A2 on the right).

2.2. PL Properties of PS, PVK, and PFO Blends

Polymeric blends are a well-known approach leading to an increase in the emission ability. Blended active layers have been demonstrated to be advantageous to fabricate optoelectronic devices such as efficient pure-color LEDs [43–45], guaranteeing easiness of fabrication, high processability, and low-cost. This situation resembles a diluted solid solution of the emitters into amorphous domains preventing aggregation caused quenching (ACQ) effect [46,47]. Different doped films were spin-coated by dissolving the dyes in a non-emissive and non-conductive (PS) or an emissive and conductive (PVK and mostly PFO) amorphous polymeric matrix. These polymers are typically employed in the construction of emissive layers. In natural light, the doped layers are homogeneous and transparent films except for A2-PFO. They display various shades of color, from yellow to red. Different dye percentages were tested in order to get better PL performance. As expected on the base of our previous study [48–51], the best emission for both samples was recorded on the more diluted blends (10 wt%), where the ACQ effect is attenuated.

The films turned out to be stable in air for three months at room temperature under natural light, showing identical optical properties both in absorption and emission. In fluorescence, broad emission bands peaked between orange and orange-red are recorded, with Stokes Shift values ranging from 83 to 183 nm. The maximum of the absorption and emission bands of the blends are reported in Table 2.

In Figure 4, the emission spectra of the blend samples are compared with the crystalline ones. The emission spectra of pure PVK and PFO are reported, for comparison, in Figure S4 of Supplementary Materials section.

A2-PVK blend is the most red-shifted emitter with CIE (0.55; 0.44) (see Table 2). Quantitatively, all A1 blends show higher PLQYs respect to A2 blends but in the case of PFO samples, clearly highlighting the difference in the electronic pattern of the two chromophores. In both cases, PLQYs of the PS blends are lower respect to the analogous PVK blends. For an azobenzene material, the PL performance of A1-PVK is a remarkable achievement. As PVK is poorly fluorescent, 40% PLQY is due mostly to the chromophore itself as the ACQ effect is suppressed. The PL performance of the A1-PFO blend is particularly interesting. Polyfluorene and its derivatives (PFs) have recently emerged as the most promising polymeric matrixes due to their PL and electroluminescence efficiencies, good thermal and chemical stability [52]. Poly (9-octylfluorene) itself emits blue light with a large bandgap. The typical approach to realize tuned emission from polyfluorene derivatives is based on the copolymerization of low-bandgap π-conjugated moieties with PFO monomers. Unfortunately, polyfluorides often show both excimer and aggregate formation during thermal annealing. The formation of excimers involves the generation of dimerized units of the polymer that emit light at energies lower than those of the polymer itself. This effect hinders the use of polyfluorenes for most applications [53].

In our case, the simple approach to dissolve a red/orange dopant in PFO produced an efficient (57% PLQY) bright orange/red (see CIE coordinates in Table 2) luminescent blend. The outstanding PL performance of A1-PFO is due to the excellent match between the chromophore and the polymeric matrix, as rationalized by DFT calculations in the Supplementary Materials part. On the contrary, the attempt to homogeneously dissolve A2 in PFO failed. Only in that case, the film was not perfectly homogeneous and transparent, and the emission of the chromophore in the red region is negligible compared to the much higher blue emission of PFO itself (see Figure 4d and Figure S4 in Supplementary Materials).

2.3. *DFT Analysis*

TDDFT approach at DFT level, using adiabatic local density approximation and ethanol as the simulated solvent, was used to run excitation energies calculations. Table 3 shows the most relevant optoelectronic properties calculated for the compounds A1 and A2. For A1, HOMO delocalization covers the entire conjugated backbone, extending to the terminal groups. LUMO is delocalized over the central rings and diazo groups. The main transition for both compounds is HOMO $\rightarrow$ LUMO, at 425 nm for A1 and 419 nm for A2. For A2, HOMO is mainly delocalized over the central ring and the oxygen atoms (Figure 5). A2 LUMO shows the electron-withdrawing effect of the nitro groups, with a higher delocalization of the orbitals over the terminal groups compared to the diethylamino derivative A1. The HOMO-LUMO gap is also higher for A2 than for A1 (Table 3). Because of the electron-withdrawing effect of the nitro groups, A2 shows a higher oxidation potential compared to A1, whereas the hole and electron reorganization energies (HRE and ERE) are typical for these class of compounds. A2 shows a decrease in the ERE due to the presence of the nitro groups. Reorganization energy (RE) is one of the parameters involved in the hopping rate, and HRE and ERE are strongly correlated to cation and anion geometries. A compound with a small RE usually shows high carrier mobility, and the energies are proportional to the deformation of the geometry during the process of charge transfer. Both derivatives show an ERE higher than their HRE, which means that there is less deformation upon electron injection compared to hole injection. Furthermore, the nitro derivative shows a smaller electron extraction potential, meaning that electron injection into A2 is easier than into

A1. The electron-withdrawing effect of the nitro groups affects the electron-transporting features of the system.

Table 3. Electro-optical properties calculated on A1 and A2 in vacuum.

Properties	A1	A2
Oxidation Potential (eV)	0.53	1.01
Reduction Potential (eV)	−1.08	−0.76
Hole Reorganization Energy (eV)	0.31	0.33
Electron Reorganization Energy (eV)	0.40	0.35
λ_{max} (nm)	425	419
E_{max} (nm)	540	538
Scaled HOMO (eV)	−4.98	−5.45
Scaled LUMO (eV)	−2.26	−2.69
HOMO-LUMO (eV)	2.72	2.76

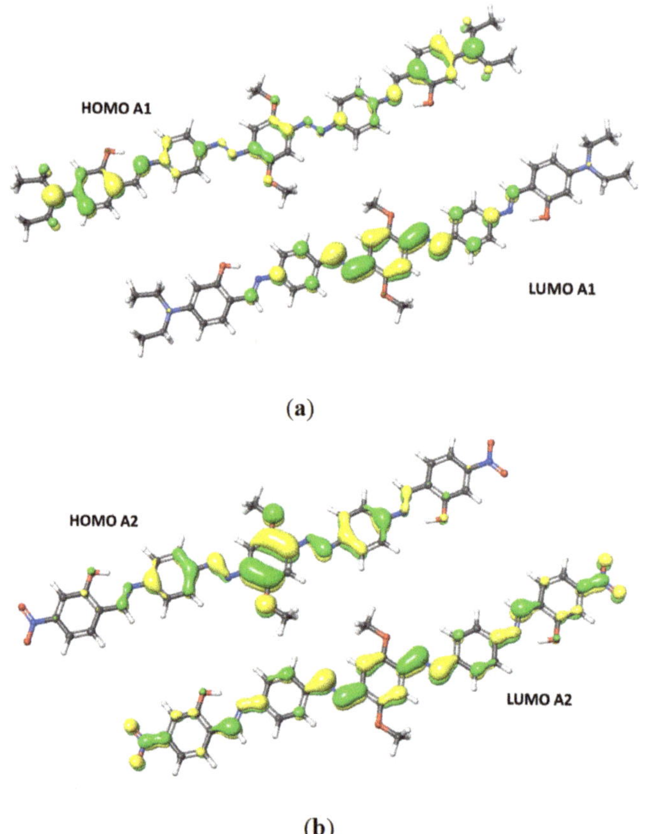

Figure 5. Frontiers orbitals HOMO and LUMO calculated for A1 (**a**) and A2 (**b**).

The significant PL performance of A1 blended in PFO with 57% PLQY must be underlined. This result appears striking, keeping in mind that PFO itself is a blue emitter and A1-PFO blend emit in the orange-red region. Moreover, this solid-state emission is due to an azobenzene dye. For this reason, we analyzed the behavior of A1 in the PFO blend, the entire discussion reported in the Supplementary Material section. The formation of stacked dimers of the azo dye and PFO resulted in modifying the electro-optical properties significantly with the lowering of HOMO and LUMO values for both systems. The new HOMO energy values are even closer to the HOMO values of the PFO (Figure S1). Overall, stacking the dyes with PFO provides a shortcut for electronic transitions and is responsible for some degree of PFO distortion. Therefore, it must be underlined the ability of the azo dyes to modulate the optoelectronic characteristics of the PFO through two effects: the steric hindrance operated by the octyl chains, and the stacking between the dyes and the polymer. Especially in the case of A1-PFO, the calculations show a significant reduction of the electronic hopping energy barrier and justify the high quantum yield observed. Analysis of the oxidation potentials of the substituted dyes revealed that the oxidation potential of A1 dyes was about 460 mV easier to oxidize (more negative potentials) than the A2. Analysis of the reduction potentials revealed that A2 was about 400 mV easier to reduce (more positive potentials) than A1 dye. The diminished ability to reduce A1 dye is attributed to the presence of electron-withdrawing groups. Compared with other azo dyes like the 4-methoxylazobenzene [54] with a E_{red} of −1.44 V, or diarylaminoazobenzenes with a reduction potential between −1.36 V and −1.50V [55], both A1 and A2 have a lower reduction tendency and higher stability (see Tables S1 and S2 in Supplementary Material).

3. Materials and Methods

Commercially available starting products were supplied by Sigma Aldrich (Sigma-Aldrich Corporation, St. Louis, MO, USA). 2-hydroxy-4-nitrobenzaldehyde was obtained as described in [56]. 4,4′-((1,1′)-(2,5-dimethoxy-1,4-phenylene)bis(diazene-2,1-diyl))dianiline (AB-NH_2) was obtained as described in [13]. ^{1}H NMR spectra were recorded in DMSO-d_6, with a Bruker Advance II 400 MHz apparatus (Bruker Corporation, Billerica, MA, USA). Mass spectrometry measurements were performed using a Q-TOF premier instrument (Waters, Milford, MA, USA) equipped by an electrospray ion source and a hybrid quadrupole-time of flight analyzer.

Zeiss Axioscope polarizing microscope (Carl Zeiss, Oberkochen, Germany) equipped with an FP90 Mettler hot stage (Mettler-Toledo, LLC-Columbus, OH, USA). DSC/TGA Perkin Elmer TGA 4000 (PerkinElmer, Inc., Waltham, MA, USA), scanning rate 10 °C/min, provided phase transition temperatures and enthalpies. The decomposition temperatures (the temperature at 5 wt.% weight loss, T_d) were measured under nitrogen flow. UV-Visible and fluorescence spectra were recorded by JASCO F-530 and FP-750 spectrometers (JASCO Inc., Mary's Court, Easton, MD, USA). Thin films of the neat samples and the polymeric blends were obtained using an SCS P6700 spin coater (Specialty Coating Systems Inc., Indianapolis, IN 46278, USA) operating at 600 rpm for 1 min (first step) and at 1200 rpm for 1 min (second step). 10 wt% NMP solution of the chromophores in commercially available PS (molecular weight 18,700 Da), in PVK (molecular weight 1100 Da) or PFO (molecular weight ≥20,000 Da), were employed.

Photoluminescence quantum yield (PLQY) measurements were conducted with a setup similar to that of de Mello et al. [57]. It considers not only the excitation laser and direct photoluminescence but also the scattering of the integrating sphere that is part of the setup. The setup consists of a 405 nm laser, whose emission does not overlap with the photoluminescence spectrum, an integrating sphere (Stellarnet Inc, Tampa, FL, USA) and a photo spectrometer (BLACK Comet Stellarnet Inc, Tampa, FL, USA). The emission was measured at five different points on the sample.

Field emission scanning electron microscopy (FESEM) images were obtained with a FEI Nova NanoSEM 450 emission SEM (Thermo Fisher Scientific Inc., Waltham, MA, USA) at an accelerating voltage of 10 kV (range of acceleration voltage: 50 V–30 kV) equipped with an Everhart Thornley detector (ETD) and a Through Lens Detector (TLD). Samples were mounted on Al specimen mounts and coated with a thin layer of Au-Pd in order to eliminate any undesirable charge effects during the SEM observations.

3.1. Synthesis of A1 and A2

The same general procedure was employed for the synthesis of A1 and A2. The synthesis of A1 is described as an example. To 3.76 g (0.01 mmol) of AB-NH_2 [39] dissolved at 70 °C in 30 mL of dry THF 3.87 g (0.02 mmol) of 4-(diethylamino)-2-hydroxybenzaldehyde was added under stirring. After 1 h, the crude product precipitated at room temperature. The compound was crystallized by dichloromethane/hexane and furtherly purified by washing with hot acetone. T_m = 280 °C; T_d = 330 °C. ^{1}H NMR (400 MHz, DMSO d_6, 25 °C, ppm): 1.12 (t, 12H), 3.46 (m, 8H), 4.00 (s, 6H), 6.21 (d, 4H), 6.44 (d, 2H), 7.46 (m, 8H), 7.65 (m, 2H), 8.89 (s, 2H), 10.65 (s, 2H). Elemental analysis calculated (%) for $C_{42}H_{46}N_8O_4$: C, 69.40; H, 6.38; N, 15.42; found: C, 69.60; H, 6.88; N, 15.88. MALDI-TOF of A1 m/z: 727.39 (M + H).

Chromophore A2 was obtained using 1,1,2,2-tetrachloroethane as the solvent and furtherly purified by washing with hot acetone and then hot dioxane: T_m = 327 °C; T_d = 330 °C. ^{1}H NMR (400 MHz, DMSO-d_6, 25 °C, ppm): 4.14 (s, 6H), 6.65 (m, 4H), 7.20–8.80 (m, 12 H), 9.20 (s, 2H), 12.55 (s, 2H). Elemental analysis calculated (%) for $C_{34}H_{26}N_8O_8$: C, 60.53; H, 3.88; N, 16.61; found: C, 60.50; H, 3.09; N, 16.80. MALDI-TOF of A2 m/z: 675.16 (M + H).

3.2. Theoretical Calculations

Quantum-mechanical calculations were performed with the Jaguar package, Schrödinger Release 2017-4 [58], on the theoretical level DFT/B3LYP, and the molecular geometry was optimized with functional B97-D3 [59]. Charges were determined using the NBO approach. Dunning's correlation-consistent triple-ζ basis set cc-pVTZ (-f), which includes a double set of polarization functions, was used for single-point calculations on optimized geometries. TD-DFT and Tamm-Dancoff [60] approximations were used to perform calculations at neutral compound geometry to extract absorption values from vertical excitation energies. The solvent was simulated using Poisson Boltzmann Solver (PBF) [61]. The stacked dimers shown in Figure S2 were constructed with the Jaguar package from the minimized structures. Each dimer was initially minimized by simulated annealing using AMBER15FB [62] forcefield and then optimized at the B3LYP level, maintaining the dimer starting structure.

Computed redox data was used to calculate "scaled" HOMO and LUMO energies, through the following equations:

$$\text{Absolute Electrode Potential} = \text{Electrode Potential} + \text{NHE Energy} \quad (1)$$

$$\text{Orbital Energy} = \text{Redox Potential} + \text{Absolute Electrode Potential} \quad (2)$$

where "NHE Energy" represents the energy of the NHE electrode in water (−4.28 V) and "Electrode Potential" represents the potential of the chosen electrode relative to NHE.

4. Conclusions

The spectroscopic behavior of two novel azobenzene chromophores with a symmetrical skeleton was systematically examined. Their coloring ability was tested in solution as neat crystalline samples and dissolved in a polymeric amorphous matrix. The doped PVC blends are an example of stable dyed polymers. In both cases, the dyes A1 and A2 are soluble up to 30 wt% in PVC matrix with CIE coordinates in the orange to red region. The films of both neat and blended dyes kept over three months under natural light at room temperature in the air perfectly retain structural and optical characteristics. No swelling nor release was detected in distilled water for 30 days.

On the other hand, the effort to obtain good steady PL response from azobenzene scaffolds is justified by the poor scientific documentation about this versatile and tunable functional group. A systematic study of the optical performance of the two differently substituted azobenzene dyes was performed, and DFT computational study gave information on HOMO-LUMO localization. The azo-based chromophores are emissive in solution and the solid-state, claiming a role as dye-dopants for emissive layers. Low-doped blends were obtained in PS, PVK, and PFO. PLQYs measured on the emissive blends are above 10% in all cases. In particular, PLQY of the orange-red A1-PFO blend (57%) represents a remarkable result for azobenzene based material, able to modulate the optoelectronic characteristics of the blue emissive PFO. The simple synthetic procedure, the affordability, solubility and processability of the dyes and the doped blends make the azo-dyes rare examples of azobenzene based materials potentially employable both as dyes and as fluorophores.

Supplementary Materials: Table S1. Electro-optical properties calculated on A1 and A2 in vacuum and associated with PFO. Table S2. Reduction potential of similar azobenzenes [2,3]. Figure S1. Energies of orbital levels of PFO, A1 and A2. The arrows show possible electronic transitions. Figure S2. Frontiers orbitals HOMO and LUMO calculated for the dimers PFO-A1 (above) and PFO-A2 (below). Figure S3. Pictures of A1 and A2 crystalline films under polarized light. Figure S4. Emission spectra of PVK (orange line) and PFO (blue line) films, excited on the absorption maxima.

Author Contributions: Conceptualization, U.C. and B.P.; data curation and formal analysis, R.D.; funding acquisition, U.C.; investigation, U.C.; methodology, R.D. and B.P.; project administration, U.C.; resources, U.C.; writing—original draft, R.D., S.P., S.C, performed the experiments; analyzed the data S.C., R.D. and R.S.; writing—review & editing, B.P. and U.C. All authors have read and agreed to the published version of the manuscript.

Funding: We gratefully acknowledge the financial aid provided by the Italian Ministry of Education, University and Research (MIUR) under grants PON PANDION 01_00375.

Acknowledgments: We thank Francesco Marrafino for assistance with TDDFT calculations and for comments that significantly improved the manuscript.

Conflicts of Interest: There are no conflicts to declare.

References

1. Ahlström, L.-H.; Sparr Eskilsson, C.; Björklund, E. Determination of banned azo dyes in consumer goods. *TrAC Trend. Anal. Chem.* **2005**, *24*, 49–56. [CrossRef]
2. Pedersen, T.G.; Johansen, P.M.; Pedersen, H.C. Characterization of azobenzene chromophores for reversible optical data storage: Molecular quantum calculations. *J. Opt. A Pure Appl. Opt.* **2000**, *2*, 272–278. [CrossRef]
3. MatoviĆ, L.; TasiĆ, N.; TriŠOviĆ, N.; LaĐAreviĆ, J.; Vitnik, V.; Vitnik, Ž.; Grgur, B.; Mijin, D. On the azo dyes derived from benzoic and cinnamic acids used as photosensitizersin dye-sensitized solar cells. *Turk. J. Chem.* **2019**, *43*, 1183–1203. [CrossRef]
4. Ding, N.; Li, Z.; Tian, X.; Zhang, J.; Guo, K.; Wang, P. Azo-based near-infrared fluorescent theranostic probe for tracking hypoxia-activated cancer chemotherapy in vivo. *Chem. Commun. Camb.* **2019**, *55*, 13172–13175. [CrossRef]
5. Tien, C.L.; Lin, R.J.; Kang, C.C.; Huang, B.Y.; Kuo, C.T.; Huang, S.Y. Electrically Controlled Diffraction Grating in Azo Dye-Doped Liquid Crystals. *Polymers* **2019**, *11*, 1051. [CrossRef]
6. Piotto, S.; Concilio, S.; Sessa, L.; Diana, R.; Torrens, G.; Juan, C.; Caruso, U.; Iannelli, P. Synthesis and Antimicrobial Studies of New Antibacterial Azo-Compounds Active against Staphylococcus aureus and Listeria monocytogenes. *Molecules* **2017**, *22*, 1372. [CrossRef]
7. Derkowska-Zielinska, B.; Matczyszyn, K.; Dudek, M.; Samoc, M.; Skowronski, L.; Chomicki, D.; Smokal, V.; Kysil, A.; Biitseva, A.; Krupka, O. Linear and nonlinear optical properties of heterocyclic azo dyes with hetaryldiazenyl substitution. *Mol. Cryst. Liq. Cryst.* **2019**, *670*, 153–159. [CrossRef]
8. Borbone, F.; Caruso, U.; Diana, R.; Panunzi, B.; Roviello, A.; Tingoli, M.; Tuzi, A. Second order nonlinear optical networks with excellent poling stability from a new trifunctional thiophene based chromophore. *Org. Electron. Phys. mater. appl.* **2009**, *10*, 53–60. [CrossRef]
9. Saeed, S.; Channar, P.A.; Saeed, A.; Larik, F.A. Fluorescence modulation of CdTe nanowire by azobenzene photochromic switches. *J. Photochem. Photobiol. A Chem.* **2019**, *369*, 159–165. [CrossRef]
10. Panunzi, B.; Borbone, F.; Capobianco, A.; Concilio, S.; Diana, R.; Peluso, A.; Piotto, S.; Tuzi, A.; Velardo, A.; Caruso, U. Synthesis, spectroscopic properties and DFT calculations of a novel multipolar azo dye and its zinc(II) complex. *Inorg. Chem. Commun.* **2017**, *84*, 103–108. [CrossRef]

11. Wei, Z.; He, L.; Chi, Z.; Ran, X.; Guo, L. Two-photon isomerization triggers two-photon-excited fluorescence of an azobenzene derivative. *Spectrochim. Acta A.* **2019**, *206*, 120–125. [CrossRef]
12. Han, M. Fluorescence Enhancement from Self-Assembled Aggregates Ii: Factors Influencing Florescence Color from Azobenzene Aggregates. *J. Mol. Eng. Mater.* **2013**, *01*, 1340008. [CrossRef]
13. Diana, R.; Panunzi, B.; Shikler, R.; Nabha, S.; Caruso, U. A symmetrical azo-based fluorophore and the derived salen multipurpose framework for emissive layers. *Inorg. Chem. Commun.* **2019**, *104*, 186–189. [CrossRef]
14. Raman, A.; Augustine, G.; Ayyadurai, N.; Easwaramoorthi, S. Gated photochromism in azobenzene-appended rhodamine cassette: Through-bond energy transfer–a universal strategy towards "Lock and Unlock" system. *J. Mater. Chem. C.* **2018**, *6*, 10497–10501. [CrossRef]
15. Borbone, F.; Caruso, U.; Palma, S.D.; Fusco, S.; Nabha, S.; Panunzi, B.; Shikler, R. High solid state photoluminescence quantum yields and effective color tuning in polyvinylpyridine based zinc(II) metallopolymers. *Macromol. Chem. Phys.* **2015**, *216*, 1516–1522. [CrossRef]
16. Satam, M.A.; Raut, R.K.; Sekar, N. Fluorescent azo disperse dyes from 3-(1,3-benzothiazol-2-yl)naphthalen-2-ol and comparison with 2-naphthol analogs. *Dyes Pigment.* **2013**, *96*, 92–103. [CrossRef]
17. Joshi, H.; Kamounah, F.S.; Gooijer, C.; van der Zwan, G.; Antonov, L. Excited state intramolecular proton transfer in some tautomeric azo dyes and schiff bases containing an intramolecular hydrogen bond. *J. Photochem. Photobiol. A Chem.* **2002**, *152*, 183–191. [CrossRef]
18. Casalboni, M.; Caruso, U.; De Maria, A.; Fusco, M.; Panunzi, B.; Quatela, A.; Roviello, A.; Sarcinelli, F.; Sirigu, A. New polyurethanes and polyesters for second-order nonlinear optical applications. *J. Polym. Sci. Part A* **2004**, *42*, 3013–3022. [CrossRef]
19. Yoshino, J.; Kano, N.; Kawashima, T. Fluorescent azobenzenes and aromatic aldimines featuring an N-B interaction. *Dalton t.* **2013**, *42*, 15826–15834. [CrossRef]
20. Tathe, A.B.; Sekar, N. Red Emitting Coumarin-Azo Dyes: Synthesis, Characterization, Linear and Non-linear Optical Properties-Experimental and Computational Approach. *J. fluoresce.* **2016**, *26*, 1279–1293. [CrossRef]
21. Mishra, V.R.; Ghanavatkar, C.W.; Sekar, N. ESIPT clubbed azo dyes as deep red emitting fluorescent molecular rotors: Photophysical properties, pH study, viscosity sensitivity, and DFT studies. *J. Lumin* **2019**, *215*, 116689. [CrossRef]
22. Nancoz, C.; Licari, G.; Beckwith, J.S.; Soederberg, M.; Dereka, B.; Rosspeintner, A.; Yushchenko, O.; Letrun, R.; Richert, S.; Lang, B.; et al. Influence of the hydrogen-bond interactions on the excited-state dynamics of a push-pull azobenzene dye: The case of Methyl Orange. *Phys. Chem. Chem. Phys. PCCP* **2018**, *20*, 7254–7264. [CrossRef]
23. Bohnke, H.; Rottger, K.; Ingle, R.A.; Marroux, H.J.B.; Bohnsack, M.; Schwalb, N.K.; Orr-Ewing, A.J.; Temps, F. Electronic Relaxation Dynamics of UV-Photoexcited 2-Aminopurine-Thymine Base Pairs in Watson-Crick and Hoogsteen Conformations. *J. Phys. Chem. B* **2019**, *123*, 2904–2914. [CrossRef]
24. Warde, U.; Sekar, N. NLOphoric mono-azo dyes with negative solvatochromism and in-built ESIPT unit from ethyl 1,3-dihydroxy-2-naphthoate: Estimation of excited state dipole moment and pH study. *Dyes Pigment.* **2017**, *137*, 384–394. [CrossRef]
25. Chen, W.; Wright, B.D.; Pang, Y. Rational design of a NIR-emitting Pd(II) sensor via oxidative cyclization to form a benzoxazole ring. *Chem. Commun. Camb.* **2012**, *48*, 3824–3826. [CrossRef]
26. Panunzi, B.; Diana, R.; Concilio, S.; Sessa, L.; Shikler, R.; Nabha, S.; Tuzi, A.; Caruso, U.; Piotto, S. Solid-state highly efficient dr mono and poly-dicyano-phenylenevinylene fluorophores. *Molecules* **2018**, *23*, 1505. [CrossRef]
27. Rauf, M.A.; Hisaindee, S.; Saleh, N. Spectroscopic studies of keto–enol tautomeric equilibrium of azo dyes. *RSC Adv.* **2015**, *5*, 18097–18110. [CrossRef]
28. Panunzi, B.; Concilio, S.; Diana, R.; Shikler, R.; Nabha, S.; Piotto, S.; Sessa, L.; Tuzi, A.; Caruso, U. Photophysical Properties of Luminescent Zinc(II)-Pyridinyloxadiazole Complexes and their Glassy Self-Assembly Networks. *Eur. J. Inorg. Chem.* **2018**, *2018*, 2709–2716. [CrossRef]
29. Borbone, F.; Caruso, U.; Concilio, S.; Nabha, S.; Piotto, S.; Shikler, R.; Tuzi, A.; Panunzi, B. From cadmium(II)-aroylhydrazone complexes to metallopolymers with enhanced photoluminescence. A structural and DFT study. *Inorg. Chim. Acta* **2017**, *458*, 129–137. [CrossRef]
30. Kasture, P.P.; Sonawane, Y.A.; Rajule, R.N.; Shankarling, G.S. Synthesis and characterisation of benzothiazole-based solid-state fluorescent azo dyes. *Coloration Technol.* **2010**, *126*, 348–352. [CrossRef]

31. Yu, H.-W.; Kim, B.-S.; Matsumoto, S. Effect of alkoxy side chain length on the solid-state fluorescence behaviour of bisazomethine dyes possessing dipropylamino terminal group. *Dyes Pigment.* **2017**, *136*, 131–139. [CrossRef]
32. Moolya, P.N.; Gadilohar, B.L.; Shankarling, G.S. Synthesis, characterisation, and study of the photophysical properties of highly stable imidazole-based novel solid-state fluorescent azo colourants. *Coloration Technol.* **2015**, *131*, 104–109. [CrossRef]
33. Goldenberg, L.M.; Lisinetskii, V.; Gritsai, Y.; Stumpe, J.; Schrader, S. Single step optical fabrication of a DFB laser device in fluorescent azobenzene-containing materials. *Adv. Mater.* **2012**, *24*, 3339–3343. [CrossRef]
34. Guo, L.; Zhang, R.; Sun, Y.; Tian, M.; Zhang, G.; Feng, R.; Li, X.; Yu, X.; He, X. Styrylpyridine salts-based red emissive two-photon turn-on probe for imaging the plasma membrane in living cells and tissues. *Analyst* **2016**, *141*, 3228–3232. [CrossRef]
35. Haidekker, M.A.; Brady, T.P.; Lichlyter, D.; Theodorakis, E.A. Effects of solvent polarity and solvent viscosity on the fluorescent properties of molecular rotors and related probes. *Bioorganic Chem.* **2005**, *33*, 415–425. [CrossRef]
36. Koenig, M.; Storti, B.; Bizzarri, R.; Guldi, D.M.; Brancato, G.; Bottari, G. A fluorescent molecular rotor showing vapochromism, aggregation-induced emission, and environmental sensing in living cells. *J. Mater. Chem. C* **2016**, *4*, 3018–3027. [CrossRef]
37. Mishra, V.R.; Ghanavatkar, C.W.; Mali, S.N.; Qureshi, S.I.; Chaudhari, H.K.; Sekar, N. Design, synthesis, antimicrobial activity and computational studies of novel azo linked substituted benzimidazole, benzoxazole and benzothiazole derivatives. *Comput. Biol. Chem.* **2019**, *78*, 330–337. [CrossRef]
38. Zampetti, A.; Minotto, A.; Squeo, B.M.; Gregoriou, V.G.; Allard, S.; Scherf, U.; Chochos, C.L.; Cacialli, F. Highly Efficient Solid-State Near-infrared Organic Light-Emitting Diodes incorporating A-D-A Dyes based on alpha, beta-unsubstituted "BODIPY" Moieties. *Sci. Rep.* **2017**, *7*, 1611. [CrossRef]
39. Diana, R.; Panunzi, B.; Concilio, S.; Marrafino, F.; Shikler, R.; Caruso, T.; Caruso, U. The Effect of Bulky Substituents on Two π-Conjugated Mesogenic Fluorophores. Their Organic Polymers and Zinc-Bridged Luminescent Networks. *Polymers* **2019**, *11*, 1379. [CrossRef]
40. Zhou, Z.; Li, W.; Hao, X.; Redshaw, C.; Chen, L.; Sun, W.-H. 6-Benzhydryl-4-methyl-2-(1H-benzoimidazol-2-yl)phenol ligands and their zinc complexes: Syntheses, characterization and photoluminescence behavior. *Inorg. Chim. Acta* **2012**, *392*, 345–353. [CrossRef]
41. Melhuish, W.H. Quantum Efficiencies of Fluorescence of Organic Substances: Effect of Solvent and Concentration of the Fluorescent Solute1. *J. Phys. Chem.* **1961**, *65*, 229–235. [CrossRef]
42. Vincett, P.; Voigt, E.; Rieckhoff, K. Phosphorescence and fluorescence of phthalocyanines. *J. Chem. Phys.* **1971**, *55*, 4131–4140. [CrossRef]
43. Berleb, S.; Brütting, W.; Schwoerer, M.; Wehrmann, R.; Elschner, A. Effect of majority carrier space charges on minority carrier injection in dye doped polymer light-emitting devices. *J. Appl. Phys.* **1998**, *83*, 4403–4409. [CrossRef]
44. Uchida, M.; Adachi, C.; Koyama, T.; Taniguchi, Y. Charge carrier trapping effect by luminescent dopant molecules in single-layer organic light emitting diodes. *J. Appl. Phys.* **1999**, *86*, 1680–1687. [CrossRef]
45. Yang, S.-H.; Wu, C.-C.; Lee, C.-F.; Liu, M.-H. Synthesis and luminescence of red MEH-PPV:P3OT polymer. *Displays* **2008**, *29*, 214–218. [CrossRef]
46. Kim, M.; Whang, D.R.; Gierschner, J.; Park, S.Y. A distyrylbenzene based highly efficient deep red/near-infrared emitting organic solid. *J. Mater. Chem. C* **2015**, *3*, 231–234. [CrossRef]
47. Wen, W.; Shi, Z.-F.; Cao, X.-P.; Xu, N.-S. Triphenylethylene-based fluorophores: Facile preparation and full-color emission in both solution and solid states. *Dyes Pigment.* **2016**, *132*, 282–290. [CrossRef]
48. Panunzi, B.; Diana, R.; Caruso, U. A Highly Efficient White Luminescent Zinc (II) Based Metallopolymer by RGB Approach. *Polymers* **2019**, *11*, 1712. [CrossRef]
49. Diana, R.; Panunzi, B.; Shikler, R.; Nabha, S.; Caruso, U. Highly efficient dicyano-phenylenevinylene fluorophore as polymer dopant or zinc-driven self-assembling building block. *Inorg. Chem. Commun.* **2019**, *104*, 145–149. [CrossRef]
50. Diana, R.; Panunzi, B.; Marrafino, F.; Piotto, S.; Caruso, U. Novel Dicyano-Phenylenevinylene Fluorophores for Low-Doped Layers: A Highly Emissive Material for Red OLEDs. *Polymers* **2019**, *11*, 1751. [CrossRef]

51. Caruso, U.; Panunzi, B.; Diana, R.; Concilio, S.; Sessa, L.; Shikler, R.; Nabha, S.; Tuzi, A.; Piotto, S. AIE/ACQ effects in two DR/NIR emitters: A structural and DFT comparative analysis. *Molecules* **2018**, *23*, 1947. [CrossRef] [PubMed]
52. Liu, J.; Chen, L.; Shao, S.; Xie, Z.; Cheng, Y.; Geng, Y.; Wang, L.; Jing, X.; Wang, F. Highly efficient red electroluminescent polymers with dopant/host system and molecular dispersion feature: Polyfluorene as the host and 2,1,3-benzothiadiazole derivatives as the red dopant. *J. Mater. Chem.* **2008**, *18*, 319–327. [CrossRef]
53. Miteva, T.; Meisel, A.; Knoll, W.; Nothofer, H.G.; Scherf, U.; Müller, D.C.; Meerholz, K.; Yasuda, A.; Neher, D. Improving the Performance of Polyfluorene-Based Organic Light-Emitting Diodes via End-capping. *Adv. Mater.* **2001**, *13*, 565–570. [CrossRef]
54. Chiu, K.Y.; Tu, Y.-J.; Lee, C.-J.; Yang, T.-F.; Lai, L.-L.; Chao, I.; Su, Y.O. Unusual spectral and electrochemical properties of azobenzene-substituted porphyrins. *Electrochim. Acta* **2012**, *62*, 51–62. [CrossRef]
55. Chiu, K.Y.; Tran, T.T.H.; Wu, C.-G.; Chang, S.-H.; Yang, T.-F.; Su, Y.O. Electrochemical studies on triarylamines featuring an azobenzene substituent and new application for small-molecule organic photovoltaics. *J. Electroanal. Chem.* **2017**, *787*, 118–124. [CrossRef]
56. Roviello, A.; Borbone, F.; Carella, A.; Diana, R.; Roviello, G.; Panunzi, B.; Ambrosio, A.; Maddalena, P. High quantum yield photoluminescence of new polyamides containing oligo-PPV amino derivatives and related oligomers. *J. Polym. Sci. Part A Polym. Chem.* **2009**, *47*, 2677–2689. [CrossRef]
57. De Mello, J.C.; Wittmann, H.F.; Friend, R.H. An improved experimental determination of external photoluminescence quantum efficiency. *Adv. Mater.* **1997**, *9*, 230–232. [CrossRef]
58. Bochevarov, A.D.; Harder, E.; Hughes, T.F.; Greenwood, J.R.; Braden, D.A.; Philipp, D.M.; Rinaldo, D.; Halls, M.D.; Zhang, J.; Friesner, R.A. Jaguar: A high-performance quantum chemistry software program with strengths in life and materials sciences. *Int. J. Quantum Chem.* **2013**, *113*, 2110–2142. [CrossRef]
59. Grimme, S.; Antony, J.; Ehrlich, S.; Krieg, H. A consistent and accurate ab initio parametrization of density functional dispersion correction (DFT-D) for the 94 elements H-Pu. *J. Chem. Phys.* **2010**, *132*, 154104. [CrossRef]
60. Fetter, A.L.; Walecka, J.D. *Quantum Theory of Many-Particle Systems*; Dover Publications: Mineola, NY, USA, 2012.
61. Marten, B.; Kim, K.; Cortis, C.; Friesner, R.A.; Murphy, R.B.; Ringnalda, M.N.; Sitkoff, D.; Honig, B. New Model for Calculation of Solvation Free Energies: Correction of Self-Consistent Reaction Field Continuum Dielectric Theory for Short-Range Hydrogen-Bonding Effects. *J. of Phys. Chem.* **1996**, *100*, 11775–11788. [CrossRef]
62. Wang, L.P.; McKiernan, K.A.; Gomes, J.; Beauchamp, K.A.; Head-Gordon, T.; Rice, J.E.; Swope, W.C.; Martinez, T.J.; Pande, V.S. Building a More Predictive Protein Force Field: A Systematic and Reproducible Route to AMBER-FB15. *J. Phys. Chem. B* **2017**, *121*, 4023–4039. [CrossRef] [PubMed]

Sample Availability: Samples of the compounds A1 and A2 are available from the authors.

Article

1,3,4-Thiadiazole-Containing Azo Dyes: Synthesis, Spectroscopic Properties and Molecular Structure

Agnieszka Kudelko [1,*], Monika Olesiejuk [1], Marcin Luczynski [1], Marcin Swiatkowski [2], Tomasz Sieranski [2] and Rafal Kruszynski [2]

1 Department of Chemical Organic Technology and Petrochemistry, The Silesian University of Technology, Krzywoustego 4, PL-44100 Gliwice, Poland; monika.olesiejuk@polsl.pl (M.O.); marcin.luczynski@polsl.pl (M.L.)

2 Department of X-ray Crystallography and Crystal Chemistry, Institute of General and Ecological Chemistry, Lodz University of Technology, Żeromskiego 116, PL-90924 Łódź, Poland; marcin.swiatkowski@p.lodz.pl (M.S.); tomasz.sieranski@p.lodz.pl (T.S.); rafal.kruszynski@p.lodz.pl (R.K.)

* Correspondence: agnieszka.kudelko@polsl.pl; Tel.: +48-32-237-17-29

Academic Editors: Jorge Bañuelos Prieto and Ugo Caruso
Received: 10 June 2020; Accepted: 13 June 2020; Published: 18 June 2020

Abstract: Three series of azo dyes derived from 2-amino-5-aryl-1,3,4-thiadiazoles and aniline, *N,N*-dimethylaniline and phenol were synthesized in high yields by a conventional diazotization-coupling sequence. The chemical structures of the prepared compounds were confirmed by ^{1}H-NMR, ^{13}C-NMR, IR, UV-Vis spectroscopy, mass spectrometry and elemental analysis. In addition, the X-ray single crystal structure of a representative azo dye was presented. For explicit determination of the influence of a substituent on radiation absorption in UV-Vis range, time-dependent density functional theory calculations were performed.

Keywords: heterocycles; 2-arylazo-5-aryl-1,3,4-thiadiazoles; azo-coupling reactions; crystal structure

1. Introduction

Thiadiazoles are five-membered heterocyclic arrangements which are rarely found in nature. Nitrogen, sulfur and carbon atoms can be arranged in several different ways in such a ring, which gives rise to several isomers: 1,2,3-thiadiazole, 1,2,4-thiadiazole, 1,2,5-thiadiazole and 1,3,4-thiadiazole [1,2]. Of these four isomers, derivatives of 1,3,4-thiadiazole seem to be the most popular among scientists. Many compounds containing such a scaffold exhibit a broad spectrum of biologic interactions and have been shown to have antifungal [3,4], antimicrobial [5], anti-inflammatory [6] and anticancer activities [7]. As well as other industrial applications, it is worth mentioning their use as viscosity stabilizers in rubber processing [8], additives for the production of lithium battery electrodes [9], dyes [10] and optoelectronic materials [11]. There are also reports of the applications of 1,3,4-thiadiazole derivatives, in particular 2,5-dimercapto-1,3,4-thiadiazoles, as lubricants [12].

Azo dyes, characterized by the presence of azo moiety (N=N) in their structure, are the most essential group of disperse dyes [13]. They have found a broad application mainly in dyestuff industry, but also in food, cosmetics and pharmaceuticals production. Generally azo dyes exhibit excellent coloring properties and offer vivid colors, starting from yellows, through oranges, reds and ending up in blues. One of the subgroups of this family are conjugated 1,3,4-thiadiazoles containing an azo group (N=N) in their structure. Such heterocyclic azo disperse dyes have received attention from the scientific community due to their brightness, clarity and affinity to various fibers [14,15]. The most common methodology to introduce this type of group into a final azo compound is a two-step transformation through appropriate diazonium salts [16]. The latter are usually produced from the reaction of primary aromatic amines with nitrites in the presence of strong mineral acids at low temperatures. Due to the extreme instability of diazonium salts [17], they are used immediately after their formation in coupling reactions with phenols or amines, substituted with electron-donating groups. Other methods used to prepare azo dyes include the condensation of nitro compounds with amines [18], reduction of nitro compounds [19,20], oxidation of amines [21,22] or condensation of nitroso compounds with amines [23]. In contrast to typical aromatic amines, diazotization and subsequent coupling of 1,3,4-thiadiazole-2-amine derivatives is rarely described in the literature [24–26], although thiadiazoles contain an important thiazole chromophoric core in their structure. This may be due to the presence of an additional electronegative nitrogen atom, which decreases the basicity of the external amino group and reduces its reactivity.

In continuation of our study on the synthesis and versatile applications of conjugated 1,3,4-thiadiazole derivatives [27–29], we attempted to synthesize and characterize the structural features of a new series of 2-phenylazo-1,3,4-thiadiazoles, functionalized with aryl substituents directly on the heterocyclic ring. Such systems combining the 1,3,4-thiadiazole ring with other aromatic compounds through an azo linker may find potential industrial applications, not only as classical synthetic dyes and pigments, laser dyes and monomers for the production of OLEDs, but also in medicine and agriculture due to the presence of a toxophoric N–C–S moiety [30].

2. Results and Discussion

2.1. Synthesis

The starting reagents in the synthesis of 1,3,4-thiadiazole-containing azo dyes were commercially available aromatic carboxylic acids substituted at position 4 with electron-donating or withdrawing groups (**1a–e**, Scheme 1). They were heated with thionyl chloride $SOCl_2$ and the resulting crude acid chlorides were transformed into derivatives of 2-benzoylhydrazinecarbothioamide (**2a–e**) in a two-phase water–toluene solvent system in the presence of base ($NaHCO_3$). The resulting acyclic intermediates underwent cyclization in concentrated sulfuric acid to give the desired 2-amino-5-aryl-1,3,4-thiadiazoles (**3a–e**). The next stages in the few-step reaction sequence were the diazotization of 2-amino-1,3,4-thiadiazole derivatives (**3a–e**) and the subsequent coupling of the diazonium salts formed from aniline, *N,N*-dimethylaniline and phenol (Scheme 1). Diazotization involved the reaction of a primary heteroaromatic amine (**3a–e**) with a nitrosating agent (nitrosyl cation) generated in situ from sodium nitrite and concentrated sulfuric acid at low temperatures. Reactions were performed in a mixture of glacial acetic acid and propionic acid in order to improve the solubility of the starting thiadiazoles. An excess of nitrous acid in the post-reaction mixture was detected with potassium iodide–starch study and eliminated by adding urea. The resulting colored diazonium salts were then used directly without purification in the coupling sequence with aromatic amines: aniline (G=NH_2) and *N,N*-dimethylaniline (G=NMe_2) and with phenol (G=OH). Due to the limited stability of diazonium salts, the transformations were carried out at low temperatures (0–5 °C) and the final products were precipitated from solution by adding base (Na_2CO_3) after reagents were combined. The reactions resulted in the formation of three series of 2-arylazo-5-aryl-1,3,4-thiadiazole dyes in various yields (**4a–e**, **5a–e**, **6a–e**, Scheme 1). The highest yields were obtained from the transformations involving phenol (**6a–e**, 75–91%, Scheme 1), while the reactions using *N,N*-dimethylaniline (**5a–e**, 56–69%, Scheme 1) and aniline (**4a–e**, 51–81%, Scheme 1) produced lower yields. Such situation may be caused by partial deactivation of anilines in acidic media. Generally, 1,3,4-thiadiazole-containing azo dyes are high-melting solids (melting point range 169–293 °C) and are insoluble in both water and hydrocarbons (hexane, toluene). They are highly soluble in DMSO, acetone and methanol. It is worth noting that among the synthesized derivatives, 12 are new compounds not yet described in the literature.

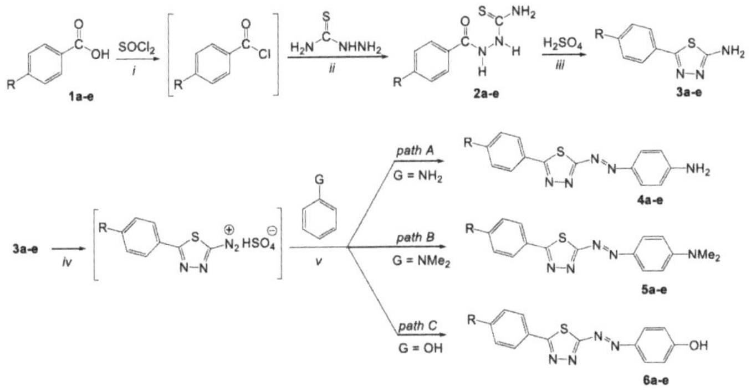

G = NH_2	R	Yield [%]	G = NMe_2	R	Yield [%]	G = OH	R	Yield [%]
4a	H	51	**5a**	H	69	**6a**	H	82
4b	MeO	53	**5b**	MeO	59	**6b**	MeO	91
4c	NO_2	70	**5c**	NO_2	60	**6c**	NO_2	80
4d	Br	68	**5d**	Br	60	**6d**	Br	75
4e	*t*-Bu	81	**5e**	*t*-Bu	56	**6e**	*t*-Bu	80

Scheme 1. Preparation of 2-amino-1,3,4-thiadiazole precursors (**3a–e**) and synthesis of 2-arylazo-5-aryl-1,3,4-thiadiazole dyes (**4a–e**, **5a–e**, **6a–e**). Reaction conditions: (i) toluene, reflux, 5–15 h; (ii) toluene, $NaHCO_3$, H_2O, rt, 24 h; (iii) conc. H_2SO_4, 24 h and then aqueous NH_3; (iv) $NaNO_2$, conc. H_2SO_4, AcOH/EtCOOH, 0–5 °C; (v) H_2O, 0–5 °C and then Na_2CO_3.

2.2. Spectral Characterization

The structures of the final diazotization products of 2-amino-5-aryl-1,3,4-thiadiazole and their subsequent coupling with aniline, *N,N*-dimethylaniline and phenol were confirmed by typical spectroscopic methods (^{1}H-NMR, ^{13}C-NMR, UV-Vis, FT-IR and HRMS). In the ^{1}H-NMR spectra of 2-arylazo-5-aryl-1,3,4-thiadiazoles (**4a–e**, **5a–e**, **6a–e**), the characteristic signals become from protons located in the benzene rings associated with the 1,3,4-thiadiazole core and appeared in the range between 6.74–8.40 ppm. Signals of the characteristic amino group (7.06–7.18 ppm), *N,N*-dimethylamino group (~3.18 ppm), *t*-butyl (~1.33 ppm) and methoxy (~3.86 ppm) were also observed. In the ^{13}C-NMR spectra the characteristic peaks from 1,3,4-thiadiazole carbon atoms C-2 and C-5 were observed at 164–166 ppm and 178–181 ppm, respectively. Signals due to methoxy group (~55.5 ppm), *t*-butyl group (30.8–34.8 ppm) and *N,N*-dimethylamino group (40.0–40.2 ppm) were observed upfield in the spectra. The formation of azo dyes **4a–e**, **5a–e**, **6a–e** was further shown by HRMS and elemental analyses.

The IR spectra of all prepared dyes were recorded from 4000 and 650 cm^{-1}. For the dyes **4a–e**, two bands, respectively at 3400–3310 cm^{-1} and 3201–3190 cm^{-1}, were visible, which were attributed to the presence of a free amino group. Such bands were not observed in the case of derivatives **5a–e**, representing a group of tertiary amines. In series **6a–e**, a broad band appeared from 3500–3100 cm^{-1}, which confirms the presence of a hydroxyl group. All dyes showed a weak band at 1505–1525 cm^{-1} for an azo group (N=N).

The UV-Vis spectra of three series of 2-arylazo-5-aryl-1,3,4-thiadiazole dyes (**4a–e**, **5a–e**, **6a–e**) measured in MeOH exhibit several (four in most cases) absorption maxima (Figure 1, Table 1). The same is observed for calculated spectra (Figure 1, Table 1) which are in good comparison with those measured. All maxima result from multiple different transitions and include both $n \rightarrow \pi^*$ and $\pi \rightarrow \pi^*$ transitions. For series **5** and **6** with compounds possessing a t-butyl group, additionally the $\sigma \rightarrow \pi^*$ transitions are observed. Due to complexity of electronic transition creating each absorption maximum, there is no one evident dependence between the substituents and absorption maximum position/absorption coefficient (Figure S1). Nevertheless, some dependences exist within groups with one kind of R or G substituent. The change of the G substituent causes the red shift of the most intense absorption maximum (appearing in the visible regions of 490–507 nm for **4a–e**, 515–530 nm for **5a–e** and 405–415 nm for **6a–e**) in order of G substituents: OH, NH_2, NMe_2. This sequence is in agreement with electron donating properties of the studied substituents. These red shifts are well reproduced in calculated spectra (Table 1). For the compounds with amino substituents (G=NH_2, NMe_2) this maximum is the most red-shifting for a compound possessing a nitro group (Table 1). For the compounds with hydroxy group (G=OH) the red shift forced by the presence of a nitro group (**6c**) is between some other red shifted compounds, and it is smaller than that shift observed for the compound **6b**. This effect is also reflected in the calculated spectra (Table 1). The simultaneous presence of a hydroxy group (G=OH) and methoxy one (R=MeO) in **6b** is an explanation of this phenomenon. For each series, the methoxy moiety induces red-shift effect (toward most of the compounds), but for amino substituents G=NH_2, NMe_2 it was smaller than for their hydroxy counterpart G=OH (influence of this group could only be noticed for calculated spectra). Increase of the red shift for stronger electron donating EDG substituents (G=NMe_2 and NH_2) is larger than caused by methoxy moiety and overrule strong electron withdrawing properties of nitro substituent. In case of the weaker EDG substituent (G=OH) with electron donating properties close to the methoxy substituent, the electron withdrawing properties of nitro substituent dominate at some point, and cause above described difference in red shifts. For each compound, the most intense (global) maximum is attributed to HOMO–LUMO (H→L) transitions (Table 1, Supplementary Materials, Figures S2–S4). For compounds possessing the nitro substituent these experimentally observed absorption maxima result also from other than H→L transitions, possessing relatively large oscillator strengths (Table 1, Figures S2–S4). In all cases, these transitions do not engage the antibonding orbitals of the nitro group (Table 1). For the second most intense absorption maximum (experimentally observed at lower wavelengths), the changes of its shift between the corresponding compounds (e.g., **4a** vs. **5a** vs. **6a**) are not as high as those observed for the most intense maxima. Those maxima appear in the ultraviolet regions of 237–267 nm for **4a–e**, 243–256 nm for **5a–e** and 243–260 nm for **6a–e** and they results from multiple $\pi\rightarrow\pi^*$ and $n\rightarrow\pi^*$ transitions involving antibonding orbitals of all substituents (i.e., NH_2, NMe_2, OH, MeO, NO_2 and Br, Table 1). The experimental and calculated spectra show some minor differences. For compounds with amino groups (G=NH_2, NMe_2), the calculated spectra do not contain absorption maxima observed experimentally in the range of 284–299 nm. This may result from the interaction of compound molecules with the solvent (i.e., formation of specific hydrogen bonds) defectively reproduced in continuous model of solvation.

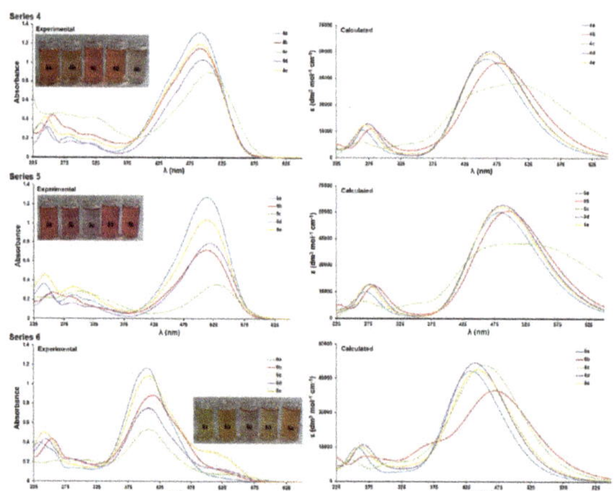

Figure 1. UV-Vis absorption spectra of the studied azo dyes in MeOH at a concentration of 4.0 × 10^{-5}-mol/L at room temperature: (**Series 4**) Absorption spectra of azo dyes **4a–e** containing 4-aminophenylazo group; (**Series 5**) absorption spectra of azo dyes **5a–e** containing 4-(*N*,*N*-dimethylamino)phenylazo group; (**Series 6**) absorption spectra of azo dyes **6a–e** containing 4-hydroxyphenylazo group. Respective calculated spectra are shown as well.

Table 1. Most important electronic transitions. H letter indicates Highest Occupied Molecular Orbital HOMO, L indicates Lowest Unoccupied Molecular Orbital LUMO and +/– (number) represent subsequent orbitals above HOMO and LUMO, respectively. The logarithm of molar absorption coefficients log ε and oscillator strengths are given in parenthesis under the wavelengths, respectively for experimental and calculated maxima. The letters **a**–**e** stand for the particular compound in a given series (**4**, **5** or **6**).

λ_{max} (nm)										The Most Important Orbitals Involved in Electronic Transitions	Character of Transition
Experimental					Calculated						
					4 (G=NH_2)						
a	b	c	d	e	a	b	c	d	e		
					261.44 (0.0909)	276.70 (0.1045)	237.50 (0.0536)	256.39 (0.0432)	270.17 (0.2800)	(**a**, e)H-2→L+1 (**b**)H→L+2 (c)H-2→L+2 (**d**)H-8→L	(**a**, **e**)n(NH_2)/π→π* (**b**)n(NH_2)/n(MeO)/π→π*(C6-rings) (c)n(NH_2)/n(NO_2)/π→π* (**d**)n(Br)/π(S-ring)→π*
237 (3.95)	256 (4.05)	267 (4.07)	249 (3.90)	248 (4.02)	262.74 (0.0776)	280.67 (0.2599)	253.28 (0.0806)	274.45 (0.3976)	273.29 (0.1360)	(a)H-6→L (**b**)H-1→L+1 (c)H-6→L+1 (**d**)H-2→L+1 (**e**) H→L+2	(a)n(NH_2)/π→π* (**b**)n(NH_2)/n(MeO)/π→π* (c)π→π* (**d**)n(Br)/n(NH_2)/π→π* (**e**)n(NH_2)/π→π*(C6-rings)
					266.87 (0.1051)		268.23 (0.0893)			(a)H→L+3 (c)H→L+4	(**a**, c)n(NH_2)/π→π*(C6-rings)
					272.57 (0.1083)					(a)H→L+2	(a)n(NH_2)/π→π*(C6-rings)
285 (3.72)	286 (3.88)		286 (3.63)	284 (3.79)							
316 (3.56)	319 (3.78)	321 (4.03)	318 (3.56)	319 (3.68)	325.66 (0.0203)	322.13 (0.0500)	312.40 (0.0522)	333.77 (0.0204)	324.73 (0.0294)	(**a**, **b**, **d**, **e**)H→L+1 (c)H-2→L+1	(**a**, **c**, **e**)n(NH_2)/π→π* (**b**)n(NH_2)/n(MeO)/π→π* (**d**)n(Br)/n(NH_2)/π→π*
							344.08 (0.1372)			(c)H-2→L	(c)n(NH_2)/π→π*
491 (4.52)	491 (4.46)	507 (4.35)	496 (4.41)	492 (4.47)		383.91 (0.1281)	428.57 (0.7203)			(b)H-1→L (c)H→L+1	(**b**)n(NH_2)/n(MeO)/π→π* (c)n(NH_2)/π→π*
					459.54 (1.4831)	478.70 (1.3233)	529.34 (0.8884)	464.44 (1.4918)	464.82 (1.4522)	H→L	(**a**, **c**, **e**)n(NH_2)/π→π* (**b**) n(NH_2)/n(MeO)/π→π* (**d**)n(Br)/n(NH_2)/π→π*

Table 1. *Cont.*

λ_{max} (nm)										The Most Important Orbitals Involved in Electronic Transitions	Character of Transition
Experimental					Calculated						
					5 (G=NMe_2)						
a	b	c	d	e	a	b	c	d	e		
					264.95 (0.1359)	278.83 (0.0592)	241.40 (0.0442)	256.30 (0.0430)	270.21 (0.0655)	(**a**)H-2→L+1 (**b**, **e**) H-1→L+1 (**c**)H-2→L+2 (**d**)H-8→L	(**a**)$\pi \rightarrow \pi^*$ (**b**) n(MeO)/n(NMe_2)/$\pi \rightarrow \pi^*$ (**c**)n(NO_2)/$\pi \rightarrow \pi^*$ (**d**)n(Br)/$\pi \rightarrow \pi^*$ (**e**)σ(*t*-Bu)/n(Me_2)/$\pi \rightarrow \pi^*$
243 (3.96)	256 (3.83)	249 (3.73)	251 (3.80)	245 (4.06)	275.87 (0.1374)	283.75 (0.1408)	261.09 (0.0967)	274.64 (0.1508)	270.38 (0.0853)	(**a**)H-6→L (**b**)H→L+2 (**c**)H-6→L+1 (**d**)H-2→L+1 (**e**)H-1→L+1	(**a**, **c**)$\pi \rightarrow \pi^*$ (**b**)n(MeO)/n(NMe_2)/$\pi \rightarrow \pi^*$(C6-ring near MeO) (**d**)n(Br)/n(NMe_2)/$\pi \rightarrow \pi^*$ (**e**)σ(*t*-Bu)/n(Me_2)/$\pi \rightarrow \pi^*$
					281.40 (0.0925)	287.77 (0.2354)	277.34 (0.1087)	282.30 (0.2530)	279.49 (0.2116)	(**a**)H→L+3 (**b**, **d**)H-5→L (**c**)H→L+4 (**e**)H-6→L	(**a**)n(NMe_2)/$\pi \rightarrow \pi^*$ (**b**)n(MeO)/$\pi \rightarrow \pi^*$ (**c**)n(NMe_2)/$\pi \rightarrow \pi^*$(C6-ring near NMe_2) (**d**)n(Br)/$\pi \rightarrow \pi^*$ (**e**)σ(*t*-Bu)/$\pi \rightarrow \pi^*$
									281.87 (0.0869)	(**e**)H→L+2	(**e**)n(NMe_2)/$\pi \rightarrow \pi^*$(C6-rings)
293 (3.79)	293 (3.73)	299 (3.85)	290 (3.61)	295 (3.92)							
320 (3.68)	334 (3.46)		325 (3.49)	323 (3.80)	337.91 (0.0197)	333.19 (0.0329)	353.80 (0.1689)	346.81 (0.0206)	336.73 (0.0263)	(**a**, **b**, **d**, **e**)H→L+1 (**c**)H-2→L	(**a**, **d**, **e**)n(NMe_2)/$\pi \rightarrow \pi^*$ (**b**)n(MeO)/n(NMe_2)/$\pi \rightarrow \pi^*$ (**c**)n(NO_2)/$\pi \rightarrow \pi^*$
						393.91 (0.0395)	453.15 (0.8353)			(**b**)H-1→L (**c**)H→L+1	(**b**)n(MeO)/n(NMe_2)/$\pi \rightarrow \pi^*$ (**c**)n(NMe_2)/$\pi \rightarrow \pi^*$
515 (4.50)	515 (4.25)	530 (3.95)	520 (4.30)	515 (4.41)	483.86 (1.4831)	497.00 (1.1993)	568.89 (0.8657)	488.69 (1.5864)	487.35 (1.5439)	H→L	(**a**, **c**, **d**, **e**)n(NMe_2)/$\pi \rightarrow \pi^*$ (**b**)n(MeO)/n(NMe_2)/$\pi \rightarrow \pi^*$

Table 1. *Cont.*

λmax (nm)										The Most Important Orbitals Involved in Electronic Transitions	Character of Transition
Experimental					Calculated						
					6 (G=OH)						
a	b	c	d	e	a	b	c	d	e		
						220.52 (0.0983)				(**b**)H-4→L+1	(**b**)π(C6-ring near OH)→π*
236 (4.00)		231 (3.68)			234.73 (0.0439)		232.01 (0.0382)			(**a**)H→L+3 (**c**)H-2→L+2	(**a**)n(OH)/π→π*(C6-ring near S-ring) (**c**)n(NO_2)/(OH)/π→π*
					251.23 (0.0758)	234.51 (0.0525)	249.20 (0.0821)	238.98 (0.0306)	234.61 (0.0464)	(**a, e**)H→L+2 (**b**)H-3→L+1 (**c**)H-8→L (**d**)H-4→L+1	(**a**)n(OH)/π→π*(C6-ring near OH) (**b**)π(C6-ring near MeO)→π* (**c**)n(NO_2)/(OH)/π→π* (**d**)π(C6-rings)→π* (**e**)σ(*t*-Bu)/n(OH)/π→π*(C6-ring near *t*-Bu)
256 (3.90)	254 (4.04)	260 (3.72)	245 (4.04)	243 (4.10)	254.98 (0.1078)	271.99 (0.1002)	264.75 (0.0369)	251.23 (0.0556)	265.07 (0.2413)	(**a**)H-2→L+1 (**b, c, e**)H-6→L (**d**)H→L+3	(**a, c**)n(OH)/π→π* (**b**)n(MeO)/n(OH)/π→π* (**d**)n(Br)/n(OH)/π→π*(C6-ring near OH) (**e**)σ(*t*-Bu)/n(OH)/π→π*
					262.06 (0.1766)			266.81 (0.1192)		(**a**)H-6→L+1 (**d**)H-2→L+1	(**a**)n(OH)/π→π* (**d**)n(Br)/n(OH)/π→π*
								270.78 (0.1692)		(**d**)H-6→L	(**d**)n(Br)/n(OH)/π→π*
298 (3.54)	291 (3.74)		290 (3.76)	290 (3.64)	297.41 (0.0427)	302.67 (0.1542)		304.17 (0.0596)	298.30 (0.0985)	(**a, b, c, e**)H→L+1	(**a**)n(OH)/π→π* (**b**)n(MeO)/n(OH)/π→π* (**d**)n(Br)/n(OH)/π→π* (**e**)σ(*t*-Bu)/n(OH)/π→π*
					325.44 (0.0438)					(**a**)H-4→L	(**a**)π(C6-ring near OH)→π*
		310 (3.74)					305.19 (0.1361)			(**c**)H-2→L+1	(**c**)n(NO_2)/(OH)/π→π*

Table 1. *Cont.*

λ_{max} (nm)										The Most Important Orbitals Involved in Electronic Transitions	Character of Transition
Experimental					Calculated						
							341.08 (0.0630)			(c)H-3→L	(c)π(C6-ring near OH)→π*
405 (4.46)	415 (4.35)	410 (4.13)	409 (4.28)	409 (3.96)	345.44 (0.0461)	370.43 (0.3618)	385.35 (0.2630)	355.56 (0.0770)	355.42 (0.1288)	(**a**, **d**, **e**)H-2→L (**b**)H-1→L (**c**)H-1→L+1	(**a**, **c**)n(OH)/π→π* (**b**)n(MeO)/n(OH)/π→π* (**d**)n(Br)/n(OH)/π→π* (**e**)σ(*t*-Bu)/n(OH)/π→π*
					432.77 (1.1904)	471.46 (0.9677)	458.02 (1.1983)	439.27 (1.2728)	444.00 (1.1983)	H→L	(**a**, **c**)n(OH)/π→π* (**b**)n(MeO)/n(OH)/π→π* (**d**)n(Br)/n(OH)/π→π* (**e**)σ(*t*-Bu)/n(OH)/π→π*

Used abbreviations: n—non-bonding orbital; S-ring—thiadiazole ring; C6-ring—benzene ring; *—antibonding orbital.

2.3. Molecular Structure

The structure of one of the obtained azo dyes was confirmed by X-ray crystal structure analysis (CCDC1946289). All atoms of the studied 2-[4-(*N*,*N*-dimethylamino)phenylazo]-5-(4-methoxyphenyl) -1,3,4-thiadiazole (**5b**) occupied general positions (Figure 2).

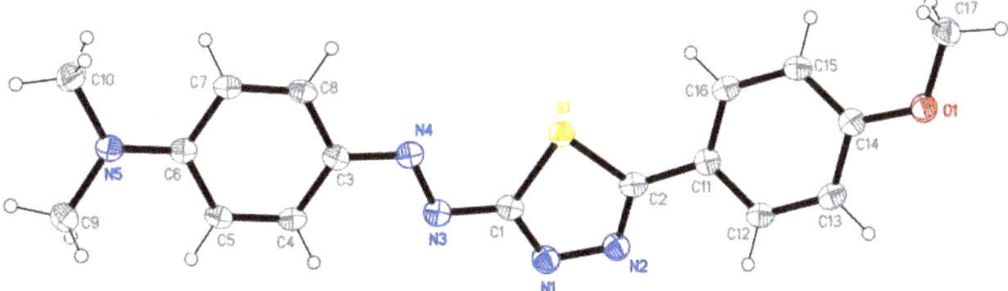

Figure 2. Molecular structure of compound **5b** plotted with 50% probability of displacement ellipsoids. Hydrogen atoms are drawn as spheres with arbitrary radii.

The five- and six- membered rings were practically planar (the largest deviating atom C2, C7, C14 stuck out 0.0060(8), 0.0079(9), 0.0073(9) Å from the weighted least squares plane containing the respective atom), but the whole molecule is considerably bent. The benzene rings containing C3 and C11 atoms were inclined at 10.11(7) and 24.90(6)° to the central 1,3,4-thiadiazole ring, respectively and mutually at 26.85(6)°. The *N*,*N*-dimethylamino group was inclined at 11.53(14)° to the neighboring benzene ring, which is a larger value than is typically observed for similar systems (4.6%–5.8°) [30]. However, it still falls within the range of 2.3%–18.8° (the angle values of median population). The C–S bonds (Table 2) were shorter than typical single C–S bonds (i.e., a mean value of 1.819 Å for a bond between an sp^3 carbon atom and a divalent sulfur) and similar to the values observed for slightly delocalized systems containing sp^2 carbon atoms (i.e., a mean value of 1.741 Å) [31]. A similar effect exists for an N–N bond (a formal bond length of 1.454 Å) and opposite effects were observed for slightly longer C=N bonds (Table 2) and double C=N bonds (i.e., a mean value of 1.279 Å [32]). Combined, these effects demonstrate small, but observable delocalization of electron density within the 1,3,4-thiadiazole ring. The azo N=N bond (Table 2) shows a transitional character between a well-localized and a completely delocalized system (mean values of 1.245 and 1.304 Å) [33]. This, along with the shortening of the neighboring C–N bonds, also demonstrates the considerable degree of delocalization of electron density within the C–N=N–C moiety. Due to the absence of classical hydrogen bond donors, the studied compounds are connected only by weak C–H•••A non-classic hydrogen bonds (where A denotes the following acceptors: N, O, S and π electrons) [34] and π•••π stacking interactions (Table 3, Table 4) [35].

Table 2. Selected structural data of studied compound **5b**.

i—j	d_{ij} [Å]	i—j—k	α_{ijk} [°]	i—j—k	α_{ijk} [°]
S1—C1	1.7462(14)	C1—S1—C2	86.22(7)	C2—C11—C16	122.25(13)
S1—C2	1.7366(14)	S1—C2—N2	114.29(11)	C5—C6—N5	121.16(12)
C1—N1	1.3092(18)	C2—N2—N1	112.80(12)	C7—C6—N5	121.27(12)
C2—N2	1.3124(18)	N2—N1—C1	112.25(12)	C6—N5—C9	120.76(12)
N1—N2	1.3757(17)	N1—C1—S1	114.42(11)	C6—N5—C10	120.34(12)
C1—N3	1.3821(18)	N1—C1—N3	120.62(13)	C9—N5—C10	118.16(12)
N3—N4	1.2907(16)	S1—C1—N3	124.91(10)	C13—C14—O1	115.86(12)
N4—C3	1.3851(18)	C1—N3—N4	111.67(11)	C15—C14—O1	124.38(13)
C2—C11	1.4651(19)	N3—N4—C3	115.29(11)	C14—O1—C17	117.14(11)
N5—C6	1.3520(18)	N4—C3—C4	125.47(12)		
N5—C9	1.4635(18)	N4—C3—C8	116.30(12)		
N5—C10	1.4606(18)	S1—C2—C11	122.98(10)		
O1—C14	1.3610(17)	N2—C2—C11	122.55(13)		
O1—C17	1.4346(17)	C2—C11—C12	118.81(13)		

Table 3. Non-classic hydrogen bonds and the first level graph motifs in the studied compound. Cg(C3) indicates centroid of ring containing C3 atom.

D-H•••A	d(D-H) [Å]	d(H•••A) [Å]	d(D•••A) [Å]	<(DHA) [°]	$G_d^a(n)$
C9—H9A•••N1 [i]	0.98	2.58	3.5351(1)	165.4	$R_2^2(22)$
C10—H10B•••O1 [ii]	0.98	2.54	3.5031(1)	166.3	C(17)
C16—H16•••S1	0.95	2.86	3.2088(1)	102.7	S(5)
C7—H7•••Cg(C3) [iii]	0.95	2.97	3.8190(1)	148.8	C(2)

Symmetry transformations used to generate equivalent atoms: (i) −x + 1, −y + 1, −z − 2; (ii) x + 1, y, z; (iii) x, −y + 1.5, z − 0.5.

The unitary graph set is composed of $R_2^2(22)C(17)S(5)C(2)$ motifs of the lowest degree. The ring motif assembles the molecules into supramolecular dimers (Figure 3), and the C(17) chain motif extends these dimers to a supramolecular ribbon extending along the crystallographic axis (Figure 4). This ribbon is constructed from alternating $R_2^2(22)R_4^4(26)$ ring motifs of a binary graph set of the lowest degree. The neighboring ribbons are interconnected by C–H•••π and π•••π interactions (Table 3, Table 4) and form a three-dimensional network assembled by non-covalent interactions.

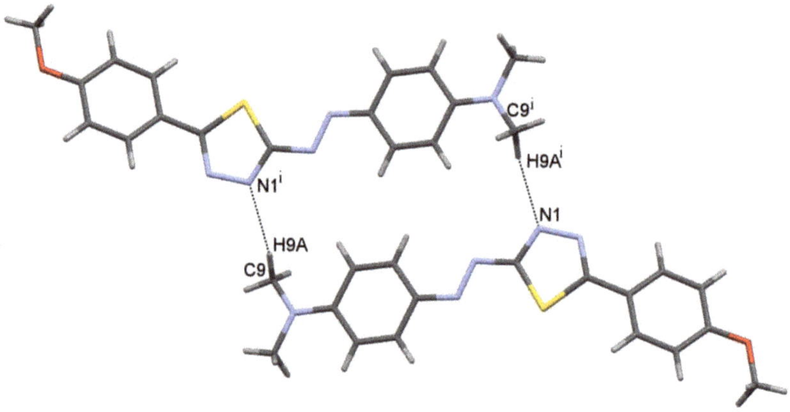

Figure 3. Supramolecular dimers forming a $R_2^2(22)$ motif of a unitary graph set. Symmetry codes as in Table 3. Non-classic hydrogen bonds are indicated by dotted lines.

Table 4. Interactions in the studied compound. Each ring is indicated by one atom, which belongs solely to this ring. α is a dihedral angle between planes I and J, β is an angle between Cg(I)-Cg(J) vector and normal to plane I and d_p is the perpendicular distance of Cg(I) on ring J plane.

R(I)•••R(J)	d(Cg•••Cg) [Å]	α [°]	β [°]	d_p [Å]
C3•••C3 [iv]	3.6217(1)	0	15.5	3.4908
C11•••C11 [v]	3.5355(1)	0	22.3	3.2703

Symmetry transformations used to generate equivalent atoms: (iv) −x+1, −y+1, −z+1;(v) −x, −y+1, −z+1.

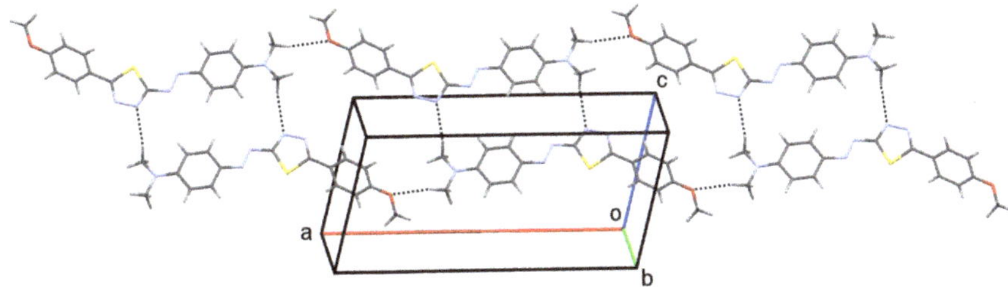

Figure 4. Part of molecular packing showing a supramolecular ribbon extending along the crystallographic axis. Hydrogen bonds indicated by dotted lines.

3. Experimental

3.1. General Information

All reagents were purchased from commercial sources and used without further purification. Melting points were measured on a Stuart SMP3 melting point apparatus. The ^{1}H-NMR and ^{13}C-NMR spectra were recorded on an Agilent 400-NMR spectrometer in DMSO-d_6 solution using TMS as the internal standard. FT-IR spectra were recorded between 4000 and 650 cm^{-1} using a FT-IR Nicolet 6700 apparatus with a Smart iTR accessory. UV-Vis spectra were recorded at room temperature in MeOH solutions (c = 4.0 × 10^{-5} mol/L) on a Jasco V-660 spectrophotometer. High-resolution mass spectra were recorded on a Waters ACQUITY UPLC/Xevo G2QT instrument. Thin-layer chromatography was performed on silica gel 60 F_{254} (Merck) TLC plates using $CHCl_3$/EtOAc (5:1 *v*/*v*) as the mobile phase. Elemental analyses were performed with a VarioEL analyzer. The excited states of the obtained compounds were calculated using TD–DFT method. All calculations (including geometry optimization) were performed utilizing Gaussian09 rev. D.01 [36] with B3LYP functional and employing 6.31 g++(2d, 2p) basis set. The geometry of all compounds was optimized and during this calculation D3 version of Grimme's dispersion with Becke–Johnson damping [37] was used. In both, the optimization process and TD–DFT calculations, the polarizable continuum model [38] was used to include overall solvation effects (with methanol selected as a solvent). The assignment of the calculated excited states to the observed experimental maxima was based on the comparison of excitation energies and the oscillator strengths/intensities of the corresponding maxima. The analysis of the character of respective orbital excitations was based on orbital contour plots.

3.2. Synthesis and Characterization

3.2.1. General Procedure for the Synthesis of 2-Benzoylhydrazinecarbothioamide Derivatives (**2a–e**)

A carboxylic acid (**1a–e**, 0.10 mol) and thionyl chloride $SOCl_2$ (22 mL, 0.30 mol) were refluxed in dry toluene (10 mL) until the acid was fully consumed (TLC; 5–15 h). After cooling, the mixture was concentrated on a rotary evaporator, washed with additional dry toluene (10 mL) and concentrated again. The crude acid chloride was then dissolved in 50 mL of dry toluene and added dropwise to a mixture of thiosemicarbazide (9.11 g, 0.10 mol), $NaHCO_3$ (8.40 g, 0.10 mol) and H_2O (150 mL). The whole solution was agitated at room temperature overnight. The precipitated solid was filtered off, dried in air and recrystallized from a mixture of EtOH–H_2O to obtain pure 2-benzoylhydrazinecarbothioamide derivatives (**2a–e**).

2-Benzoylhydrazinecarbothioamide (**2a**). The product was obtained as a white solid (8.60 g, 44%); mp 197–198 °C (196–198 °C [39]).

2-(4-Methoxybenzoylhydrazinecarbothioamide (**2b**). The product was obtained as a white solid (15.32 g, 68%); mp 225–227 °C (226 °C [40]).

2-(4-Nitrobenzoylhydrazinecarbothioamide (**2c**). The product was obtained as a yellow solid (18.02 g, 75%); mp 212–214 °C (214 °C [41]).

2-(4-Bromobenzoylhydrazinecarbothioamide (**2d**). The product was obtained as a white solid (16.72 g, 75%); mp 216–218 °C. ^{1}H-NMR (400 MHz, DMSO-d_6): δ 7.68–7.72 (m, 3H, Ar: H-3, H-5, NH), 7.82–7.88 (m, 3H, Ar: H-2, H-6, NH), 9.34 (s, 1H, NH), 10.45 (s, 1H, NH); ^{13}C-NMR (100 MHz, DMSO-d_6): δ 125.5, 129.5, 129.9, 131.1, 131.7, 164.9 (C=O), 182.0 (C=S); Anal. Calcd for $C_8H_8BrN_3OS$: C, 35.05; H, 2.94; N, 15.33. Found: C, 34.95; H, 2.89; N, 15.40.

2-(4-t-Butylbenzoylhydrazinecarbothioamide (**2e**). The product was obtained as a beige solid (12.06 g, 48%); mp 210–211 °C. ^{1}H-NMR (400 MHz, DMSO-d_6): δ 1.33 (s, 9H, C(CH_3)$_3$), 7.48 (d, 2H, *J* = 8.4 Hz, Ar: H-3, H-5), 7.52–7.60 (m, 2H, NH_2), 7.84 (d, 2H, *J* = 8.4 Hz, Ar: H-2, H-6), 9.31 (s, 1H, NH), 10.39 (s, 1H, NH); ^{13}C-NMR (100 MHz, DMSO-d_6): δ 30.8, 34.8, 124.9, 126.1, 128.3, 154.6, 165.6 (C=O), 182.0 (C=S); Anal. Calcd for $C_{12}H_{17}N_3OS$: C, 57.34; H, 6.82; N, 16.72. Found: C, 57.21; H, 6.88; N, 16.63.

3.2.2. General Procedure for the Synthesis of 2-Amino-1,3,4-thiadiazole Derivatives (**3a–e**)

Derivatives of 2-benzoylhydrazinecarbothioamide **2a–e** (0.05 mol) were dissolved in 50 mL of concentrated H_2SO_4 and agitated at room temperature overnight. Then, the mixture was alkalized with 25% ammonia and the precipitated product was filtered off, dried in air and recrystallized from a mixture of EtOH–H_2O to yield the corresponding 2-amino-1,3,4-thiadiazoles **3a–e**.

2-Amino-5-phenyl-1,3,4-thiadiazole (**3a**). The product was obtained as a white solid (3.89 g, 45%); mp 232–233 °C (230 °C [42]).

2-Amino-5-(4-methoxyphenyl)-1,3,4-thiadiazole (**3b**). The product was obtained as a white solid (3.42 g, 33%); mp 184–185 °C (185–187 °C [43]).

2-Amino-5-(4-nitrophenyl)-1,3,4-thiadiazole (**3c**). The product was obtained as a yellow solid (8.33 g, 75%); mp 254–256 °C (250–252 °C [44]).

2-Amino-5-(4-bromophenyl)-1,3,4-thiadiazole (**3d**). The product was obtained as a white solid (8.96 g, 70%); mp 226 °C (222–224 °C [41]).

2-Amino-5-(4-t-butylphenyl)-1,3,4-thiadiazole (**3e**). The product was obtained as a white solid (4.78 g, 41%); mp 251–253 °C. ^{1}H-NMR (400 MHz, DMSO-d_6): δ 1.29 (s, 9H, C(CH_3)$_3$), 7.35 (s, 2H, NH_2), 7.48 (d, 2H, *J* = 8.4 Hz, Ar: H-3, H-5), 7.67 (d, 2H, *J* = 8.4 Hz, Ar: H-2, H-6); ^{13}C-NMR (100 MHz, DMSO-d_6): δ 30.8, 34.5, 125.8, 126.0, 128.3, 152.2, 156.3, 168.1; Anal. Calcd for $C_{12}H_{15}N_3S$: C, 61.77; H, 6.48; N, 18.01. Found: C, 61.70; H, 6.40; N, 18.12.

3.2.3. General Procedure for the Synthesis of 2-(4-Aminophenylazo)-5-phenyl-1,3,4-thiadiazole Derivatives (**4a–e**)

Sodium nitrite (1.31 g, 0.019 mol) was introduced portion wise into concentrated sulfuric acid (20 mL). The solution was heated to 50 °C in a water bath until complete dissolution and then rapidly cooled in an ice/salt bath to 0 °C. In the meantime, the solution of the appropriate 2-amino-5-aryl-1,3,4-thiadiazole **3a–e** (0.015 mol) in glacial acetic acid (30 mL) and propionic acid (15 mL) was prepared and added dropwise to an agitated solution of sodium nitrite in concentrated sulfuric acid at 0–5 °C. Then, the mixture was stirred for 24 h, and excess nitrous acid was decomposed by the addition of urea. The resulting diazonium salt solution was slowly introduced into the mixture of aniline (1.37 mL, 1.39 g, 0.015 mol) in 15 mL of water at 0–5 °C. The colored mixture was stirred at room temperature for the next 24 h and finally neutralized with saturated sodium carbonate solution (75 mL). The solid was filtered off, washed twice with hot water (2 × 25 mL) and dried in air. The crude product (**4a–e**) was purified by column chromatography on silica gel ($CHCl_3$/EtOAc, 5:1 *v/v*).

2-(4-Aminophenylazo)-5-phenyl-1,3,4-thiadiazole (**4a**). The product was obtained as a brown reddish solid (2.11 g, 51%); mp 236–238 °C; R_f ($CHCl_3$/EtOAc, 5:1 *v/v*) 0.18. ^{1}H-NMR (400 MHz, DMSO-d_6): δ 6.75 (d, 2H, *J* = 8.8 Hz, 2-Ar: H-3, H-5), 7.13 (s, 2H, NH_2), 7.57–7.59 (m, 3H, 5-Ar: H-3′, H-4′, H-5′), 7.77 (d, 2H, *J* = 8.8 Hz, 2-Ar: H-2, H-6), 8.04 (d, 2H, *J* = 8.4 Hz, 5-Ar: H-2′, H-6′); ^{13}C-NMR (100 MHz, DMSO-d_6): δ 114.1, 127.5, 127.8, 129.4, 130.1, 131.5, 142.3, 156.6, 165.7, 180.2. IR (ATR) ν: 3311, 1636,

1507 cm^{-1}; HRMS calcd for ($C_{14}H_{11}N_5S + H^+$): 282.0813; found: 282.0818; Anal. Calcd for $C_{14}H_{11}N_5S$: C, 59.77; H, 3.94; N, 24.89. Found: C, 59.83; H, 3.97; N, 24.93.

2-(4-Aminophenylazo)-5-(4-methoxyphenyl)-1,3,4-thiadiazole (**4b**). The product was obtained as a brown reddish solid (2.47 g, 53%); mp 269–270 °C; R_f ($CHCl_3$/EtOAc, 5:1 *v/v*) 0.16. ^{1}H-NMR (400 MHz, DMSO-d_6): δ 3.86 (s, 3H, OCH_3), 6.74 (d, 2H, *J* = 8.8 Hz, 2-Ar: H-3, H-5), 7.06 (s, 2H, NH_2), 7.11 (d, 2H, *J* = 8.8 Hz, 5-Ar: H-3′, H-5′), 7.75 (d, 2H, *J* = 8.8 Hz, 2-Ar: H-2, H-6), 7.98 (d, 2H, *J* = 8.8 Hz, 5-Ar: H-2′, H-6′); ^{13}C-NMR (100 MHz, DMSO-d_6): δ 55.5, 114.1, 114.8, 122.6, 127.6, 129.2, 142.3, 156.3, 161.8, 165.6, 179.5. IR (ATR) ν: 3422, 3333, 1651, 1508 cm^{-1}; HRMS calcd for ($C_{15}H_{13}N_5SO + H^+$): 312.0919; found: 312.0912; Anal. Calcd for $C_{15}H_{13}N_5SO$: C, 57.86; H, 4.21; N, 22.49. Found: C, 57.73; H, 4.24; N, 22.43.

2-(4-Aminophenylazo)-5-(4-nitrophenyl)-1,3,4-thiadiazole (**4c**). The product was obtained as a dark brown solid (3.42 g, 70%); mp 176–178 °C (178–179 °C [24]); R_f ($CHCl_3$/EtOAc, 5:1 *v/v*) 0.23. IR (ATR) ν: 3403, 1637, 1513 cm^{-1}; Anal. Calcd for $C_{14}H_{10}N_6O_2S$: C, 51.53; H, 3.09; N, 25.75. Found: C, 51.47; H, 3.05; N, 25.77.

2-(4-Aminophenylazo)-5-(4-bromophenyl)-1,3,4-thiadiazole (**4d**). The product was obtained as a dark violet solid (3.67 g, 68%); mp 261–263 °C; R_f ($CHCl_3$/EtOAc, 3:1 *v/v*) 0.23. ^{1}H-NMR (400 MHz DMSO-d_6): δ 6.75 (d, 2H, *J* = 9.2 Hz, 2-Ar: H-3, H-5), 7.18 (s, 2H, NH_2), 7.75–7.78 (m, 4H, 2-Ar, 5-Ar: H-2, H-6, H-3′, H-5′), 7.97 (d, 2H, *J* = 8.8 Hz, 5-Ar: H-2′, H-6′); ^{13}C-NMR (100 MHz, DMSO-d_6): δ 114.2, 124.9, 129.2, 129.3, 130.4, 132.4, 142.3, 156.7, 164.6, 180.5. IR (ATR) ν: 3321, 1636, 1507 cm^{-1}; HRMS calcd for ($C_{14}H_{10}N_5SBr$+H^+): 359.9919; found: 359.9914; Anal. Calcd for $C_{14}H_{10}N_5SBr$: C, 46.68; H, 2.80; N, 19.44. Found: C, 46.81; H, 2.79; N, 19.49.

2-(4-Aminophenylazo)-5-(4-t-butylphenyl)-1,3,4-thiadiazole (**4e**). The product was obtained as a brown solid (2.43 g, 81%); mp 171–173 °C; R_f ($CHCl_3$/EtOAc, 5:1 *v/v*) 0.27. ^{1}H-NMR (400 MHz, DMSO-d_6): δ 1.33 (s, 9H, $C(CH_3)_3$), 6.74 (d, 2H, *J* = 8.4 Hz, 2-Ar: H-3, H-5), 7.10 (s, 2H, NH_2), 7.58 (d, 2H, *J* = 8.8 Hz, 5-Ar: H-3′, H-5′), 7.76 (d, 2H, *J* = 8.4 Hz, 2-Ar: H-2, H-6), 7.96 (d, 2H, *J* = 8.8 Hz, 5-Ar: H-2′, H-6′); ^{13}C-NMR (100 MHz, DMSO-d_6): δ 30.8, 34.8, 114.1, 126.2, 127.3, 127.4, 130.2, 142.3, 154.5, 156.5, 165.7, 179.9. IR (ATR) ν: 3325, 1621, 1507 cm^{-1}; HRMS calcd for ($C_{18}H_{19}N_5S$+H^+): 338.1440; found: 338.1438; Anal. Calcd for $C_{18}H_{19}N_5S$: C, 64.07; H, 5.68; N, 20.75. Found: C, 64.21; H, 5.70; N, 20.71.

3.2.4. General procedure for the synthesis of 2-[4-(*N,N*-dimethylamino)phenylazo]-5-phenyl-1,3,4-thiadiazole derivatives (**5a–e**)

Sodium nitrite (1.31 g, 0.019 mol) was introduced portion wise into concentrated sulfuric acid (20 mL). The solution was heated to 50 °C in a water bath until complete dissolution and then rapidly cooled in an ice/salt bath to 0 °C. In the meantime, a solution of the appropriate 2-amino-5-aryl-1,3,4-thiadiazole **3a–e** (0.015 mol) in glacial acetic acid (30 mL) and propionic acid (15 mL) was prepared and added dropwise to an agitated solution of sodium nitrite in concentrated sulfuric acid at 0–5 °C. Then, the mixture was stirred for 24 h and excess nitrous acid was decomposed by the addition of urea. The resulting diazonium salt solution was slowly introduced into the mixture of N,N-dimethylaniline (1.90 mL, 1.82 g, 0.015 mol) in 15 mL of water at 0–5 °C. The colored mixture was stirred at room temperature for the next 24 h and finally neutralized with saturated sodium carbonate solution (75 mL). The solid was filtered off, washed twice with hot water (2 × 25 mL) and dried in air. The crude product (**5a–e**) was purified by column chromatography on silica gel ($CHCl_3$/EtOAc, 5:1 *v/v*).

2-[4-(N,N-Dimethylamino)phenylazo]-5-phenyl-1,3,4-thiadiazole (**5a**). The product was obtained as a brown solid (3.20 g, 69%); mp 194–195 °C (191 °C [26]); R_f ($CHCl_3$/EtOAc, 5:1 *v/v*) 0.32. IR (ATR) ν: 1603, 1525 cm^{-1}; Anal. Calcd for $C_{16}H_{15}N_5S$: C, 62.11; H, 4.89; N, 22.64. Found: C, 62.08; H, 4.86; N, 22.67.

2-[4-(N,N-Dimethylamino)phenylazo]-5-(4-methoxyphenyl)-1,3,4-thiadiazole (**5b**). The product was obtained as a dark red solid (3.00 g, 59%); mp 219–221 °C; R_f ($CHCl_3$/EtOAc, 5:1 *v/v*) 0.27. ^{1}H-NMR (400 MHz, DMSO-d_6): δ 3.17 (s, 6H, $N(CH_3)_2$), 3.86 (s, 3H, OCH_3), 6.92 (d, 2H, *J* = 9.2 Hz, 2-Ar: H-3, H-5), 7.12 (d, 2H, *J* = 8.8 Hz, 5-Ar: H-3′, H-5′), 7.85 (d, 2H, *J* = 9.2 Hz, 2-Ar: H-2, H-6), 7.99 (d, 2H, *J* = 8.8 Hz, 5-Ar: H-2′, H-6′); ^{13}C-NMR (100 MHz, DMSO-d_6): δ 40.2, 55.5, 112.2, 114.8, 122.5, 126.0, 129.2, 129.4, 142.0, 154.6, 164.8, 179.4. IR (ATR) ν: 1603, 1516 cm^{-1}; HRMS calcd for ($C_{17}H_{17}N_5SO$+H^+):

340.1232; found: 340.1233; Anal. Calcd for $C_{17}H_{17}N_5SO$: C, 60.16; H, 5.05; N, 20.63. Found: C, 60.27; H, 5.03; N, 20.61.

2-[4-(N,N-Dimethylamino)phenylazo]-5-(4-nitrophenyl)-1,3,4-thiadiazole (**5c**). The product was obtained as a brown solid (3.19 g, 60%); mp 169–171 °C; R_f ($CHCl_3$/EtOAc, 5:1 *v*/*v*) 0.38. ^{1}H-NMR (400 MHz, DMSO-d_6): δ 3.18 (s, 6H, $N(CH_3)_2$), 7.04 (d, 2H, *J* = 8.8 Hz, 2-Ar: H-3, H-5), 7.96 (d, 2H, *J* = 9.2 Hz, 5-Ar: H-3′, H-5′), 8.37–8.40 (m, 4H, 2-Ar, 5-Ar: H-2, H-6, H-2′, H-6′); ^{13}C-NMR (100 MHz, DMSO-d_6): δ 40.0, 112.3, 124.4, 124.5, 126.9, 129.1, 130.1, 142.1, 155.5, 164.8, 179.8. IR (ATR) ν: 1597, 1516 cm^{-1}; HRMS calcd for ($C_{16}H_{14}N_6SO_2$ + H^+): 355.0977; found: 355.0979; Anal. Calcd for $C_{16}H_{14}N_6SO_2$: C, 54.23; H, 3.98; N, 23.71. Found: C, 54.40; H, 3.99; N, 23.76.

5-(4-Bromophenyl)-2-[4-(N,N-dimethylamino)phenylazo]-1,3,4-thiadiazole (**5d**). The product was obtained as a dark violet solid (3.49 g, 60%); mp 201–203 °C; R_f ($CHCl_3$/EtOAc, 5:1 *v*/*v*) 0.40. ^{1}H-NMR (400 MHz, DMSO-d_6): δ 3.18 (s, 6H, $N(CH_3)_2$), 6.91 (d, 2H, *J* = 9.2 Hz, 2-Ar: H-3, H-5), 7.77 (d, 2H, *J* = 8.4 Hz, 5-Ar: H-3′, H-5′), 7.84 (d, 2H, *J* = 9.2 Hz, 2-Ar: H-2, H-6), 7.97 (d, 2H, *J* = 8.4 Hz, 5-Ar: H-2′, H-6′); ^{13}C-NMR (100 MHz, DMSO-d_6): δ 40.0, 112.3, 124.9, 128.8, 129.2, 129.3, 132.4, 142.1, 154.8, 164.7, 180.5. IR (ATR) ν: 1597, 1522 cm^{-1}; HRMS calcd for ($C_{16}H_{14}N_5SBr$+H^+): 388.0232; found: 388.0238; Anal. Calcd for $C_{16}H_{14}N_5SBr$: C, 49.49; H, 3.63; N, 18.04. Found: C, 49.37; H, 3.66; N, 18.09.

5-(4-t-Butylphenyl)-2-[4-(N,N-dimethylamino)phenylazo]-1,3,4-thiadiazole (**5e**). The product was obtained as a dark violet solid (3.07 g, 56%); mp 225–227 °C; R_f ($CHCl_3$/EtOAc, 5:1 *v*/*v*) 0.39. ^{1}H-NMR (400 MHz, DMSO-d_6): δ 1.33 (s, 9H, $C(CH_3)_3$), 3.18 (s, 6H, $N(CH_3)_2$), 6.93 (d, 2H, *J* = 9.2 Hz, 2-Ar: H-3, H-5), 7.59 (d, 2H, *J* = 8.4 Hz, 5-Ar: H-3′, H-5′), 7.86 (d, 2H, *J* = 9.2 Hz, 2-Ar: H-2, H-6), 7.97 (d, 2H, *J* = 8.4 Hz, 5-Ar: H-2′, H-6′); ^{13}C-NMR (100 MHz, DMSO-d_6): δ 30.8, 34.8, 40.1, 112.3, 126.2, 127.4, 127.6, 142.1, 153.4, 154.5, 154.7, 165.8, 179.8. IR (ATR) ν: 1604, 1524 cm^{-1}; Anal. Calcd for $C_{20}H_{23}N_5S$: HRMS calcd for ($C_{20}H_{23}N_5S$+H^+): 366.1752; found: 366.1754; Anal. Calcd for $C_{20}H_{23}N_5S$: C, 65.72; H, 6.34; N, 19.16. Found: C, 65.80; H, 6.33; N, 19.18.

3.2.5. General Procedure for the Synthesis of 2-(4-Hydroxyphenylazo)-5-phenyl-1,3,4-thiadiazole Derivatives (6a–e)

Sodium nitrite (1.31 g, 0.019 mol) was introduced portion wise into concentrated sulfuric acid (20 mL). The solution was heated to 50 °C in a water bath until complete dissolution and then rapidly cooled in an ice/salt bath to 0 °C. In the meantime, a solution of the appropriate 2-amino-5-aryl-1,3,4-thiadiazole **3a–e** (0.015 mol) in glacial acetic acid (30 mL) and propionic acid (15 mL) was prepared and added dropwise to an agitated solution of sodium nitrite in concentrated sulfuric acid at 0–5 °C. Then, the mixture was stirred for 24 h and excess nitrous acid was decomposed by the addition of urea. The resulting diazonium salt solution was slowly introduced into the mixture of phenol (1.41 g, 0.015 mol) in 15 mL of water at 0–5 °C. The colored mixture was stirred at room temperature for 24 h and finally neutralized with saturated sodium carbonate solution (75 mL). The solid was filtered off, washed twice with hot water (2 × 25 mL) and dried in air. The crude product (**6a–e**) was purified by column chromatography on silica gel ($CHCl_3$/EtOAc, 5:1 *v*/*v*).

2-(4-Hydroxyphenylazo)-5-phenyl-1,3,4-thiadiazole (**6a**). The product was obtained as a red solid (3.47 g, 82%); mp 200–202 °C (196–198 °C [25]); R_f ($CHCl_3$/EtOAc, 5:1 *v*/*v*) 0.25. IR (ATR) ν: 3060, 1608, 1512 cm^{-1}; Anal. Calcd for $C_{14}H_{10}N_4OS$: C, 59.56; H, 3.57; N, 19.85. Found: C, 59.50; H, 3.52; N, 19.87.

2-(4-Hydroxyphenylazo)-5-(4-methoxyphenyl)-1,3,4-thiadiazole (**6b**). The product was obtained as an orange solid (4.26 g, 91%); mp 271–273 °C; R_f ($CHCl_3$/EtOAc, 5:1 *v*/*v*) 0.23. ^{1}H-NMR (400 MHz, DMSO-d_6): δ 3.86 (s, 3H, OCH_3), 7.02 (d, 2H, *J* = 9.2 Hz, 2-Ar: H-3, H-5), 7.13 (d, 2H, *J* = 8.8 Hz, 5-Ar: H-3′, H-5′), 7.92 (d, 2H, *J* = 9.2 Hz, 2-Ar: H-2, H-6), 8.03 (d, 2H, *J* = 8.8 Hz, 5-Ar: H-2′, H-6′); ^{13}C-NMR (100 MHz, DMSO-d_6): δ 55.5, 114.9, 116.7, 122.1, 126.8, 129.6, 144.7, 162.2, 164.1, 167.3, 178.4. IR (ATR) ν: 3099, 1601, 1518 cm^{-1}; HRMS calcd for ($C_{15}H_{12}N_4SO_2$+H^+): 313.0759; found: 313.0755; Anal. Calcd for $C_{15}H_{12}N_4SO_2$: C, 57.68; H, 3.87; N, 17.94. Found: C, 57.59; H, 3.85; N, 17.99.

2-(4-Hydroxyphenylazo)-5-(4-nitrophenyl)-1,3,4-thiadiazole (**6c**). The product was obtained as a dark brown solid (3.92 g, 80%); mp 258–260 °C; R_f ($CHCl_3$/EtOAc, 5:1 *v*/*v*) 0.35. ^{1}H-NMR (400 MHz,

DMSO-d_6): δ 6.97 (d, 2H, *J* = 8.8 Hz, 2-Ar: H-3, H-5), 7.92 (d, 2H, *J* = 8.8 Hz, 5-Ar: H-3′, H-5′), 8.28–8.40 (m, 4H, 2-Ar, 5-Ar: H-2, H-6, H-2′, H-6′); ^{13}C-NMR (100 MHz, DMSO-d_6): δ 117.4, 124.4, 124.6, 127.5, 129.0, 144.7, 157.3, 164.1, 167.4, 179.1. IR (ATR) ν: 3080, 1598, 1515 cm^{-1}; HRMS calcd for ($C_{14}H_9N_5SO_3$ + H^+): 328.0504; found: 328.0502; Anal. Calcd for $C_{14}H_9N_5SO_3$: C, 51.37; H, 2.77; N, 21.40. Found: C, 51.50; H, 2.78; N, 21.47.

5-(4-Bromophenyl)-2-(4-hydroxyphenylazo)-1,3,4-thiadiazole (**6d**). The product was obtained as a brown solid (4.06 g, 75%); mp 263–265 °C; R_f ($CHCl_3$/EtOAc, 5:1 *v*/*v*) 0.33. ^{1}H-NMR (400 MHz, DMSO-d_6): δ 7.03 (d, 2H, J= 9.2 Hz, 2-Ar: H-3, H-5), 7.80 (d, 2H, *J* = 8.8 Hz, 5-Ar: H-3′, H-5′), 7.94 (d, 2H, *J* = 9.2 Hz, 2-Ar: H-2, H-6), 8.03 (d, 2H, *J* = 8.8 Hz, 5-Ar: H-2′, H-6′); ^{13}C-NMR (100 MHz, DMSO-d_6): δ 116.8, 125.5, 127.1, 129.6, 131.7, 132.5, 144.7, 164.4, 166.4, 179.4. IR (ATR) ν: 3047, 1587, 1507 cm^{-1}; HRMS calcd for ($C_{14}H_9N_4SOBr$ + H^+): 360.9759; found: 360.9757; Anal. Calcd for $C_{14}H_9N_4SOBr$: C, 46.55; H, 2.51; N, 15.51. Found: C, 46.61; H, 2.50; N, 15.59.

5-(4-t-Butylphenyl)-2-(4-hydroxyphenylazo)-1,3,4-thiadiazole (**6e**). The product was obtained as an orange solid (4.06 g, 80%); mp 291–293 °C; R_f ($CHCl_3$/EtOAc, 5:1 *v*/*v*) 0.37. ^{1}H-NMR (400 MHz, DMSO-d_6): δ 1.33 (s, 9H, $C(CH_3)_3$), 7.03 (d, 2H, *J* = 9.2 Hz, 2-Ar: H-3; H-5), 7.60 (d, 2H, *J* = 8.8 Hz, 5-Ar: H-3′, H-5′), 7.94 (d, 2H, *J* = 9.2 Hz, 2-Ar: H-2, H-6), 8.01 (d, 2H, *J* = 8.8 Hz, 5-Ar: H-2′, H-6′); ^{13}C-NMR (100 MHz, DMSO-d_6): δ 30.8, 34.8, 116.7, 126.3, 126.9, 127.6, 129.6, 144.8, 155.1, 164.1, 167.5, 178.8. IR (ATR) ν: 3100, 1582, 1507 cm^{-1}; HRMS calcd for ($C_{18}H_{18}N_4SO$ + H^+): 339.1280; found: 339.1281; Anal. Calcd for $C_{18}H_{18}N_4SO$: C, 63.88; H, 5.36; N, 16.56. Found: C, 63.76; H, 5.39; N, 16.51.

4. Conclusions

An efficient synthetic strategy for the preparation of new azo dyes using weak heteroaromatic amines containing a 1,3,4-thiadiazole scaffold as the diazo component and aniline, *N*,*N*-dimethylaniline and phenol as simultaneous coupling precursors, was presented. Three series of azo dyes, the derivatives of 2-phenylazo-5-phenyl-1,3,4-thiadiazole, were obtained in good to excellent yields. The new dyes obtained—demonstrating a broad color spectrum and good solubility in some polar solvents—may serve as potential candidates for the dye industry and will be tested in the near future.

Supplementary Materials: Copies of the 1H-NMR, 13C-NMR, UV-Vis and IR spectra of the compounds as well as figures showing the calculated spectra and the contour plots of the most important molecular orbitals are available in the online Supplementary Materials.

Author Contributions: A.K., M.O. and M.L. conceived and designed the experiments, performed the experiments and analyzed the data. M.S., T.S. and R.K. performed emission measurements, X-ray structural analysis and time-dependent density functional theory calculations. A.K. wrote the manuscript with the help of M.O. and R.K. All authors read and approved the final manuscript.

Funding: The synthetic part of the project was financially supported by the EU under the project RPO WL RPLU.01.02.00-06-1116/16. The crystallographic part was financed by funds allocated by the Ministry of Science and Higher Education to the Institute of General and Ecological Chemistry, Lodz University of Technology in Poland.

Conflicts of Interest: The authors declare no conflicts of interest.

References

1. Koutentis, P.A.; Constantinides, C.P. 1,3,4-Thiadiazoles. In *Comprehensive Heterocyclic Chemistry III*; Katritzky, A., Ramsden, C.A., Scriven, E.F.V., Taylor, R.J.K., Eds.; Elsevier Science Ltd.: Oxford, UK, 2008; Volume 5, pp. 567–605.
2. Hu, Y.; Li, C.; Wang, X.; Yang, Y.; Zhu, H. 1,3,4-Thiadiazole: Synthesis, reactions, and applications in medicinal, agricultural, and materials chemistry. *Chem. Rev.* **2014**, *114*, 5572–5610. [CrossRef] [PubMed]
3. Li, Y.; Geng, J.; Liu, Y.; Yu, S.; Zhao, G. Thiadiazole—A promising structure in medicinal chemistry. *Chem. Med. Chem.* **2013**, *8*, 27–41. [CrossRef] [PubMed]
4. Karaburun, A.; Acar Cevik, U.; Osmaniye, D.; Saglik, B.; Kaya Cavusoglu, B.; Levent, S.; Ozkay, Y.; Koparal, A.; Behcet, M.; Kaplancikli, Z.A. Synthesis and evaluation of new 1,3,4-thiadiazole derivatives as potent antifungal agents. *Molecules* **2018**, *23*, 3129. [CrossRef] [PubMed]

5. Gur, M.; Sener, N.; Kastas, C.A.; Ozkan, O.E.; Muglu, H.; Elmaswaria, M.A.M. Synthesis and characterization of some new heteroaromatic compounds having chirality adjacent to a 1,3,4-thiadiazole moiety and their antimicrobial activities. *J. Heterocycl. Chem.* **2017**, *54*, 3578–3590. [CrossRef]
6. Hafez, H.N.; Hegab, M.I.; Ahmed-Farag, I.S.; El-Gazzar, A.B.A. A facile regioselective synthesis of novel spiro-thioxanthene and spiro-xanthene-9′,2-[1,3,4]thiadiazole derivatives as potential analgesic and anti-inflammatory agents. *Bioorg. Med. Chem.* **2008**, *18*, 4538–4543. [CrossRef]
7. Altintop, M.D.; Sever, B.; Ozdemir, A.; Iglin, S.; Alti, O.; Turan-Zitouni, G.; Kaplancikli, Z.A. Synthesis and evaluation of a series of 1,3,4-thiadiazole derivatives as potential anticancer agents. *Anticancer Agents Med. Chem.* **2019**, *18*, 1606–1616. [CrossRef]
8. Richwine, J.R. 2,5-Dimercapto-1,3,4-thiadiazole as a Cross-Linker for Saturated Halogen-Containing Polymers. Inventor: Hercules Incorporated Wilmington, Assignee. U.S. Patent 4288576, 8 September 1981.
9. Jin, L.; Wang, G.; Li, X. Poly(2,5-dimercapto-1,3,4-thiadiazole)/sulfonated graphene composite as cathode material for rechargeable lithium batteries. *J. Appl. Electrochem.* **2011**, *41*, 377–382. [CrossRef]
10. Maradiya, H.R. Monoazo disperse dyes based on 2-amino-1,3,4-thiadiazole derivatives. *J. Serb. Chem. Soc.* **2002**, *67*, 709–718. [CrossRef]
11. Yan, H.; Su, H.; Tian, D.; Miao, F.; Li, H. Synthesis of triazolo-thiadiazole fluorescent organic nanoparticles as primary sensor toward Ag^+ and the complex of Ag^+ as secondary sensor toward cysteine. *Sens. Actuators B Chem.* **2011**, *160*, 656–661. [CrossRef]
12. Hipler, F.; Fischer, R.A.; Muller, J. Matrix-isolation pyrolysis investigation of mercapto-functionalized 1,3,4-thiadiazoles: Thermal stability of thiadiazole lubricant additives. *Phys. Chem. Chem. Phys.* **2005**, *7*, 731–737. [CrossRef]
13. Dawson, J.F. Developments in disperse dyes. *Color. Technol.* **1978**, *9*, 25–35. [CrossRef]
14. Arcoria, A.; De Giorgi, M.R.; Fatuzzo, F.; Longo, M.L. Dyeing properties of basic azo-dyes from 2-amino thiadiazole. *Dyes Pigment.* **1993**, *21*, 67–74. [CrossRef]
15. De Giorgi, M.R.; Carpignano, R.; Cerniani, A. Structure optimization in a series of thiadiazole disperse dyes using a chemometric approach. *Dyes Pigment.* **1998**, *37*, 187–196. [CrossRef]
16. Zollinger, H. Azo dyes and Pigments. In *Color Chemistry: Syntheses, Properties, and Applications of Organic Dyes and Pigments*, 3rd ed.; Viley-VCH: Zurich, Switzerland, 2003; pp. 165–254.
17. Ullrich, R.; Grewer, T. Decomposition of aromatic diazonium compounds. *Thermochim. Acta* **1993**, *225*, 201–211. [CrossRef]
18. Zhao, R.; Tan, C.; Xie, Y.; Gao, C.; Liu, H.; Jiang, Y. One step synthesis of azo compounds from nitroaromatics and anilines. *Tetrahedron Lett.* **2011**, *52*, 3805–3809. [CrossRef]
19. Chung, T.F.; Wu, Y.M.; Cheng, C.H. Reduction of aromatic nitro compounds by ethylenediamine. A new selective reagent for the synthesis of symmetric azo compounds. *J. Org. Chem.* **1984**, *49*, 1215–1217. [CrossRef]
20. Srinivasa, G.R.; Abiraj, K.; Channe Gowda, D. Lead-catalyzed synthesis of azo compounds by ammonium acetate reduction of aromatic nitro compounds. *Synth. Commun.* **2003**, *33*, 4221–4227. [CrossRef]
21. Noureldin, N.A.; Bellegarde, J.W. A novel method. The synthesis of ketones and azobenzenes using supported permanganate. *Synthesis* **1999**, *6*, 939–942. [CrossRef]
22. Bhatnagar, I.; George, M.V. Oxidation with metal oxides. III. Oxidation of diamines and hydrazines with manganese dioxide. *J. Org. Chem.* **1968**, *33*, 2407–2411. [CrossRef]
23. Faustino, H.; El-Shisthawy, R.M.; Reis, L.V.; Santos, P.F.; Almeida, P. 2-Nitrosobenzothiazoles: Useful synthons for new azobenzothiazole dyes. *Tetrahedron Lett.* **2008**, *49*, 6907–6909. [CrossRef]
24. Tambe, S.M.; Tasaganva, R.G.; Inamdar, S.R.; Kariduraganavar, M.Y. Synthesis and characterization of nonlinear optical side-chain polyimides containing the thiadiazole chromophores. *J. Appl. Pol. Sci.* **2012**, *125*, 1049–1058. [CrossRef]
25. Tomi, I.H.R.; Al-Daraji, A.H.R.; Al-Qaysi, R.R.T.; Hasson, M.M.; Al-Dulaimy, K.H.D. Synthesis, characterization and biological activities of some azo derivatives of aminothiadiazole derived from nicotinic and isonicotinic acids. *Arab. J. Chem.* **2014**, *7*, 687–694. [CrossRef]
26. Kumar, C.T.K.; Keshavayya, J.; Rajesk, T.N.; Peethambar, S.K.; Ali, A.R.S. Synthesis, characterization, and biological activity of 5-phenyl-1,3,4-thiadiazole-2-amine incorporated azo dye derivatives. *Org. Chem. Int.* **2013**, 370626. [CrossRef]

27. Kedzia, A.; Kudelko, A.; Swiatkowski, M.; Kruszynski, R. Microwave-promoted synthesis of highly luminescent s-tetrazine-1,3,4-oxadiazole and s-tetrazine-1,3,4-thiadiazole hybrids. *Dyes Pigment.* **2020**, *172*, 107865. [CrossRef]
28. Wróblowska, M.; Kudelko, A.; Kuźnik, N.; Łaba, K.; Łapkowski, M. Synthesis of Extended 1,3,4-Oxadiazole and 1,3,4-Thiadiazole Derivatives in the Suzuki Cross-Coupling Reactions. *J. Heterocycl. Chem.* **2017**, *54*, 1550–1557. [CrossRef]
29. Wróblowska, M.; Kudelko, A.; Łapkowski, M. Efficient Synthesis of Conjugated 1,3,4-Thiadiazole Hybrids through Palladium-Catalyzed Cross Coupling of 2,5-Bis(4-bromophenyl)-1,3,4-thiadiazole with Boronic Acids. *Synlett* **2015**, *26*, 2127–2130. [CrossRef]
30. Mavrova, A.T.; Wesselinova, D.; Tsenov, Y.A.; Denkova, P. Synthesis, cytotoxity and effects of some 1,2,4-triazole and 1,3,4-thiadiazole derivatives on immunocompetent cells. *Eur. J. Med. Chem.* **2009**, *44*, 63–69. [CrossRef]
31. Groom, C.R.; Bruno, I.J.; Lightfoot, M.P.; Ward, S.C. The Cambridge Structural Database. *Acta Crystallogr. B* **2016**, *72*, 171–179. [CrossRef]
32. Cowley, J.M. Scattering factors for the diffraction of electrons by crystalline solids. In *International Tables for Crystallography, Volume C: Mathematical, Physical and Chemical Tables*, 3rd ed.; Prince, E., Ed.; Kluwer Academic Publishers: Dordrecht, The Netherlands, 2004; pp. 259–262.
33. Sheldrick, G.M. Crystal structure refinement with SHELXL. *Acta Crystallogr.* **2015**, *C71*, 3–8.
34. Desiraju, G.R.; Steiner, T. *The Weak Hydrogen Bond in Structural Chemistry and Biology*; Oxford University Press: Oxford, UK, 1999; pp. 293–342.
35. Kruszynski, R.; Sieranski, T. Can stacking interactions exist beyond the commonly accepted limits? *Cryst. Growth Des.* **2016**, *16*, 587–595. [CrossRef]
36. Frisch, M.J.; Trucks, G.W.; Schlegel, H.B.; Scuseria, G.E.; Robb, M.A.; Cheeseman, J.R.; Scalmani, G.; Barone, V.; Mennucci, B.; Petersson, G.A.; et al. *Gaussian 09, Revision, D.01*; Gaussian, Inc.: Wallingford, CT, USA, 2009.
37. Grimme, S.; Ehrlich, S.; Goerigk, L. Effect of the damping function in dispersion corrected density functional theory. *J. Comp. Chem.* **2011**, *32*, 1456–1465. [CrossRef] [PubMed]
38. Tomasi, J.; Mennucci, B.; Cammi, R. Quantum mechanical continuum solvation models. *Chem. Rev.* **2005**, *105*, 2999–3093. [CrossRef]
39. Campaigne, E.; Selby, T.P. Thiazoles and thiadiazines. The condensation of ethyl 4-chloroacetoacetate with thiosemicarbazide. *J. Heterocycl. Chem.* **1978**, *15*, 401–411. [CrossRef]
40. Plumitallo, A.; Cardia, M.C.; Distinto, S.; DeLogu, A.; Maccioni, E. Synthesis and anti-microbial activity evaluation of some new 1-benzoyl-isothiosemicarbazides. *Farmaco* **2004**, *59*, 945–952. [CrossRef] [PubMed]
41. Malbec, F.; Milcent, R.; Barbier, G. Dérivés de la dihydro-2,4 triazole-1,2,4 thione-3 et de l'amino-2 thiadiazole-1,3,4 à partir de nouvelles thiosemicarbazones d'esters. *J. Heterocycl. Chem.* **1984**, *21*, 1689–1698. [CrossRef]
42. Giri, S.; Nizamuddin Srivastava, U.C. Synthesis of some N-(5-aryl/aryloxymethyl-1,3,4-thiadiazol-2-yl) glyoxylamide thiosemicarbazones as potential antiviral and antifungal agents. *Agric. Biol. Chem.* **1983**, *47*, 103–105. [CrossRef]
43. Jatav, V.; Mishra, P.; Kashaw, S.; Stables, J.P. Synthesis and CNS depressant activity of some novel 3-[5-substituted 1,3,4-thiadiazole-2-yl]-2-styryl quinazoline-4(3*H*)-ones. *Eur. J. Med. Chem.* **2008**, *43*, 135–141. [CrossRef] [PubMed]
44. Carvalho, A.S.; Gibaldi, D.; Pinto, A.C.; Bozza, M.; Boechat, N. Synthesis and trypanocidal evaluation of news 5-[N-(3-(5-substituted)-1,3,4-thiadiazolyl)]amino-1-methyl-4-nitroimidazoles. *Lett. Drug Des. Discov.* **2006**, *3*, 98–101. [CrossRef]

Sample Availability: Samples of the compounds **4a–e**, **5a–e**, **6a–e** are available from the authors.

Article

A Flavone-Based Solvatochromic Probe with A Low Expected Perturbation Impact on the Membrane Physical State

Simona Concilio [1,*], Miriam Di Martino [2], Anna Maria Nardiello [2], Barbara Panunzi [3], Lucia Sessa [2], Ylenia Miele [4], Federico Rossi [5] and Stefano Piotto [2,*]

1. Department of Industrial Engineering, University of Salerno, 84084 Fisciano, Italy
2. Department of Pharmacy, University of Salerno, 84084 Fisciano, Italy; miriamdimartino@hotmail.it (M.D.M.); annardiello@unisa.it (A.M.N.); lucsessa@unisa.it (L.S.)
3. Department of Agriculture, University of Napoli Federico II, 80055 Portici, Italy; barbara.panunzi@unina.it
4. Department of Chemistry and Biology "A. Zambelli", University of Salerno, 84084 Fisciano, Italy; ymiele@unisa.it
5. Department of Earth, Environmental and Physical Sciences "DEEP Sciences", University of Siena, 53100 Siena, Italy; federico.rossi2@unisi.it

* Correspondence: sconcilio@unisa.it (S.C.); piotto@unisa.it (S.P.); Tel.: +39-089-964115 (S.C.); +39-089-969795 (S.P.)

Academic Editors: Prieto Jorge Bañuelos and Ugo Caruso
Received: 7 July 2020; Accepted: 28 July 2020; Published: 29 July 2020

Abstract: The study of the cell membrane is an ambitious and arduous objective since its physical state is regulated by a series of processes that guarantee its regular functionality. Among the different methods of analysis, fluorescence spectroscopy is a technique of election, non-invasive, and easy to use. Besides, molecular dynamics analysis (MD) on model membranes provides useful information on the possibility of using a new probe, following its positioning in the membrane, and evaluating the possible perturbation of the double layer. In this work, we report the rational design and the synthesis of a new fluorescent solvatochromic probe and its characterization in model membranes. The probe consists of a fluorescent aromatic nucleus of a 3-hydroxyflavone moiety, provided with a saturated chain of 18 carbon atoms and a zwitterionic head so to facilitate the anchoring to the polar heads of the lipid bilayer and avoid the complete internalization. It was possible to study the behavior of the probe in GUV model membranes by MD analysis and fluorescence microscopy, demonstrating that the new probe can efficiently be incorporated in the lipid bilayer, and give a color response, thanks to is solvatochromic properties. Moreover, MD simulation of the probe in the membrane supports the hypothesis of a reduced perturbation of the membrane physical state.

Keywords: flavone; solvatochromic probe; membrane

---

1. Introduction

Biological membranes play an essential role in almost every cellular process and are characterized by a variety of lipid species, which differ in their physicochemical properties. The immiscibility of these components causes heterogeneity of the membrane and the formation of so-called "lipid rafts". Rafts represent liquid-ordered domains (Lo). They consist mainly of saturated hydrocarbons (sphingolipids and phospholipids) and are free to move in the liquid disordered bilayer (Ld) of the membrane constituted by unsaturated fatty acids [1]. In recent years, there has been a growing interest in the study of membrane heterogeneity. Fluorescence techniques are one of the most widely used methods of characterization for membrane analysis, providing excellent tools for this purpose. These methods require the use of fluorescent probes, able to penetrate the membranes allowing a precise and immediate analysis [2,3]. The number of probes developed for the study of raft membranes is continuously increasing; those so far reported have restrictions related to the type of membrane and the fluorescence technique used [4–6].

The most recently identified class of probes corresponds to the class of environment-sensitive probes, which can directly distinguish the Lo and Ld phases due to differences in their intrinsic properties. These probes can change their fluorescence properties, fluorescence intensity, and even the emission color in response to changes in the environment [5,7–9]. This class includes solvatochromic fluorescent probes that show variations in their fluorescence properties depending on the polarity and hydration of the environment. Very promising in this context are the probes of the family of 3-hydroxyflavones (3-HF), in which an electron-donating moiety is present in position 4′. They show an excited state intramolecular proton transfer (ESIPT) reaction between the normal excited state N* and the tautomeric proton transfer product T*. The ESIPT process takes place by intramolecular proton transfer from the 3-hydroxy group to the 4-carbonyl (Scheme 1), thus resulting in two intense and well-separated emission bands [10,11].

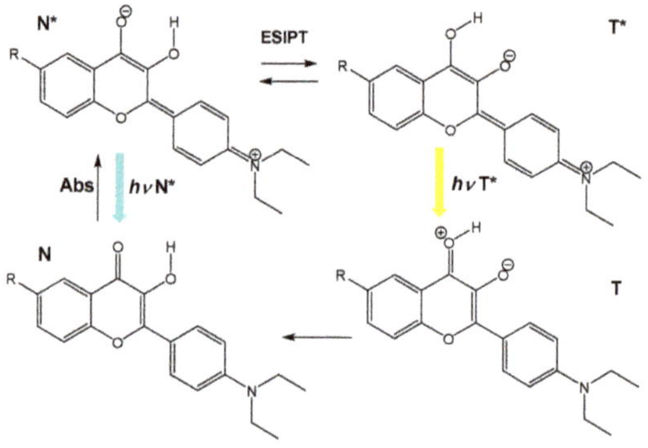

Scheme 1. Diagram of ESIPT Reaction in 3HF.

The intensity and the position of the two emission bands depend on the solvent polarity. The states N* and T* have different charge distributions and interact differently with their environment. The N* form shows a high charge separation and, therefore, a higher dipole moment than the T* state. Proton transfer in the ESIPT process from the 3-hydroxy to the 4-carbonyl group causes a shift of the negative charge, reducing the polar character of the T* species. On the other hand, the presence of the electron-donating N,N-diethylamino group, contributes to the increase of the dipole moment of the species N* (this is not observed for the T* state) [12]. This difference in the distribution of charge results in a higher dependence of the N* band, located at shorter wavelengths, on the solvent polarity. An important parameter is the intensity ratio of emission maxima of the two bands I^{N^*}/I^{T^*}, which is directly correlated to the energy of the two states and provides an indication of the dependence on the solvent-polarity [13]. In aprotic solvents, the ESIPT process is speedy, because it is barrier-free, and this results in a decrease in intensity of the N* state band. In protic solvents, the solvation and formation of H-intermolecular bonds impose a high activation barrier that leads to a redshift of the N* band with an increase in intensity. In contrast, the fluorescence of the T* state is almost indifferent to the solvent polarity and remains practically unchanged [14,15]. This common effect in protic solvents, such as methanol and ethanol, sometimes results in the presence of a single band in the emission spectra, because of the overlap of the N* band, shifted at longer wavelengths, and T* band poorly resolved [16,17].

It is well known that the insertion of a probe into a bilayer can result in a, sometimes drastic, perturbation of the membrane physical state. [18–20]. This perturbation can impact the experimental results, so it is of primary importance to verify that the probe is capable to fit into the double layer without significantly changing its characteristics.

In this work, we report the design, synthesis, characterization, and the study in lipid vesicles and model membranes of a new solvatochromic probe for membrane analysis. The new probe, called 3HF18 and bearing the 3-hydroxyflavone moiety, has been provided with a saturated chain of 18 carbon atoms and a zwitterionic head so to facilitate the anchoring to the polar heads of the lipid bilayer and avoid the complete internalization, compared to similar existing probes [21]. The presence of the electron-donating N,N-diethylamino group on the 3-hydroxyflavone moiety, contributes to the increase of the dipole moment of the N* form, and ensures a high solvatochromic effect. Also, absorption and fluorescence analyses in organic solvents with different polarities were carried out to investigate the solvatochromic properties. Finally, the behavior of the probe 3HF18 in the model membrane was studied through DFT analysis and by fluorescence microscopy on GUVs.

2. Results and Discussion

2.1. *Probe Design and Synthesis*

The synthetic scheme of the probe 3HF18 (**4**) is illustrated in Figure 1.

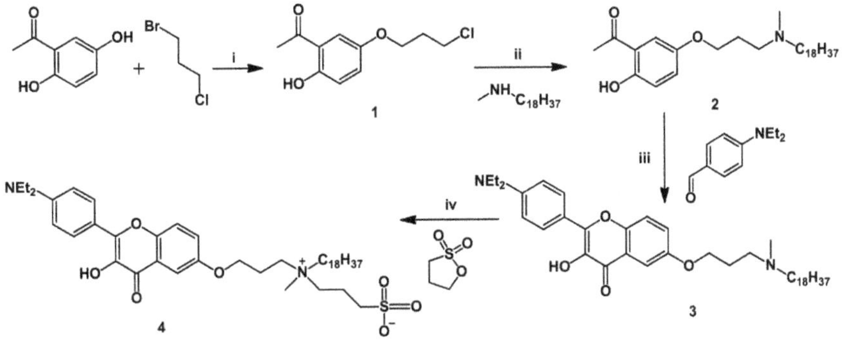

Figure 1. Synthetic scheme of the 3HF18 probe (**4**). *Reagents and Conditions*: (i) K_2CO_3, DMF dry, 80 °C, 4 h; (ii) KI, K_2CO_3, CH_3CN dry, reflux, 48 h; (iii) (a) NaOMe, DMF dry, RT, 24 h; (b) EtOH, NaOMe, H_2O_2, reflux, 30 min; (iv) K_2CO_3, CH_3CN dry, reflux 48 h.

Compound **1** was prepared according to a reported method [16]. The introduction of the C_{18} hydrocarbon chain was obtained by reaction with N-methyloctadecylamine in the presence of a base (step ii). Compound **2** was condensed with N,N-diethylaminobenzaldehyde under basic conditions, and then oxidized in the presence of hydrogen peroxide (step iii), so affording the fluorescent 3-hydroxyflavone core by oxidative cyclization (Algar–Flynn–Oyamada modified reaction) [22]. The quaternization of the amine by reaction with 1,3-propanesultone gave the fluorescent probe product 3HF18 (**4**) with a zwitterionic group. The products of all steps were purified by column chromatography or preparative TLC and fully characterized. NMR and mass spectrometry have confirmed the structure of the final product.

2.2. *Optical Characterization*

The probe 3HF18 was analyzed by UV-Vis spectrophotometry and spectrofluorimetry in solvents with different polarity (Table 1, Figures 2 and 3). The UV-Vis spectra show absorption maxima between 400 nm and 450 nm, with a redshift of more than 50 nm and an increase in absorbance as the polarity of the solvent increases. Emission spectra confirm the solvatochromic property of the probe in emission rather than in absorption. The probe has been shown to detect even slight changes in the polarity of the environment. In polar solvents both N* and T* are energy stabilized and this stabilization is more pronounced for N* due to the higher electrical dipole moment. For very polar protic solvents, N* is at energy lower than T*, and therefore only one transition is observed (Figure 3). Moreover, as reported in the literature for analogous compounds, the probe's core is poorly soluble and nonfluorescent in water [23]. The solubility increases in water/methanol mixtures. In these cases, the fluorescence spectra show a single fluorescence peak around 520 nm (Figure S5 in Supplementary Materials). As can be observed in Figure 3, the maximum emission ranges from 475 nm in dioxane to 551 nm ethylene glycolfor the N* band. While the T* band, located at longer wavelengths, shows almost no influence with the solvent polarity because the stabilization effect of polar solvents acts on both T and T* states. The I^{N^*}/I^{T^*} ratio represents the dependence of the fluorescence intensity on the solvent polarity. It is possible to observe an increase in the intensity of fluorescence, which accompanies the redshift, changing from apolar solvents such as dioxane to a single band (N* state) in protic polar solvents such as methanol, propan-2-ol and ethylene glycol.

Table 1. Absorption and emission data of 3HF18 in solvents with different polarities.

Solvent	λ_{Abs} (nm) [1]	ε (M^{-1} cm^{-1}) [2]	$\lambda^{N^*}_{Em}$ (nm)	$\lambda^{T^*}_{Em}$ (nm)	I^{N^*}/I^{T^*}
Dioxane	405	$1.1{\cdot}10^4$	475	571	0.48
AcOet	406	$1.3{\cdot}10^4$	483	568	0.94
DCM	416	$2.4{\cdot}10^4$	502	564	1.8
Acetone	426	$3.2{\cdot}10^4$	502	567	1.9
Propan-2-ol	440	$3.4{\cdot}10^4$	515	- [3]	- [3]
MeOH	440	$3.4{\cdot}10^4$	534	- [3]	- [3]
DMF	438	$4.4{\cdot}10^4$	509	576	2.05
DMSO	440	$4.2{\cdot}10^4$	519	- [4]	- [4]
Ethylene Glycol	453	$4.4{\cdot}10^4$	551	- [3]	- [3]
H_2O [5]	-	-	-	-	-

[1] Maximum absorption wavelength; [2] Molar extinction coefficients, from Lambert-Beer Law; [3] A single N* peak is observed; [4] The T* band, poorly resolved, appears as a shoulder; [5] The probe is poorly soluble in water. For the emission spectra, all the solutions were excited at 398 nm.

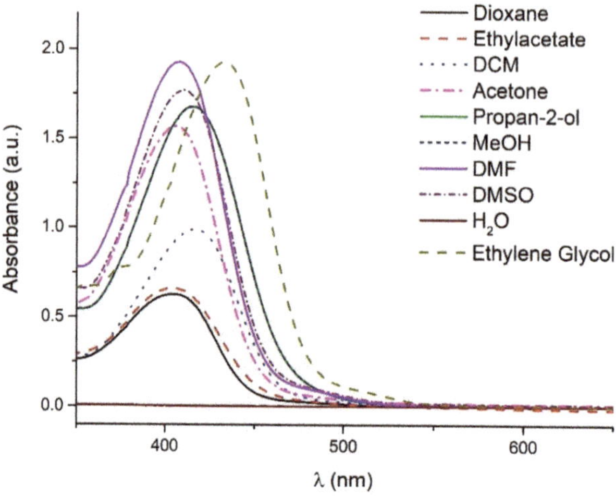

Figure 2. Absorption spectra of 3HF18 in solvents of different polarity.

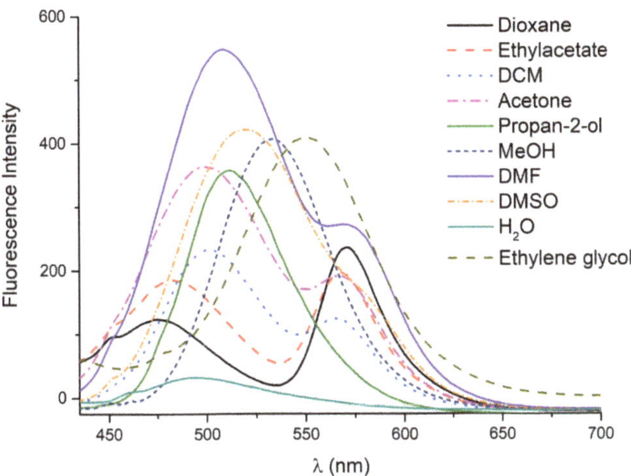

Figure 3. Emission spectra of 3HF18 in solvents of different polarity, excited at 398 nm.

In Figure 4, a picture of different solutions of the probe 3HF18, in representative solvents with increasing polarity is reported, under white (a) and UV (b) light.

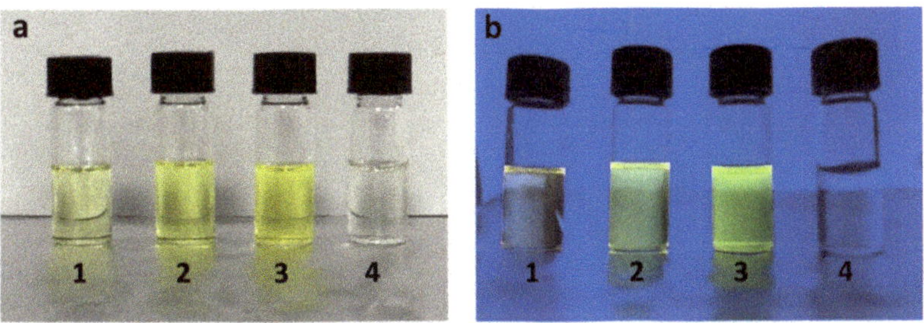

Figure 4. Picture of 3HF18 dissolved in different representative solvents: (1) Dioxane, (2) DCM, (3) MeOH, (4) H_2O: (**a**) under white light and (**b**) under UV light (365 nm).

From the image in Figure 4, the solvatochromism of the 3HF18 molecule is visible to the naked eye. The emission color of the solutions ranges from orange to yellow-green, as the polarity of the solvent increases. The aqueous solution appears colorless due to the poor solubility of the compound.

2.3. *Analysis of Fluorescence in Lipid Membranes*

The heterogeneity and dynamism of cell membranes make it challenging to study their structure because they are incredibly complex entities. For this reason, the simplest model for membranes is represented by unilamellar vesicles. In particular, giant unilamellar vesicles, GUVs, with dimensions comparable to the sizes of a cell (in an interval of about 1–100 μm), are particularly useful as model membranes. Over the years, several methods have been developed for the design of GUVs. Many methods have been proposed to control the size and composition of the vesicles [24,25]. One of the most recently developed methods is the "droplet transfer" or "reverse emulsion" [26], characterized by the stratification of an aqueous phase and a water-in-oil emulsion with the formation of the bilayer at the interface of the two phases. This method permits to easily encapsulate hydrophilic compounds in the aqueous lumen [27,28] and blend hydrophobic species in the membrane, such as fatty acids [29,30] or polymers [31,32]. In this work, the "droplet transfer" method was used for the formation of vesicles of POPC and a ternary mixture of POPC:DPPC:Chol to simulate the two phases

Lo and Ld of a membrane. The solvatochromism of the probe 3HF18 can be used to characterize cell membranes or their simplified models consisting of lipid vesicles. The probe has been successfully encapsulated both in pure POPC vesicles in the Ld phase, Figure 5A,C and in mixed POPC:DPPC:Chol membranes (Lo phases, Figure 5B,D). Fluorescence has been recorded at two different wavelength intervals, where the emission maxima do not overlap: in (Figure 5A,B) λ_{exc} = 405 nm, the emission range is λ = 477–527 nm; in Figure 5C,D λ_{exc} = 405 nm and the emission range is λ = 550–600 nm. Cholesterol is reported to increase polarity along the membrane surface, while in the hydrophobic core it decreases polarity [33]. In the case of mixed vesicles, therefore, due to the effect of Chol, the surface polarity may be higher than in vesicles of POPC alone. In a previous work [21], we have shown that membrane probes with a zwitterionic head can be easily placed with the chromophore beneath the membrane surface and with the hydrophobic tail inserted between the tails of phospholipids. In this case, a higher I^{N^*}/I^{T^*} emission ratio in GUVs of POPC/DPPC/Chol is observed (Figure 5E,F) which supports the notion that the chromophore is positioned below the surface of the membrane.

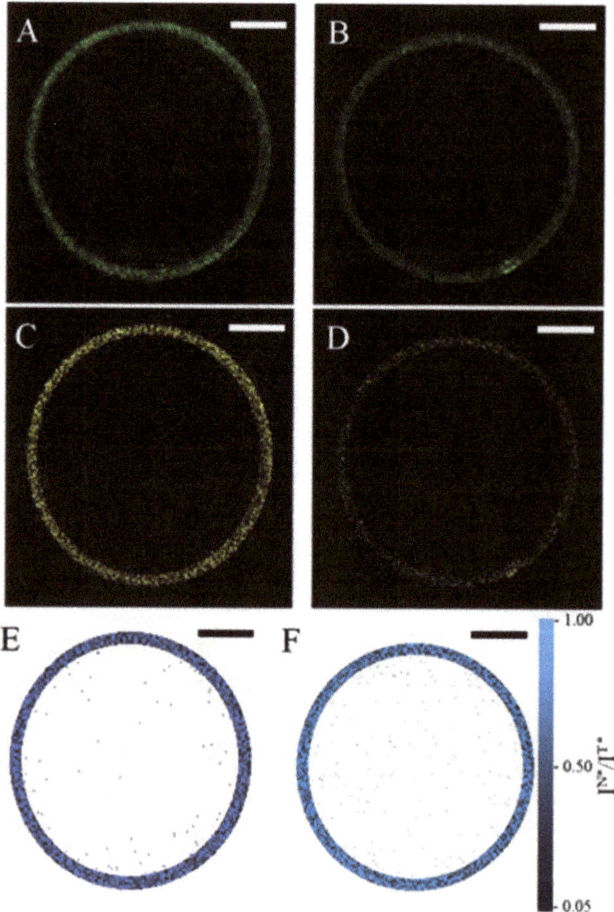

Figure 5. Fluorescence confocal images (**A–D**) and intensity ratio (**E**,**F**) of vesicles made of POPC (**A**,**C**,**E**) and POPC/DPPC/Chol 1/0.5/0.7 (**B**,**D**,**F**). Excitation at λ = 405 nm, Emission at λ = 477–527 nm (**A**,**B**) and λ = 550–600 nm (**C**,**D**). Bars scale 5 µm. Contrast of the images has been adjusted to improve the quality of the pictures.

The fluorescence of the probe is limited to the surface of the vesicles, thus confirming the anchorage of the probe to the double layer. Molecular dynamics studies indicate that the flip flop mechanism of the probe is significantly lower than that of fluorophore alone and also than that of POPC and DPPC. In fact, the zwitterionic head of the probe has a dipole inverted compared to that of phospholipids and increases its membrane cohesion.

2.4. Molecular Dynamics in Model Membranes. Thickness and Area Per Lipid

Through molecular dynamics (MD) analysis, double layer thickness and area per lipid values were calculated. The experimental values of double layer thickness and area per lipid (A_l) are commonly used for the validation of simulations [34,35]. The average lipid bilayer thickness, calculated as the average distance between the phosphorus atoms of the two double-layer sheets, showed that the insertion of the probe, in both excited N* and T* forms, does not cause membrane perturbation (Table 2).

Table 2. Thickness values (in Å) of the two membranes in the presence and absence of the probe.

POPC [1]	DPPC-Chol [1]	POPC [2]	DPPC-Chol [2]
37.3 ± 0.3	45.9 ± 0.2	39.8 ± 0.8	44.8
POPC + 3HF18(N*) [3]	**POPC + 3HF18(T*) [4]**	**DPPC-Chol + 3HF18(N*) [5]**	**DPPC-Chol + 3HF18(T*) [6]**
37.4 ± 0.2	37.8 ± 0.2	46.4 ± 0.3	46.1 ± 0.2

[1] Thickness of simulated POPC and DPPC-Chol membranes; [2] Experimental thickness of POPC and DPPC-Chol membranes; [3] Thickness of POPC membrane with N* 3HF18; [4] Thickness of POPC membrane with the T* 3HF18; [5] Thickness of DPPC-Chol membrane with N* 3HF18; [6] Thickness of DPPC-Chol membrane with T* 3HF18.

The thickness values obtained from pure membrane simulations are 37.3 Å for POPC and 45.9 Å for DPPC-Chol. The calculated thickness values are in good agreement with the experimental values of 39.8 ± 0.8 Å [35] for POPC and 44.8 Å for DPPC-Chol [34], respectively. From the values shown in Table 2, it can be seen that the inclusion of the 3HF18 molecule does not cause a change in the thickness of the analyzed membranes. Table 3 shows the area per lipid (A_l) data obtained from simulations on different systems compared with the experimental data obtained from CHARMM-GUI [36].

Table 3. Area per Lipid (A_l) values of the two membranes with and without the probe 3HF18.

POPC [1]	DPPC-Chol [1]	POPC + 3HF18 (N*) [2]	POPC + 3HF18 (T*) [3]	DPPC-Chol + 3HF18 (N*) [4]	DPPC-Chol + 3HF18 (T*) [5]
68.3	43.2-48.6	69.9 ± 0.5	69.2 ± 0.5	43.2 ± 0.2	43.6 ± 0.2

[1] Experimental A_l of POPC e DPPC-Chol membranes; [2] A_l of POPC membrane with N* 3HF18; [3] A_l of POPC membrane with T* 3HF18; [4] A_l of DPPC-Chol membrane with N* 3HF18; [5] A_l of DPPC-Chol membrane with T* 3HF18.

When the cholesterol concentration inside the membrane is high (>20%), it is necessary to decrease the area per lipid values by 10–20%. Therefore DPPC-Chol displays a range of values. Also, for this analysis, the areas per lipid in the presence and absence of the probe are comparable, so we can conclude that the presence of the probe does not cause perturbation in the membranes.

2.5. Deuterium Order Parameter SCD

To assess whether the probe causes a change in membrane fluidity, we performed an analysis of the deuterium order parameter (SCD. The SCD parameter is a measure of the motor anisotropy of the C-D bond analyzed and provides its orientation over time. It is derived from NMR experiments. SCD analysis provides comprehensive information on membrane fluidity and allows us to predict changes in the fluidity of the lipophilic chains of phospholipids near a guest molecule (in this case, the 3HF18 probe). The fluidity of the POPC and DPPC-Chol chains was evaluated in the presence and absence of the 3HF18 probe in the forms N, N* and T* (Figures 6 and 7). The results of the N-form calculations are reported in the Supplementary Materials section. In Figure 6, the SCDs for POPC phospholipid chains around 5 Å from the guest molecule is shown. In Figure 6a, the unsaturated chain is shown, and in Figure 6b, the saturated one.

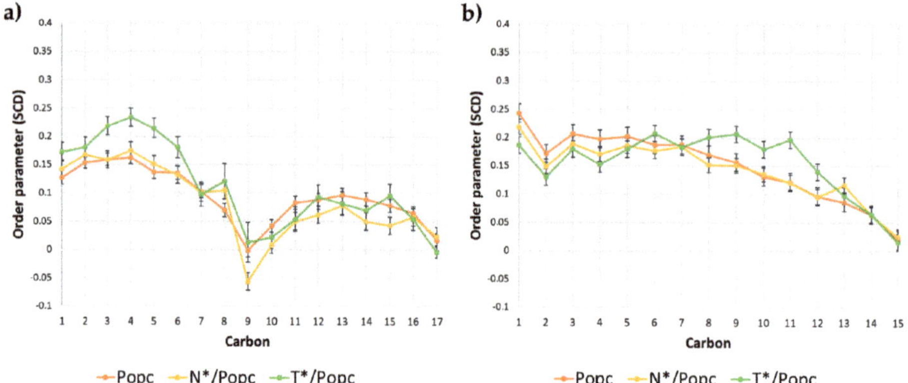

Figure 6. Order parameter SCD for (**a**) the unsaturated oleic and (**b**) the saturated palmitoyl acyl chains of POPC phospholipid in POPC pure membrane (orange curve) used as control and in POPC/3HF18 membrane (N*, yellow curve; T* green curve). Notes: On the *Y*-axis, the SCD is indicated; on the *X*-axis, the carbon atom position is reported, starting from the first (1) alpha carbon atom in the chains.

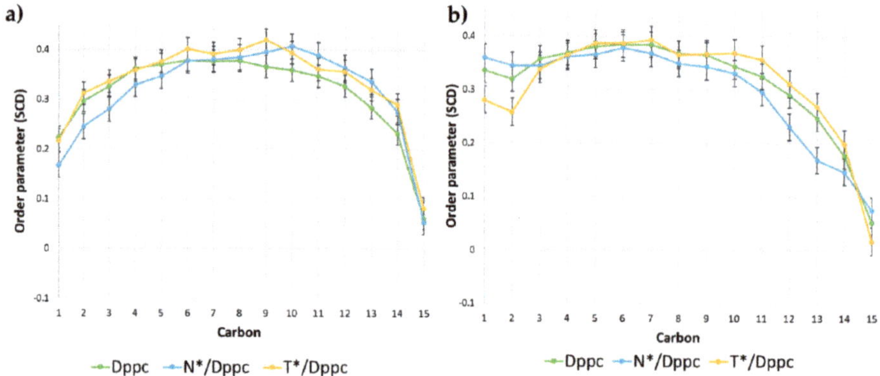

Figure 7. Order parameter SCD for (**a**) the saturated Sn-1 palmitoyl chain and (**b**) the Sn-2 saturated palmitoyl chains of DPPC phospholipid in DPPC pure membrane (green curve) used as control and in DPPC/3HF18 membrane (N*, blue curve; T* yellow curve). Notes: On the *Y*-axis, the SCD is indicated; on the *X*-axis, the carbon atom position is reported, starting from the first (1) alpha carbon atom in the chains.

The SCD decreases sharply in the region near the double bond (carbon atom 9) and approaches 0 in the tail region, indicating the highest disorder within the double layer (Figure 6). Similar SCD profiles were observed experimentally using NMR studies [37]. Figure 7 shows the SCD values of the two saturated chains of DPPC. Even for the DPPC, in the final region of the tail, the SCD decreases towards 0, thus showing a greater disorder.

From Figures 6 and 7, we can see that the probe (both tautomeric forms of 3HF18) interacts with the aliphatic chains of phospholipids and does not cause a significant change in the order of the membranes.

2.6. Pseudo-Semantic Analysis

A new approach to evaluate membrane fluidity is to evaluate SCD by pseudo-semantic analysis. The SCD parameter, obtained from MD analysis, is converted into alphabetical strings, which are compared and clustered by similarity. The advantage compared to the previous analysis, is the possibility to compare the different membrane/probe systems and to analyze the fluidity of the individual frames during a simulation. The result, displayed through a dispersion graph, is shown in Figure 8.

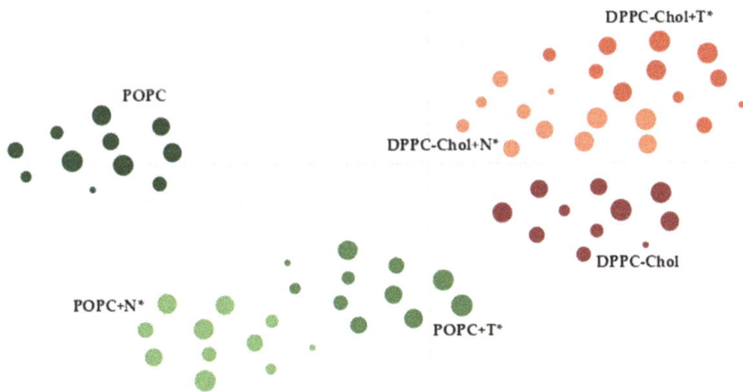

Figure 8. t-SNE dispersion graph. Each point corresponds to a frame of molecular dynamics. Circle dimension is related to the time (ns). Green circles: POPC membranes; red/orange: DPPC-Chol systems.

In Figure 8, the six systems analyzed are shown. Each simulation of MD generates 11 frames, whose dimensions increase with the time of the simulation (ns). Each circle corresponds to one of the 11 frames of the trajectory. In different shades of green are shown the systems in POPC, and in red/orange the systems in DPPC-Chol. The analysis is capable to capture the similarity among different membranes at different times. Figure 8 shows that DPPC-Chol membranes are rather similar, forming almost a single cluster (top right). This fact indicates that the presence of the probe does not significantly perturb the membrane and that the order of the aliphatic chains during the whole dynamics remains similar. On the contrary, POPC membrane (and reasonably any kind of Ld membranes), are slightly influenced by the presence of the probe. The ordering effect of the probe is revealed by the change in the position of the POPC + T* cluster toward the DPPC-Chol cluster.

2.7. The Binding Energy Between Probe and Membrane

In order to check whether either tautomeric form has a preference for the Lo or Ld phase of the membranes, the binding energy with the membrane/water system has been calculated. The binding energies for the last 5 ns of the simulation are shown in Table 4.

Table 4. Binding energy values for the 3HF18 (form N* and T*) in POPC and DPPC-Chol membrane.

	POPC + 3HF18(N*) [1]	**POPC + 3HF18(T*)** [2]	**DPPC-Chol + 3HF18(N*)** [3]	**DPPC-Chol + 3HF18(T*)** [4]
Binding energy (kcal/mol)	286 ± 14	249 ± 18	273 ± 12	281 ± 16

[1] The binding energy of the 3HF18 probe in POPC membrane N* form; [2] Binding energy of the 3HF18 probe in POPC membrane T* form; [3] Binding energy of the 3HF18 probe in DPPC-Chol membrane N* form; [4] Binding energy of the 3HF18 probe in DPPC-Chol membrane T* form.

As we can see from Table 4 and Figure 9, each tautomeric form has a preference for a membrane: N* has better interaction with the POPC membrane while T* interacts better with DPPC-Chol. To validate the energy binding results, a t-test statistical analysis was performed for the different POPC and DPPC systems with the two forms of the 3HF18 probe, and the results are reported in the Supplementary Materials section. The analysis shows that the energies are significantly different, with a probability of statistical significance of over 98%.

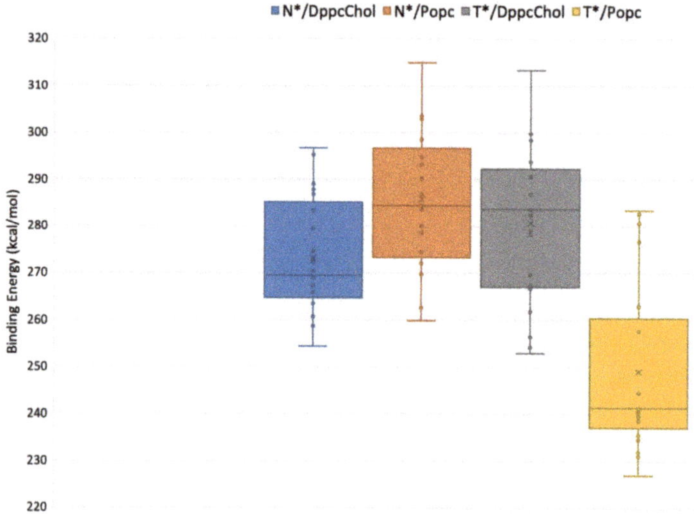

Figure 9. Box plot of binding energies on four systems: the 3HF18 probe in the N* form in DPPC-Chol membrane (blue); the 3HF18 probe in the N* form in POPC membrane (orange); the 3HF18 probe in the T* form in DPPC-Chol membrane (grey); the 3HF18 probe in the T* form in POPC membrane (yellow). The binding energy expressed in kcal/mol is shown on the *y*-axis.

3. Materials and Methods

All the reagents and solvents were purchased from Sigma Aldrich (Milan, Italy) and used without further purification. Optical observations of the probe in GUVs were performed by using a TCS SP8 Confocal laser (Leica, Wetzlar, Germany), UV laser λ = 405 nm, emission range for the N* transition λ = 477–527 nm, emission range for the T* transition λ = 550–600 nm). The images were acquired with a sequential scan between lines and an average of three frames. UV–vis absorption spectra of the samples were recorded at 25 °C in acetonitrile solution, on a Lambda 800 spectrophotometer (Perkin Elmer, Rodgau, Germany). The spectral region 650–240 nm was investigated by using a cell path length of 1.0 cm. Probe concentration of about 3.0×10^{-5} mol L^{-1} was used. Fluorescence spectra were performed at 25 °C in acetonitrile solution, on a Jasco FP-750 Spectrofluorometer (JASCO Europe, Cremella (LC) Italy), at a concentration of about 3.0×10^{-7} mol L^{-1}. ^{1}H-NMR spectra were recorded with a DRX/400 spectrometer (Bruker, Billerica, MA, USA). Chemical shifts are reported relative to the residual solvent peak (chloroform-d: H = 7.26 ppm). The following abbreviations are used to express spin multiplicities in ^{1}H NMR spectra: s = singlet; d = doublet; dd = double doublet; t = triplet; m = multiplet. High-resolution mass spectra were acquired on an LTQ-Orbitrap instrument (Thermo-Fisher, Waltham, MA, USA) operating in positive ion mode. The probe was dissolved in methanol at a concentration of 0.1 mg/mL and injected into the MS ion source. Spectra were acquired in the 150–800 *m*/*z* range.

3.1. Synthesis

The probe 3HF18 was synthesized according to the following procedure.

3.1.1. Synthesis of 5-(3-chloropropoxy)-2-hydroxy-acetophenone (**1**)

2,5-Dihydroxyacetophenone (2.0 g, 0.013 mol) was dissolved in dry DMF (13 mL). Then, K_2CO_3 (1 eq, 1.8 g, 0.013 mol) and 1-bromo-3-chloropropane (1.1 eq, 1.4 mL, 0.0145 mol) were added. The reaction mixture was stirred under a nitrogen atmosphere heating at 80 °C for 4 h. The solution became dark. After cooling, it was diluted with water and extracted with ethyl acetate (three times). The combined organic layers were dried over anhydrous sodium sulfate, filtered and the solvent evaporated under reduced pressure. The obtained dark oil was purified by flash column chromatography in silica gel (hexane: ethyl acetate, 98:2) to produce compound **1** as a light-yellow solid (1.13 g, 0.005 mol). Yield 38%. ^{1}H-NMR (400 MHz, $CDCl_3$): δ 2.16 (m, 2H), 2.55 (s, 3H), 3.70 (t, 2H), 4.03 (t, 2H), 6.86 (1, 2H), 7.05 (dd, 1H), 7.14 (d, 1H).

3.1.2. Synthesis of 5-(3-(methyl(octadecyl)amino)propoxy)-2-hydroxyacetophenone (**2**)

5-(3-Chloropropoxy)-2-hydroxyacetophenone (**1**, 1.0 g, 0.0044 mol) was dissolved in dry acetonitrile (55 mL) and KI (1.5 eq, 1.095 g, 0.0066 mol), K_2CO_3 (2.5 eq, 1.52 g, 0.011 mol) and N-methyloctadecylamine (1.2 eq, 1.50 g, 0.0053 mol) were added to the solution. The reaction mixture refluxed under stirring in a nitrogen atmosphere for 48 h. After cooling, the solution was extracted with chloroform three times. The organic layer was dried over anhydrous sodium sulfate, filtered, and the solvent was evaporated under reduced pressure. The residue was purified by flash column chromatography in silica gel (dichloromethane: methanol, 95:5) to produce the desired product 2 as a brown solid (1.530 g, 0.0033 mol). Yield 74%. ^{1}H-NMR (400 MHz, $CDCl_3$): δ 0.91 (t, 3 H), 1.27 (s, 32 H), 2.10 (q, 2 H), 2.43 (s, 3 H), 2.58 (t, 2 H), 2.63 (s, 3H), 2.76 (t, 2 H), 4.04 (t, 2 H), 6.92 (d, 1 H), 7.11 (dd, 1 H), 7.21 (d, 1 H).

3.1.3. Synthesis of 4′-diethylamino-6-(3- methyloctadecylaminopropoxy)-3-hydroxyflavone (**3**)

5-(3-(Methyl(octadecyl)amino)propoxy)-2-hydroxyacetophenone (**2**, 800 mg, 1.68 mmol) was dissolved in DMF (10.0 mL) and then 4-diethylaminobenzaldehyde (1.05 eq, 313 mg, 1.76 mmol) and sodium methoxide (4.0 eq, 363.0 mg, 6.72 mmol) were added. The obtained mixture was stirred for 24 h to give dark red chalcone, which was used in the next step without isolation. This mixture was then diluted with 10.0 mL of ethanol and then sodium methoxide (12 eq., 1.088 g, 20.16 mmol) was added. The temperature was reduced to 0 °C to add hydrogen peroxide (10 eq, 1.9 mL). This mixture was kept refluxing for 30 min under stirring and the dark red solution turned orange. The solution was cooled at room temperature, diluted with water and neutralized with 10% hydrochloric acid. The product was then extracted with ethyl acetate. The solvent was removed under vacuum and the crude product was purified by flash column chromatography in silica gel (dichloromethane: methanol, 90:10) to give the desired product **3** as a yellow solid (148 mg, 0.228 mmol). Yield 13.5%. ^{1}H-NMR (400 MHz, $CDCl_3$): δ 0.8 (t, 3H), 1.05 (t, 6 H), 1.2 (s, 30 H), 1.95 (q, 2 H), 2.35 (q, 2 H), 2.75 (s, 3 H), 2.95 (t, 2 H), 3.2 (t, 2 H), 3.38 (q, 4 H), 4.1 (t, 2 H), 6.7 (d, 2 H), 7.15 (dd, 1H), 7.41 (d, 1 H), 7.45 (d, 1 H), 8.07 (d, 2 H).

3.1.4. Synthesis of N-[3-(4′-diethylamino-3-hydroxyflavonyl-6-oxy)propyl]-N,N-(methyloctadecyl ammonium) propane-1-sulfonate (**4**)

4′-Diethylamino-6-(3- methyloctadecylaminopropoxy)-3-hydroxyflavone (**3**, 90 mg, 0.138 mmol) was dissolved in dry acetonitrile (10 mL) and 1,3-propanesultone (2.5 eq, 42.0 mg, 0.345 mmol) and K_2CO_3 (1 eq.,19.0 mg, 0.138 mmol) were added. The reaction was stirred at reflux under nitrogen atmosphere for 48 h. The solvent was evaporated under reduced pressure and the crude was purified by preparative TLC (dichloromethane: methanol, 85:15) to give product **4** as a yellow solid (25 mg; 0.032 mmol). Yield 24%. ^{1}H-NMR (400 MHz, MeOD): δ 0.91 (6 H, t), 1.23-1.36 (33 H, m), 1.79 (2 H, m),

2.25 (2 H, m), 2.34 (2 H, m), 2.91 (2 H, t), 3.13 (3 H, s), 3.52 (4 H, m), 3.59 (4 H, m), 4.27 (2 H, t), 6.85 (2 H, d), 7.41 (1 H, dd), 7.59 (1 H, d), 7.65 (1 H, d), 8.22 (2 H, d). ^{13}C-NMR (600 MHz, MeOD): δ 11.50, 13.01, 18.17, 21.26, 21.87, 22.31, 25.94, 28.78, 29.05, 29.13, 29.24, 29.37, 31.65, 48.16, 58.46, 60.39, 61.64, 64.70, 78.04, 104.71, 119.46, 121.76, 123.05, 129.44, 136.87, 150.11, 155.06, 172.16. TOF-MS (ES+) calcd for $C_{44}H_{70}N_2O_7S + Na^+$: 793.48; found $[M + Na]^+$: 793.58.

3.2. Preparation of GUVs

GUVs were made either with pure POPC or with a mixture POPC:DPPC:Chol = 1:0.5:0.7, in both cases the preparation method was the same. Here we report only the one for the mixed membrane. In brief, an Eppendorf tube (1.5 mL) was filled with: (*i*) an "outer solution" (0.5 mL of water, 200 mM glucose); (*ii*) an interfacial phase (0.3 mL, 0.3 mM POPC in mineral oil, 0.15 mM of DPPC in mineral oil, 0.21 mM of Chol in mineral oil); (iii) a water/oil emulsion (0.6 mL) prepared by pipetting up and down the "inner phase" made mixing 20 µL of water (200 mM sucrose) with an hydrophobic phase. The hydrophobic phase was prepared by mixing POPC (0.3 mM), DPPC (0.15 mM), Chol (0.21 mM) in mineral oil and adding the 3HF18 probe (4 µM) first dissolved in chloroform and subsequently dried after the evaporation of the solvent. The tube was immediately centrifuged (6000 rpm, 10 min, RT). After centrifugation, mineral oil was removed, and GUVs were washed twice to remove free solutes by pelleting/resuspension. 30 µL of pellet was resuspended in 20 µL of glucose (200 mM). 20 µL of solution was put on a microscope glass slide and observed with confocal microscopy. Typically, GUVs had diameters between 12–50 µm [26,27,29,38].

3.3. Molecular Dynamics Simulations

All MD simulations were performed with YASARA Structure v19.12.14 [39]. The force-field used is AMBER15FB under NPT (normal pressure and temperature) conditions, coupling the system to a Berendsen thermostat and using a combined solvent pressure and density control. The applied cut-off is 8 Å, and the Particle Mesh Ewald (PME) model has been used to calculate the electrostatic interactions, as it has been demonstrated that the correct treatment of electrostatic is essential in biological membrane systems. The systems are fully hydrated (water density 0.997 g/mL). The simulation cell was neutralized with NaCl, with a final concentration of 0.9%. After membrane equilibration, the probe was incorporated by positioning the aliphatic chain perpendicular to the bilayer [40–42]. MD simulations lasted 15 ns, of which only the last 5 ns were used for analysis. With the same MD parameters, the binding energy between the 3HF18 probe and the hydrated membranes was also calculated.

3.4. Model Membrane Structural Parameters

The structural parameters of the membrane are the thickness, the fluidity of the aliphatic chains, and the area per lipid. The last parameter has been calculated by dividing the surface area of the membrane by the number of lipids that compose it. The lipid area was then compared with the experimental membrane values reported in CHARMM-GUI [36]. The fluidity was measured indirectly by calculating the deuterium order parameter (SCD) with the VMD MembPlugin software [43,44].

3.5. Pseudo-Semantic Approach for the Study of Membrane Fluidity

A new approach for the study of membrane fluidity [45,46] consists of converting the results obtained from the dynamics into alphabetical strings, which are then compared and clustered by similarity. The parameter examined, in this case, is the SCD calculated using VMD MembPlugin [44]. The last 5 ns of simulation correspond to 11 frames. For each frame, the SCD values of phospholipids within 5 Å of the probe are taken into account. Each SCD value is assigned a letter. A total of 66 alphabetical strings are produced (11 for each system). In each string, the SCD values of each carbon making up the aliphatic chain of phospholipids under investigation are given in letters. The strings were then processed with Sysa (www.softmining.it/sysa), an online tool able to analyze the similarities between them. Sysa pre-processes the input files in order to eliminate redundant data and then extract user-defined k-mer length sub-strings. The metric used in this study is Szymkiew Simpson Coefficient with k-mer between 2 and 8. At the end of the computation, Sysa generated a matrix of distances. Due to the difficulty of displaying a matrix with more than three dimensions, we decided to process it using t-SNE (t-distributed stochastic neighbor embedding). t-SNE is a two-dimensional matrix dimensionality reduction algorithm [47].The algorithm allows keeping objects that were already close to each other in the original space at reduced dimensionality.

4. Conclusions

A novel fluorescent molecule was designed, synthesized, and optically characterized as a probe for lipid membranes. From the optical and confocal microscopy analysis of GUVs loaded with the probe, the probe is able to anchor to the lipid double layer, without internalizing. In addition, the fluorescence of the 3HF18 probe is visible in both Ld and Lo vesicles. MD simulations show that the insertion of the probe does not disturb the membrane, since both the membrane thickness and the order parameter remain essentially unchanged. Furthermore, thanks to a new semi-semantic approach, it was possible to assess that Ld membranes are more sensitive to probe insertion, and that the 3HF18 molecule shows a preference for the Lo or Ld phases of a model membrane, depending on the most stable tautomeric form.

Supplementary Materials: The following are available online, Table S1: Thickness values (in Å) of the two membranes with and without the probe, Figure S1: SCD of the unsaturated POPC chain with (blue) and without (green) the probe, Figure S2: SCD of POPC saturated chain with (blue) and without (green) probe, Figure S3: SCD of the DPPC Sn-1 palmitoyl chain (DPPC-Chol) with and without probe, Figure S4: SCD of the DPPC Sn-2 palmitoyl chain (DPPC-Chol) with and without probe, Figure S5: Fluorescence spectra of probe 3HF18 in binary mixture MeOH/H_2O in different ratios, Table S2: Results of t-Student test.

Author Contributions: Conceptualization, S.C. and S.P.; methodology, L.S. and F.R.; investigation, M.D.M., B.P., A.M.N. and Y.M.; data curation, L.S. and A.M.N.; writing—original draft preparation, M.D.M.; writing—review and editing, S.C. and S.P.; supervision, F.R. and B.P.; funding acquisition, S.P. All authors have read and agreed to the published version of the manuscript.

Funding: This research was funded by the Italian Ministry of Education, University and Research MIUR (grant number 300395FRB18), and by the project Clinglio, within the European Commission H2020 Framework Programme (agreement number 755179).

Acknowledgments: Jlenia Brunetti is gratefully acknowledged for data acquisition at the confocal microscope.

Conflicts of Interest: The authors declare no conflict of interest.

References

1. Simons, K.; Vaz, W.L.C. Model Systems, Lipid Rafts, and Cell Membranes 1. *Annu. Rev. Biophys. Biomol. Struct.* **2004**, *33*, 269–295. [CrossRef] [PubMed]
2. Vaz, W.L.C.; Melo, E. Fluorescence Spectroscopic Studies on Phase Heterogeneity in Lipid Bilayer Membranes. *J. Fluoresc.* **2001**, *11*, 255–271. [CrossRef]
3. Di Martino, M.; Marrafino, F.; Diana, R.; Iannelli, P.; Concilio, S. Fluorescent Probes for Applications in Bioimaging. In *Advances in Bionanomaterials II*; Piotto, S., Concilio, S., Sessa, L., Rossi, F., Eds.; Lecture Notes in Bioengineering; Springer International Publishing: Cham, Switzerland, 2020; pp. 243–258. [CrossRef]
4. Klymchenko, A.S.; Kreder, R. Fluorescent Probes for Lipid Rafts: From Model Membranes to Living Cells. *Chem. Boil.* **2014**, *21*, 97–113. [CrossRef] [PubMed]
5. Diana, R.; Panunzi, B.; Tuzi, A.; Piotto, S.; Concilio, S.; Caruso, U. An Amphiphilic Pyridinoyl-hydrazone Probe for Colorimetric and Fluorescence pH Sensing. *Molecules* **2019**, *24*, 3833. [CrossRef] [PubMed]
6. Borbone, F.; Caruso, U.; Concilio, S.; Nabha, S.; Piotto, S.; Shikler, R.; Tuzi, A.; Panunzi, B. From cadmium (II)-aroylhydrazone complexes to metallopolymers with enhanced photoluminescence. A structural and DFT study. *Inorg. Chim. Acta.* **2017**, *458*, 129–137. [CrossRef]
7. Baumgart, T.; Hunt, G.; Farkas, E.R.; Webb, W.W.; Feigenson, G.W. Fluorescence probe partitioning between Lo/Ld phases in lipid membranes. *Biochim. Biophys. Acta (BBA)-Biomembr.* **2007**, *1768*, 2182–2194. [CrossRef]
8. Demchenko, A.P.; Mely, Y.; Duportail, G.; Klymchenko, A.S. Monitoring Biophysical Properties of Lipid Membranes by Environment-Sensitive Fluorescent Probes. *Biophys. J.* **2009**, *96*, 3461–3470. [CrossRef]

9. Panunzi, B.; Borbone, F.; Capobianco, A.; Concilio, S.; Diana, R.; Peluso, A.; Piotto, S.; Tuzi, A.; Velardo, A.; Caruso, U. Synthesis, spectroscopic properties and DFT calculations of a novel multipolar azo dye and its zinc(II) complex. *Inorg. Chem. Commun.* **2017**, *84*, 103–108. [CrossRef]
10. Bansal, M.; Kaur, R. Electromeric effect of substitution at 6 th position in 2-(Furan-2-yl)-3-hydroxy-4 H-chromen-4-one (FHC) on the absorption and emission spectra. *J. Chem. Sci.* **2015**, *127*, 405–412. [CrossRef]
11. Klymchenko, A.S.; Demchenko, A.P. Electrochromic Modulation of Excited-State Intramolecular Proton Transfer: The New Principle in Design of Fluorescence Sensors. *J. Am. Chem. Soc.* **2002**, *124*, 12372–12379. [CrossRef]
12. Klymchenko, A.S.; Demchenko, A.P. Multiparametric probing of intermolecular interactions with fluorescent dye exhibiting excited state intramolecular proton transfer. *Phys. Chem. Chem. Phys.* **2003**, *5*, 461–468. [CrossRef]
13. Klymchenko, A.S.; Duportail, G.; Ozturk, T.; Pivovarenko, V.; Mély, Y.; Demchenko, A.P. Novel two-band ratiometric fluorescence probes with different location and orientation in phospholipid membranes. *Chem. Boil.* **2002**, *9*, 1199–1208. [CrossRef]
14. Ghosh, D.; Batuta, S.; Das, S.; Begum, N.A.; Mandal, D. Proton Transfer Dynamics of 4′-N,N-Dimethylamino-3-hydroxyflavone Observed in Hydrogen-Bonding Solvents and Aqueous Micelles. *J. Phys. Chem. B* **2015**, *119*, 5650–5661. [CrossRef] [PubMed]
15. Shynkar, V.V.; Mely, Y.; Duportail, G.; Piemont, E.; Klymchenko, A.S.; Demchenko, A.P. Picosecond time-resolved fluorescence studies are consistent with reversible excited-state intramolecular proton transfer in 4 ′-(dialkylamino)-3-hydroxyflavones. *J. Phys. Chem. A* **2003**, *107*, 9522–9529. [CrossRef]
16. Darwich, Z.; Kucherak, O.; Kreder, R.; Richert, L.; Vauchelles, R.; Mély, Y.; Klymchenko, A.S. Rational design of fluorescent membrane probes for apoptosis based on 3-hydroxyflavone. *Methods Appl. Fluoresc.* **2013**, *1*, 25002. [CrossRef] [PubMed]
17. Ozturk, T.; Klymchenko, A.S.; Capan, A.; Oncul, S.; Cikrikci, S.; Taskiran, S.; Tasan, B.; Kaynak, F.B.; Ozbey, S.; Demchenko, A.P. New 3-hydroxyflavone derivatives for probing hydrophobic sites in microheterogeneous systems. *Tetrahedron* **2007**, *63*, 10290–10299. [CrossRef]
18. Vigh, L.; Escribá, P.V.; Sonnleitner, A.; Sonnleitner, M.; Piotto, S.; Maresca, B.; Horvàth, I.; Harwood, J.L. The significance of lipid composition for membrane activity: New concepts and ways of assessing function. *Prog. Lipid Res.* **2005**, *44*, 303–344. [CrossRef]
19. Török, Z.; Crul, T.; Maresca, B.; Schütz, G.J.; Viana, F.; Dindia, L.; Piotto, S.; Brameshuber, M.; Balogh, G.; Péter, M.; et al. Plasma membranes as heat stress sensors: From lipid-controlled molecular switches to therapeutic applications. *Biochim. Biophys. Acta (BBA)-Biomembr.* **2014**, *1838*, 1594–1618. [CrossRef]
20. Piotto, S.; Trapani, A.; Bianchino, E.; Ibarguren, M.; Lopez, D.J.; Busquets, X.; Concilio, S. The effect of hydroxylated fatty acid-containing phospholipids in the remodeling of lipid membranes. *Biochim. Biophys. Acta (BBA)-Biomembr.* **2014**, *1838*, 1509–1517. [CrossRef]
21. Concilio, S.; Ferrentino, I.; Sessa, L.; Massa, A.; Iannelli, P.; Diana, R.; Panunzi, B.; Rella, A.; Piotto, S. A novel fluorescent solvatochromic probe for lipid bilayers. *Supramol. Chem.* **2017**, *29*, 1–9. [CrossRef]
22. Kreder, R.; Oncul, S.; Kucherak, O.; Pyrshev, K.A.; Klymchenko, A.S.; Real, E.; Mély, Y. Blue fluorogenic probes for cell plasma membranes fill the gap in multicolour imaging. *RSC Adv.* **2015**, *5*, 22899–22905. [CrossRef]
23. Shynkar, V.V.; Klymchenko, A.S.; Kunzelmann, C.; Duportail, G.; Muller, C.; Demchenko, A.P.; Freyssinet, J.-M.; Mely, Y. Fluorescent Biomembrane Probe for Ratiometric Detection of Apoptosis. *J. Am. Chem. Soc.* **2007**, *129*, 2187–2193. [CrossRef] [PubMed]
24. Fenz, S.F.; Sengupta, K. Giant vesicles as cell models. *Integr. Boil.* **2012**, *4*, 982. [CrossRef] [PubMed]
25. Walde, P.; Cosentino, K.; Engel, H.; Stano, P. Giant Vesicles: Preparations and Applications. *ChemBioChem* **2010**, *11*, 848–865. [CrossRef] [PubMed]
26. Pautot, S.; Frisken, B.J.; Weitz, D.A. Production of Unilamellar Vesicles Using an Inverted Emulsion. *Langmuir* **2003**, *19*, 2870–2879. [CrossRef]
27. Miele, Y.; Bánsági, T.; Taylor, A.; Stano, P.; Rossi, F. Engineering Enzyme-Driven Dynamic Behaviour in Lipid Vesicles. In *Advances in Artificial Life, Evolutionary Computation and Systems Chemistry*; Rossi, F., Mavelli, F., Stano, P., Caivano, D., Eds.; Communications in Computer and Information Science; Springer International Publishing: Cham, Switzerland, 2016; pp. 197–208. [CrossRef]

28. Moga, A.; Yandrapalli, N.; Dimova, R.; Robinson, T. Optimization of the Inverted Emulsion Method for High-Yield Production of Biomimetic Giant Unilamellar Vesicles. *ChemBioChem* **2019**, *20*, 2674–2682. [CrossRef]
29. Stano, P.; De Souza, T.P.; Carrara, P.; Altamura, E.; Caputo, M.; Luisi, P.L.; Mavelli, F.; D'Aguanno, E. Recent Biophysical Issues About the Preparation of Solute-Filled Lipid Vesicles. *Mech. Adv. Mater. Struct.* **2014**, *22*, 748–759. [CrossRef]
30. Miele, Y.; Medveczky, Z.; Holló, G.; Tegze, B.; Derényi, I.; Hórvölgyi, Z.; Altamura, E.; Lagzi, I.; Rossi, F. Self-division of giant vesicles driven by an internal enzymatic reaction. *Chem. Sci.* **2020**, *11*, 3228–3235. [CrossRef]
31. Peyret, A.; Ibarboure, E.; Le Meins, J.; Lecommandoux, S. Asymmetric Hybrid Polymer–Lipid Giant Vesicles as Cell Membrane Mimics. *Adv. Sci.* **2017**, *5*, 1700453. [CrossRef]
32. Miele, Y.; Mingotaud, A.-F.; Caruso, E.; Malacarne, M.C.; Izzo, L.; Lonetti, B.; Rossi, F. Hybrid giant lipid vesicles incorporating a PMMA-based copolymer. *Biochim. et Biophys. Acta (BBA)-Gen. Subj.* **2020**, 129611. [CrossRef]
33. Subczynski, W.K.; Pasenkiewicz-Gierula, M.; Widomska, J.; Mainali, L.; Raguz, M. High Cholesterol/Low Cholesterol: Effects in Biological Membranes: A Review. *Cell Biophys.* **2017**, *75*, 369–385. [CrossRef] [PubMed]
34. Drolle, E.; Kučerka, N.; Hoopes, M.; Choi, Y.; Katsaras, J.; Karttunen, M.; Leonenko, Z. Effect of melatonin and cholesterol on the structure of DOPC and DPPC membranes. *Biochim. Biophys. Acta (BBA)-Biomembr.* **2013**, *1828*, 2247–2254. [CrossRef] [PubMed]
35. Kučerka, N.; Nieh, M.-P.; Katsaras, J. Fluid phase lipid areas and bilayer thicknesses of commonly used phosphatidylcholines as a function of temperature. *Biochim. Biophys. Acta (BBA)-Biomembr.* **2011**, *1808*, 2761–2771. [CrossRef]
36. Wu, E.L.; Cheng, X.; Jo, S.; Rui, H.; Song, K.C.; Dávila-Contreras, E.M.; Qi, Y.; Lee, J.; Monje-Galvan, V.; Venable, R.M. CHARMM-GUI membrane builder toward realistic biological membrane simulations. *J. Comput. Chem.* **2014**, *35*, 1997–2004. [CrossRef] [PubMed]
37. Warschawski, D.; Devaux, P.F. Order parameters of unsaturated phospholipids in membranes and the effect of cholesterol: A 1H–13C solid-state NMR study at natural abundance. *Eur. Biophys. J.* **2005**, *34*, 987–996. [CrossRef] [PubMed]
38. Stano, P.; Wodlei, F.; Carrara, P.; Ristori, S.; Marchettini, N.; Rossi, F. Approaches to Molecular Communication Between Synthetic Compartments Based on Encapsulated Chemical Oscillators. In *Advances in Artificial Life and Evolutionary Computation*; Pizzuti, C., Spezzano, G., Eds.; Communications in Computer and Information Science; Springer International Publishing: Cham, Switzerland, 2014; pp. 58–74. [CrossRef]
39. Krieger, E.; Vriend, G. YASARA View–molecular graphics for all devices–from smartphones to workstations. *Bioinformatics* **2014**, *30*, 2981–2982. [CrossRef]
40. Casas, J.; Ibarguren, M.; Álvarez, R.; Terés, S.; Lladó, V.; Piotto, S.; Concilio, S.; Busquets, X.; Lopez, D.J.; Escribá, P.V.; et al. G protein-membrane interactions II: Effect of G protein-linked lipids on membrane structure and G protein-membrane interactions. *Biochim. et Biophys. Acta (BBA)–Biomembr.* **2017**, *1859*, 1526–1535. [CrossRef]
41. Piotto, S.; Concilio, S.; Bianchino, E.; Iannelli, P.; Lopez, D.J.; Terés, S.; Ibarguren, M.; Barceló-Coblijn, G.; Martin, M.L.; Guardiola-Serrano, F.; et al. Differential effect of 2-hydroxyoleic acid enantiomers on protein (sphingomyelin synthase) and lipid (membrane) targets. *Biochim. et Biophys. Acta (BBA)–Biomembr.* **2014**, *1838*, 1628–1637. [CrossRef]
42. Mario, S.; Di Marino, S.; Grimaldi, M.; Campana, F.; Vitiello, G.; Piotto, S.; D'Errico, G.; D'Ursi, A.M. Structural features of the C8 antiviral peptide in a membrane-mimicking environment. *Biochim. Biophys. Acta (BBA)–Biomembr.* **2014**, *1838*, 1010–1018. [CrossRef]
43. Humphrey, W.; Dalke, A.; Schulten, K. VMD: Visual molecular dynamics. *J. Mol. Graph.* **1996**, *14*, 33–38. [CrossRef]
44. Guixà-González, R.; Rodríguez-Espigares, I.; Ramirez-Anguita, J.M.; Carrió-Gaspar, P.; Monne, H.M.-S.; Giorgino, T.; Selent, J. MEMBPLUGIN: Studying membrane complexity in VMD. *Bioinform.* **2014**, *30*, 1478–1480. [CrossRef] [PubMed]
45. Nardiello, A.M.; Piotto, S.; Di Biasi, L.; Sessa, L. Pseudo-semantic Approach to Study Model Membranes. In *Advances in Bionanomaterials II*; Piotto, S., Concilio, S., Sessa, L., Rossi, F., Eds.; Lecture Notes in Bioengineering; Springer International Publishing: Cham, Switzerland, 2020; pp. 120–127. [CrossRef]

46. Piotto, S.; Nardiello, A.M.; Di Biasi, L.; Sessa, L. Encoding Materials Dynamics for Machine Learning Applications. In *Advances in Bionanomaterials II*; Piotto, S., Concilio, S., Sessa, L., Rossi, F., Eds.; Lecture Notes in Bioengineering; Springer International Publishing: Cham, Switzerland, 2020; pp. 128–136. [CrossRef]
47. Maaten, L.v.d.; Hinton, G. Visualizing data using t-SNE. *J. Mach. Learn. Res.* **2008**, *9*, 2579–2605.

Sample Availability: Samples of the compound 3HF18 are available from the authors.

MDPI

Article

Sulfanyl Porphyrazines with Morpholinylethyl Periphery—Synthesis, Electrochemistry, and Photocatalytic Studies after Deposition on Titanium(IV) Oxide P25 Nanoparticles

Tomasz Koczorowski [1,*], Wojciech Szczolko [1], Anna Teubert [2] and Tomasz Goslinski [1]

[1] Chair and Department of Chemical Technology of Drugs, Poznan University of Medical Sciences, Grunwaldzka 6, 60-780 Poznan, Poland; wszczolko@wp.pl (W.S.); tomasz.goslinski@ump.edu.pl (T.G.)

[2] Institute of Bioorganic Chemistry, Polish Academy of Sciences, Z. Noskowskiego 12, 61-704 Poznan, Poland; ateubert@ibch.poznan.pl

* Correspondence: tkoczorowski@ump.edu.pl

Abstract: The syntheses, spectral UV–Vis, NMR, and electrochemical as well as photocatalytic properties of novel magnesium(II) and zinc(II) symmetrical sulfanyl porphyrazines with 2-(morpholin-4-yl)ethylsulfanyl peripheral substituents are presented. Both porphyrazine derivatives were synthesized in cyclotetramerization reactions and subsequently embedded on the surface of commercially available P25 titanium(IV) oxide nanoparticles. The obtained macrocyclic compounds were broadly characterized by ESI MS spectrometry, 1D and 2D NMR techniques, UV–Vis spectroscopy, and subjected to electrochemical studies. Both hybrid materials, consisting of porphyrazine derivatives embedded on the titanium(IV) oxide nanoparticles' surface, were characterized in terms of particle size and distribution. Next, they were subjected to photocatalytic studies with 1,3-diphenylisobenzofuran, a known singlet oxygen quencher. The applicability of the obtained hybrid material consisting of titanium(IV) oxide P25 nanoparticles and magnesium(II) porphyrazine derivative was assessed in photocatalytic studies with selected active pharmaceutical ingredients, such as diclofenac sodium salt and ibuprofen.

Keywords: catalysis; electrochemistry; morpholine; porphyrazine; titanium(IV) oxide

Citation: Koczorowski, T.; Szczolko, W.; Teubert, A.; Goslinski, T. Sulfanyl Porphyrazines with Morpholinylethyl Periphery—Synthesis, Electrochemistry, and Photocatalytic Studies after Deposition on Titanium(IV) Oxide P25 Nanoparticles. *Molecules* **2021**, *26*, 2280. https://doi.org/10.3390/molecules26082280

Academic Editor: Ana Margarida Gomes da Silva

Received: 29 March 2021
Accepted: 13 April 2021
Published: 15 April 2021

1. Introduction

Porphyrazines (Pzs) are synthetic tetrapyrrole macrocyclic molecules, known as aza-analogues of porphyrins. Their physicochemical properties can be tuned by the exchange of the central metal cation or by peripheral substitution [1]. Substituted porphyrazines reveal high absorption in the UV–Vis region and good effectiveness for singlet oxygen generation. In addition, they are usually soluble in organic solvents [2–5]. Their unique physicochemical properties, including optical and electrochemical ones, make them useful in biosensing [6], photocatalysis [7], nonlinear optics [8], and biomedicine, where they can be considered as photosensitizers for photodynamic therapy [9]. Amino and sulfanyl porphyrazines can be obtained from a cyclotetramerization reaction starting from diaminomaleonitrile or dimercaptomaleonitrile disodium salt derivatives, respectively.

Porphyrazines peripherally substituted with sulfanyl moieties have been studied widely over the last twenty years. They are well-soluble in common organic solvents [10–12], and present interesting optical [13,14] and electrochemical properties, and have therefore been applied as sensing materials in technology [15–21]. Other applications of sulfanyl Pzs concern photodynamic therapy (PDT) [22–24], wastewater treatment [25–27], and catalysis [28–31]. Symmetrical octa-substituted sulfanyl porphyrazines, unlike their unsymmetrical derivatives, usually present low singlet oxygen generation quantum yields, and therefore their biological activities are limited [32,33]. Recently, many symmetrical and unsymmetrical magnesium(II) and zinc(II) sulfanyl porphyrazines with bulky dendrimeric periphery were synthesized and evaluated in terms of their suitability for photodynamic therapy (PDT) [34–38]. In addition, some sulfanyl porphyrazines and phthalocyanines with morpholinyl moieties were studied as photosensitizers for PDT and photodynamic antimicrobial chemotherapy (PACT) and revealed promising potential [39–41].

Due to the presence of an expanded aromatic system in porphyrazines, they are often highly hydrophobic, and therefore less soluble or even insoluble in water. Thus, diverse methods, using specific carriers, have been employed to allow porphyrazines and related compounds to form stable suspensions in aqueous solutions. The carriers most widely applied in medicine and technology are liposomes [24,42,43], metal and metal oxide nanoparticles [44–46], and polymeric nanomaterials [47]. Among the metal oxide nanoparticles, one of the most interesting is titanium(IV) oxide nanoparticles (TiO_2), which have unique photochemical features. Uncoated TiO_2 nanoparticles are photoactive only when irradiated with UV light. However, TiO_2 nanocarriers coated with photoactive compounds, like porphyrinoid macrocycles, when irradiated with light of an appropriate energy, can absorb light and participate in energy transfer, and thus more effectively take part in photochemical reactions [48–50].

Herein, we present the synthesis, spectral UV–Vis, NMR, and electrochemical as well as photocatalytic properties of novel magnesium(II) and zinc(II) symmetrical sulfanyl porphyrazines with 2-(morpholin-4-yl)ethylsulfanyl peripheral substituents. The synthesized macrocyclic compounds were embedded on the surface of commercially available P25 titanium(IV) oxide nanoparticles. The obtained grafted hybrid material was subjected to photocatalytic studies with 1,3-diphenylisobenzofuran, a known singlet oxygen quencher, and with diclofenac sodium salt and ibuprofen as examples of active pharmaceutical ingredients.

2. Results and Discussion

2.1. Synthesis and Physicochemical Characterization

The synthetic pathway was based on the alkylation reaction of commercially available mercaptomaleonitrile disodium salt (**1**) with 4-(2-chloroethyl)morpholine hydrochloride (**2**) in dimethylformamide (DMF), and with potassium carbonate as a base, which led to compound **3** (Scheme 1) [20,51]. Next, the Linstead macrocyclization reaction of **3** with magnesium butanolate as a base in *n*-butanol led to novel symmetric sulfanyl magnesium(II) porphyrazine **4** [52]. Simultaneously, compound **3** was also used in the macrocyclization reaction with $Zn(OAc)_2$ and 1,8-diazabicyclo[5.4.0]undec-7-ene (DBU) in *n*-pentanol to give zinc(II) porphyrazine **5** [53].

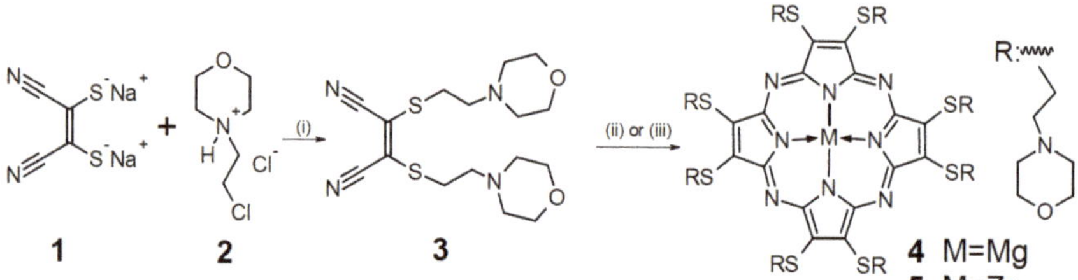

Scheme 1. Reagents and conditions: (i) K_2CO_3, DMF, 60 °C, 24 h; (ii) $Mg(n\text{-}BuO)_2$, *n*-butanol, reflux, 24 h; (iii) $Zn(OAc)_2$, DBU, *n*-pentanol, reflux, 24 h; DBU—1,8-diazabicyclo[5.4.0]undec-7-ene, DMF—dimethylformamide.

All synthesized novel macrocyclic compounds were characterized using various analytical techniques, including high-resolution mass spectra (ESI), one- and two-dimensional NMR spectroscopy, and UV–Vis spectrophotometry. Notably, Pzs **4** and **5** have very low melting points at 113–116 °C and 121–123 °C, respectively, which is relatively low compared with other sulfanyl and amino porphyrazines that melt over 300 °C [20,54].

In the UV–Vis spectra of porphyrazines **4** and **5** in dichloromethane, two intensive bands were found: a Soret or B band in the range of 250–400 nm, and a Q band between 600 and 800 nm (Figure 1). Magnesium(II) Pz **4** revealed strong absorption Soret and Q bands with maxima at 375 m and 671 nm, whereas zinc(II) Pz **5** maxima appeared at 373 and 668 nm, respectively. The only difference between the UV–Vis spectra of **4** and **5** was noted in the region between 450 and 550 nm, where, for Pz **4,** only a weak absorption maximum at 497 nm appeared. A similar effect was noted before for sulfanyl porphyrazines with peripheral phthalimide motifs [38].

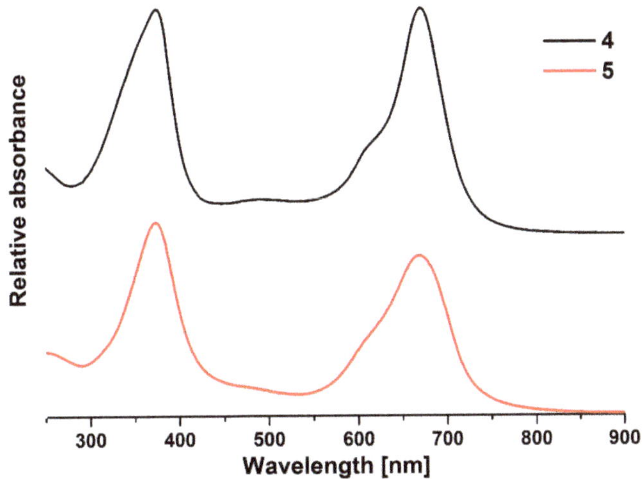

Figure 1. The UV–Vis spectrum of Pz **4** (black line) and Pz **5** (red line) in dichloromethane.

In the NMR spectra of porphyrazines **4** and **5**, which were recorded in pyridine-d_5, four distinguishable signals in the aliphatic region were noted in the ^{1}H NMR, whereas there were six signals in the ^{13}C NMR (four aliphatic and two aromatic, originating from pyrrolyl rings of macrocycle, see Supplementary Materials). The two-dimensional technique ^{1}H-^{1}H COSY NMR was used to assist allocation and analysis of protons within ethylsulfanyl and morpholinyl substituents. Signal shift analyses performed for Pz **4** and Pz **5** indicated similarities for two morpholinyl proton signals and one of two methylene signals within the ethylene linker. A slight difference in the shifts of signals was noted for another methylene group within the ethylene linker. These methylene protons appeared as a singlet at 4.64 ppm for Pz **4**, and as a multiplet in the range of 3.81–3.89 ppm for Pz **5**. In addition, in the ^{1}H NMR spectra of Pz **4** and **5**, the proton signals were generally up-field shifted in comparison with those in the structure of symmetrical magnesium(II) phthalocyanine with eight 2-(morpholin-4-yl)ethoxy substituents [39]. This finding could be explained by a higher electronegativity of oxygen atoms than sulfur present in the peripheries of both groups of macrocycles.

2.2. Electrochemistry

The electrochemical study was performed to assess the electrochemical properties of the obtained porphyrazines, aiming to propose their prospective potential applicabilities. The cyclic (CV) and differential pulse (DPV) voltammetry measurements were conducted in organic solvent (dichloromethane), with the addition of supporting electrolyte: 0.1 M tetrabutylammonium perchlorate. The classic three-electrode system was employed with the glassy carbon working electrode. Due to the use of Ag wire as a pseudo-reference electrode, the ferrocene was added as an internal standard, and all results were adjusted to the ferrocene/ferrocenium peak potential. In the CV experiments, the scan potential range was set between 50 and 250 mV/s. The obtained results are presented in Figures 2 and 3 and Table 1.

In the voltammograms recorded for both porphyrazines **4** and **5**, four redox peak potentials were noted. Almost all redox peaks observed in the CV voltammograms are irreversible due to the aggregation-disaggregation behavior of porphyrazines in dichloromethane; thus, redox pairs can be observed only in the DPV measurements. A similar phenomenon was previously observed for iron(II) porphyrazine bearing identical periphery to both herein studied Pzs [20]. However, the oxidation peaks (**IV**) at 0.62 V for Pz **4** and 0.53 V for Pz **5** were noted in the DPV voltammogram only when the applied potential was increasing over time from −2.0 to 0.7 V. Conversely, when the applied potential was decreasing over time, oxidation peaks were not observed (Figures 2 and 3). The oxidation peak currents were at least five times higher than the highest reduction peak (**I**) in the case of both porphyrazines (Figures 2 and 3). Such a high oxidation peak current could also be a result of an aggregation phenomenon, which was previously observed for other sulfanyl porphyrazines [15]. Notably, the peak potentials of zinc(II) porphyrazine **5** shifted to more negative potentials in comparison with magnesium(II) complex **4** (Table 1), which is the result of different metal cations present inside their cores [55].

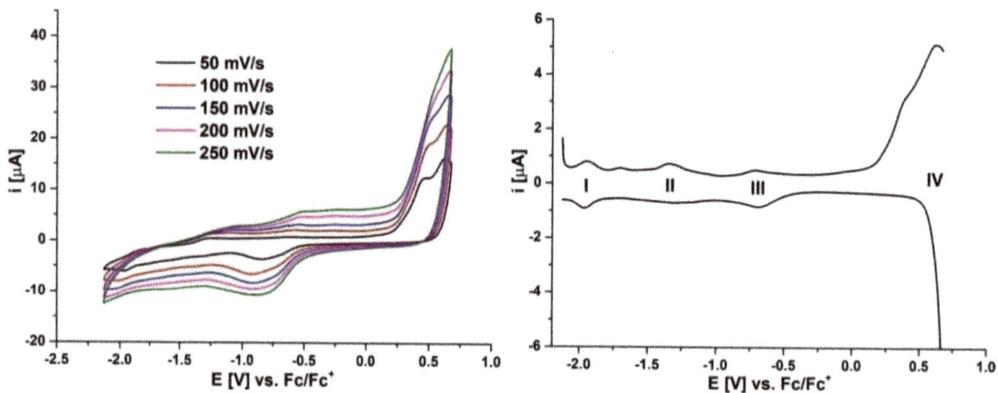

Figure 2. The cyclic and differential pulse voltammograms of porphyrazine **4** in 0.1 M TBAP/DCM. The DPV parameters: modulation amplitude 20 mV and step rate 5 mV·s^{-1}.

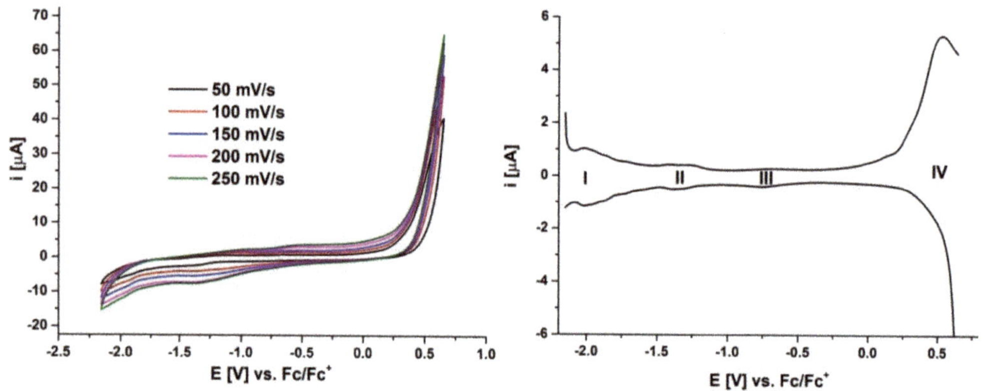

Figure 3. The cyclic and differential pulse voltammograms of porphyrazine **5** in 0.1 M TBAP/DCM. The DPV parameters: modulation amplitude 20 mV and step rate 5 mV·s^{-1}.

Measurements performed for amino porphyrazines in dichloromethane indicated that the first oxidation peak appears below 0 V vs. Fc/Fc^+ [56,57]. A different situation was observed when alkyl- or phenylsulfanyl substituents were present in the macrocyclic periphery. Then, peaks shifted toward positive values due to a strong electron-withdrawing effect [55,58]. The oxidation peak potentials (IV) of Pzs **4** and **5** comply with this rule.

Table 1. The electrochemical data of porphyrazines **4** and **5**.

Pz		$MPz^{(-2)}/MPz^{(-3)}$ I	$MPz^{(-1)}/MPz^{(-2)}$ II	$MPz^{(0)}/MPz^{(-1)}$ III	$MPz^{(0)}/MPz^{(+1)}$ IV
4	$E_{1/2}$ [V] vs. Fc/Fc^+	−1.95	−1.33	−0.70	0.62
5	$E_{1/2}$ [V] vs. Fc/Fc^+	−2.02	−1.34	−0.74	0.53

In order to calculate the HOMO-LUMO energy levels in compounds such as porphyrinoid macrocycles, electrochemical measurements were used. The electrochemical energy gap ($E_{gap\ el}$) was calculated by determining the onset potentials of first oxidation and first reduction processes originating from the porphyrazine ring with the use of the following equations:

$$E_{HOMO} = -(V_{onset\ ox} - V_{FOC} + 4.8)\ eV,$$

$$E_{LUMO} = -(V_{onset\ red} - V_{FOC} + 4.8)\ eV,$$

$$E_{gap\ el} = (E_{LUMO} - E_{HOMO})\ eV$$

In the above equations, V_{FOC} stands for the ferrocene half-wave potential, $V_{onset\ ox}$ is assigned as the Pz oxidation onset, and $V_{onset\ red}$ represents the Pz reduction onset. For all these values (eV), the calculations of the first oxidation and the first reduction were adjusted to the ferrocene's energy level at −4.8 eV. The value of 4.8 eV refers to a standard electrode potential for normal hydrogen electrode (NHE) at −4.6 eV on the zero vacuum level scale, and a value of 0.2 eV versus NHE for the potential of ferrocene standard [59,60]. The calculated electrochemical energy gap was slightly higher for magnesium(II) porphyrazine **4** than for zinc(II) complex **5** (Figure 4 and Table 2).

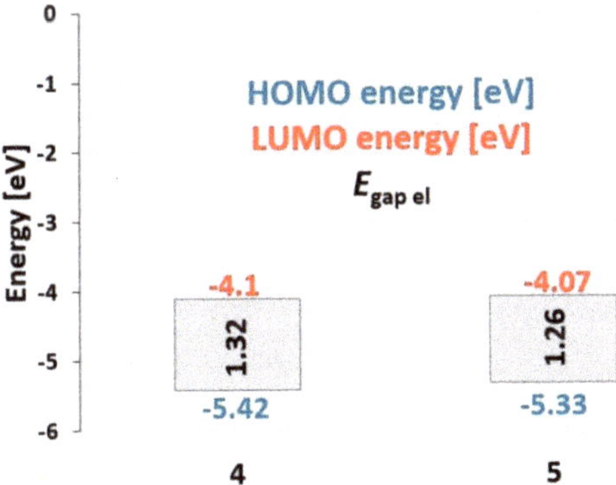

Figure 4. The HOMO-LUMO energy levels estimated for electrochemical data for porphyrazines **4** and **5**.

At the same time as the previous calculations, we calculated the optical band gaps ($E_{gap\ opt}$) based on the UV–Vis spectra Q band onsets, according to the equation $E = hc/\lambda_{onset}$ [61]. Both electrochemical and optical energy gaps, which we obtained, were subsequently compared and are presented in Table 2. The optical band gaps ($E_{gap\ opt}$)

for porphyrazines **4** and **5** were found to be in agreement with those obtained from electrochemical measurements within approx. 0.2 eV.

Table 2. The optical and electrochemical HOMO-LUMO band gaps for porphyrazines **4** and **5**.

	λ_{max} [nm] Soret	λ_{max} [nm] Q Band	λ_{onset} [nm]	Optical Band Gap $E_{gap\ opt}$ [eV]	Electrochemical Band Gap $E_{gap\ el}$ [eV]
4	375	671	820	1.51	1.32
5	373	668	850	1.46	1.26

2.3. TiO_2 Deposition and Characterization

Titanium(IV) oxide (TiO_2) nanoparticles have often been utilized in diverse studies as carriers for photosensitizers [62,63]. Among many titanium(IV) oxide types, the commercially available P25, consisting of a mixture of crystal phases of anatase and rutile, is the most popular. Herein, the hybrid materials, type Pz@P25, were prepared by depositing porphyrazine **4** or **5** on the surface of TiO_2 nanoparticles. The solutions of macrocycles were added to titania suspension, sonicated, and mixed for 72 h, yielding **4**@P25 and **5**@P25, respectively. The resulting hybrid materials contained 5% (*w*/*w*) of the macrocycle. The sizes and the dispersities of the obtained nanomaterials were subjected to the detailed analyses using a NanoSight LM10 instrument (sCMOS camera, 405 nm laser), equipped with a nanoparticle tracking analysis system. The diameters of the obtained hybrid materials were assessed and compared with the pure TiO_2 nanoparticles. The results are presented in Table 3.

Table 3. The particle size distribution of P25, **4**@P25, and **5**@P25.

Material	Measured Particle Size (nm)	Polydispersity Index [a]
P25	74.8 ± 7.7	0.17
4@P25	327.4 ± 15.5	0.13
5@P25	322.5 ± 28.2	0.15

[a] Calculated according to the formula PDI = $(SD/\text{mean diameter})^2$ [64].

Considering the measured particle size values, strong agglomeration of the hybrid nanoparticles was observed. The mean particle sizes of the **4**@P25 and **5**@P25 were four times higher than the unmodified P25 (74.8 ± 7.7 nm). This could suggest that the deposition of the macrocycles strongly influences the titanium(IV) oxide nanoparticles. In addition, in both cases, the calculated polydispersity indices were below 0.2, which indicated that the distributions of nanoparticles within the studied hybrid materials are monodisperse. It also seems that the presence of sulfanyl porphyrazines on the surface of P25 nanoparticles hampers the electrostatic interactions between Pz@P25 nanoparticles and allows obtaining Pz@P25 of specific diameters.

2.4. Photocatalysis

All hybrid materials were assessed for their photocatalytic oxidation abilities. These properties were studied with the use of a known singlet oxygen quencher, 1,3-diphenylisobenzofurane (DPBF), according to previously presented procedures [20,65]. In the UV–Vis spectrum, the decrease in the DPBF absorption band at 413 nm over a period of time results from the transformation of DPBF towards the new product, which is 1,2-dibenzoylbenzene (Scheme 2).

Measurements were performed in DMF at ambient temperature, and with the use of red-light LED lamps (665 nm). The intensity of light was adjusted to 10 mW/cm^2. The irradiations were conducted in a 10 mm quartz cuvette equipped with a magnetic stirrer. The results of photocatalytic reactions were evaluated with the use of UV–Vis spectrophotometry.

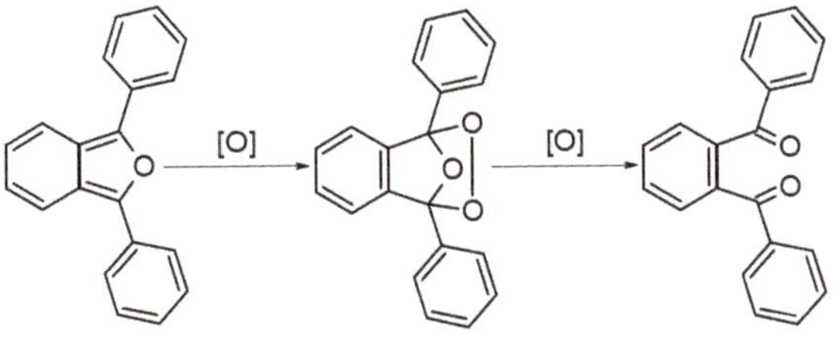

Scheme 2. The oxidation of 1,3-diphenylisobenzofurane to 1,2-dibenzoylbenzene.

The photocatalytic oxidations of DPBF were performed using three types of materials: **4**@P25, **5**@P25, and unmodified P25. Their catalytic activity was assessed by recording the UV–Vis scans within the range of 250–800 nm for 8 min, and every 2 min. The **4**@P25 hybrid material containing magnesium(II) sulfanyl porphyrazine deposited on the surface of TiO_2 nanoparticles revealed the highest photocatalytic activity. Moderate activity was noted for **5**@P25 and the lowest activity for the unmodified P25 (Figure 5). In the case of **5**@P25 nanoparticles, the linear plots of DPBF absorbance were decreasing with time, which indicates that the photooxidation process follows the first-order kinetics (Figure 5D). In the parallel study, the R^2 value measured for **4**@P25 hybrid material and P25 nanoparticles slightly deviated from unity.

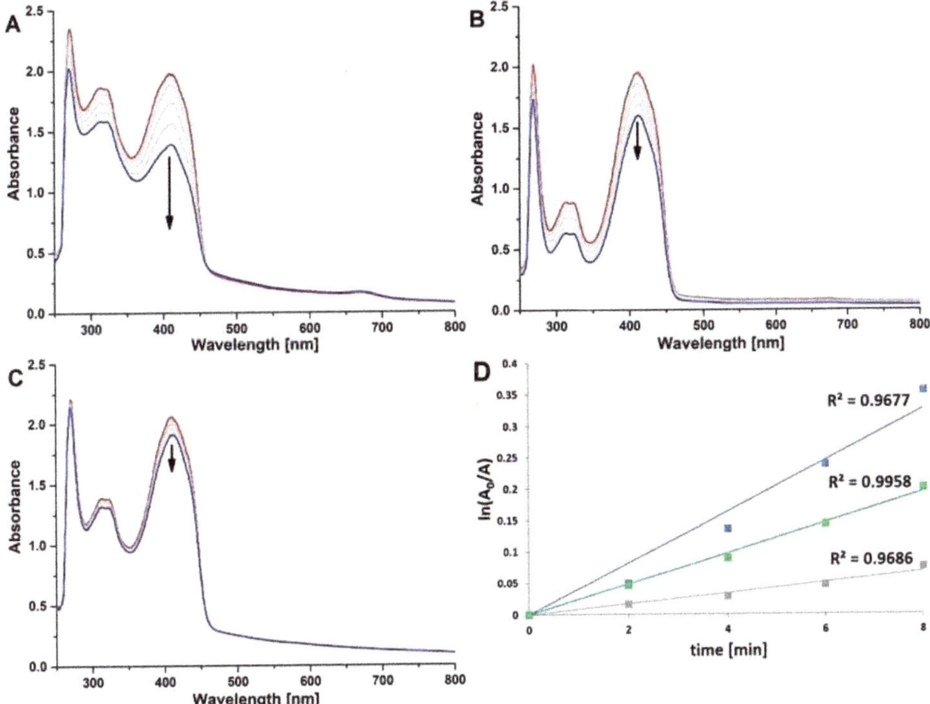

Figure 5. The UV–Vis spectra for the oxidation of DPBF in dimethylformamide in the presence of **4**@P25 (**A**), **5**@P25 (**B**), and P25 (**C**) as catalysts over a period of time. (**D**) the plots of DPBF absorbance in time in the presence of **4**@P25 (blue), **5**@P25 (green), and P25 (grey); DPBF—1,3-diphenylisobenzofurane.

The results obtained in the photocatalytic oxidation study with DPBF indicated the **4**@P25 hybrid material as a candidate for further photocatalytic study with selected active pharmaceutical ingredients (APIs): diclofenac sodium salt and ibuprofen. Both APIs are common non-steroidal anti-inflammatory drugs, and therefore constitute an important component of drug-related pollutants in water. What is essential is that the UV–Vis method can be employed for the photodegradation study of both APIs. The photodegradation study was performed in the same conditions as previously established for the photooxidation of DPBF. The results are presented in Figure 6. In the UV–Vis spectra, decreases in both APIs absorbances over a period of time were observed, which indicates the photodegradation of the studied compounds. For this reason, the obtained hybrid material **4@P25** can be considered an efficient heterogenic catalyst for further photooxidation studies of diverse organic compounds.

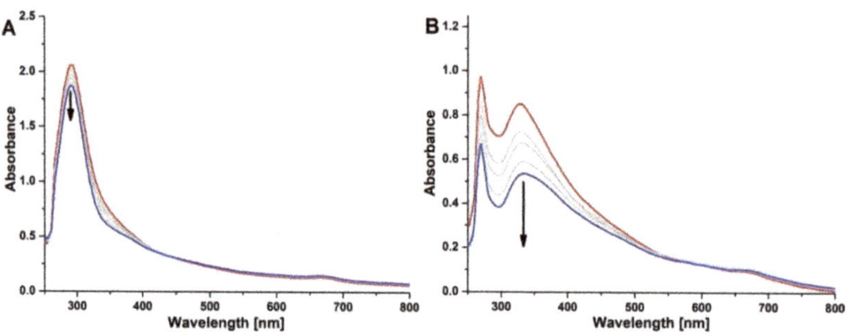

Figure 6. The UV–Vis spectra of photooxidation of diclofenac sodium salt (**A**) and ibuprofen (**B**) in dimethylformamide with **4@P25** as a catalyst over a period of time.

3. Materials and Methods

3.1. Materials and Instruments

All the reactions described in this paper were conducted under argon. Before attempting the reaction, the glassware was oven-dried (at 140 °C). All solvents were rotary evaporated under vacuum at or below 40 °C. All reaction temperatures reported in the experimental section refer to the external bath temperatures. The reactions were performed on a Heidolph MR Hei-Tec, equipped with Radleys Heat-On heating mantle. All solvents and reagents were obtained from commercial suppliers (Merck, Darmstadt, Germany; TCI, Zwijndrecht, Belgium, and Fluorochem, Hadfield, UK), and used without any further purification, unless otherwise stated. Melting points were measured with the use of a Stuart Bibby apparatus (Triad Scientific, Staffordshire, UK) and are uncorrected. Flash column chromatography was performed on a Merck neutral aluminum oxide gel, whereas thin-layer chromatography (TLC) was performed on aluminum oxide F254 plates (Merck) and visualized with a UV lamp (λ_{max} 254 or 365 nm). UV–Vis spectra were recorded with the use of an Ocean Optics USB 2000+ spectrometer (Ocean Opitics Inc., Largo, FL, USA). ^{1}H NMR and ^{13}C NMR spectra were recorded using Bruker Avance 400 and 500 (Bruker, Karlsruhe, Germany) spectrometers. Chemical shifts (δ) are specified in parts per million (ppm) and are referenced against a residual solvent peak (pyridine-d_5), whereas coupling constants (*J*) are calculated in Hertz (Hz). The abbreviations *s*, *t*, and *m* refer to singlet, triplet, and multiplet, respectively. Mass spectra (ESI MS) were performed in the Wielkopolska Centre of Advanced Technologies, Adam Mickiewicz University in Poznan, Poland.

3.2. Synthesis

2,3-Bis[2-(morpholin-4-ylo)ethylsulfanyl]-(2*Z*)-butene-1,4-dinitrile (3) is a known compound synthesized and characterized earlier in our group [20]: dimercaptomaleonitrile disodium salt (558 mg; 3.0 mmol) (**1**), 4-(2-chloroethyl)morpholine hydrochloride (1.396 g; 7.5 mmol) (**2**) and K_2CO_3 (4.140 g; 30.0 mmol) were mixed in DMF (30 mL) at 60 °C for 24 h under argon. Next, the reaction mixture was cooled to room temperature and filtered through Celite, then the filtrate was evaporated with toluene to a dry solid residue. Crude solid was subjected to a flash column chromatography with Al_2O_3 (DCM:CH_3OH; 50:1) to give **3** as yellow crystals (810 mg; 71% yield).

{2,3,7,8,12,13,17,18-Octakis[2-(morpholin-4-yl)ethylsulfanyl]porphyrazinato} magn esium(II) (4): magnesium turnings (45 mg; 1.88 mmol) and iodide (1 crystal) were suspended in *n*-butanol (15 mL) and refluxed for 6 h in an inert atmosphere. After the reaction mixture was cooled to room temperature, compound **3** (692 mg; 1.88 mmol) was added and the mixture was refluxed for 18 h. Next the mixture was filtrated through Celite and the solvents were evaporated with toluene to a dry solid. Crude product was purified by column chromatography on Al_2O_3 (DCM:MeOH, 50:1) to give porphyrazine **4** (110 mg; 16% yield) as a green-blue solid: mp 113-116 °C; R_f (DCM:MeOH:$N(C_2H_5)_3$, 10:1:0.1) 0.42. UV–Vis (DCM): λ_{max}, nm (logε) 375 (4.68), 497 (3.91), 671 (4.71). ^1H NMR (400 MHz; pyridine-d_5): δH, ppm 2.60 (*s*, 32H, morph-CH_2), 3.11 (*s*, 16H, CH_2), 3.67 (*s*, 32H, morph-CH_2), 4.64 (*t*, ^3J = 10.0 Hz, 16H, CH_2). ^{13}C NMR (100 MHz; pyridine-d_5): δC, ppm 33.5; 54.5; 60.0; 67.5; 141.9; 158.4. HRMS ESI (pos): calc. for $C_{64}H_{97}N_{16}O_8S_8Mg$ m/z 1497.5291 $[M + H]^+$; found m/z 1497.5340 $[M + H]^+$.

{2,3,7,8,12,13,17,18-Octakis[2-(morpholin-4-yl)ethylsulfanyl]porphyrazinato} zinc(II) (5): dimercaptomaleonitrile **3** (700 mg; 1.9 mmol), $Zn(OAc)_2$ (175 mg, 0.95 mmol) and DBU (142 μL, 0.95 mmol) in *n*-pentanol (4 mL) were refluxed in an inert atmosphere for 18 h. Next, the reaction mixture was filtrated through Celite and the filtrate was evaporated with toluene to dryness. Crude solid was purified by column chromatography with Al_2O_3 (DCM:MeOH; 50:1→10:1) to give porphyrazine **5** (90 mg; 12%) as a dark green solid: mp 121–123 °C; R_f (DCM:MeOH:$N(C_2H_5)_3$, 10:1:0.1) 0.57. UV–Vis (DCM): λ_{max}, nm (logε) 373 (4.79); 668 (4.71). ^1H NMR (500 MHz; pyridine-d_5): δH, ppm 2.54–2.63 (*m*, 32H, morph-CH_2), 3.09 (*s*, 16H, CH_2), 3.63–3.69 (*m*, 32H, morph-CH_2), 3.81–3.89 (*m*, 16H, CH_2). ^{13}C NMR (125 MHz; pyridine-d_5): δC, ppm 33.34; 54.37; 59.88; 67.38; 141.92; 157.34. HRMS ESI (pos): calc. for $C_{64}H_{97}N_{16}O_8S_8Zn$ m/z 1539.4721 $[M+H]^+$; found m/z 1539.4756 $[M + H]^+$.

3.3. Electrochemical Studies

The electrochemical studies were performed with a Metrohm Autolab PGSTAT128N potentiostat (Metrohm, Herisau, Switzerland). The data acquisition and storage were driven by Metrohm Nova 2.1.4 software (Metrohm). The measurements were obtained with the use of a glassy carbon (GC) working electrode (area = 0.071 cm^2), Ag wire (pseudo-reference electrode), and a platinum wire (counter electrode). Before each procedure, the GC electrode was polished with aqueous 50 nm Al_2O_3 slurry (purchased from Sigma-Aldrich) using a polishing cloth and was subsequently washed in an ultrasonic bath with deionized water for 10 min to remove inorganic impurities. Ferrocene/ferrocenium couple (Fc/Fc^+) was applied as an internal standard. The solvent (dichloromethane) containing a supporting electrolyte (0.1 M tetrabutylammonium perchlorate (TBAP)) in a glass cell (volume 10 mL) was deoxygenated by purging nitrogen gas for 10 min prior to each experiment. All electrochemical experiments were carried out at 22 °C. The solvent and reagent were purchased from Sigma-Aldrich Chemie GmbH, Steinheim, Germany.

3.4. Deposition of Porphyrazines on TiO_2 P25 Nanoparticles

Studied porphyrazines were deposited on P25 Aeroxide®® titanium(IV) oxide (TiO_2) nanoparticles using the chemical deposition method [50]. In general, porphyrazine **4** or **5** in the amount of 5 mg was added to a dispersion of 100 mg P25 nanoparticles (sized approx. 21 nm) in 20 mL of dichloromethane:methanol mixture (1:1, *v*/*v*). After the reaction mixture had been stirred for 72 h, the solvents were evaporated on a rotary evaporator. Next, the obtained hybrid material was air dried for 24 h. The ratio of the macrocycle to the P25 TiO_2 was 1:20 (*w*/*w*).

The hybrid materials were subjected to nanoparticle size measurements using a Malvern Panalytical NanoSight LM10 instrument (Malvern, UK), equipped with sCMOS camera, and 405 nm laser. The data acquisition and storage were provided by Nanoparticle Tracking Analysis (NTA) 3.2 Dev Build 3.2.16 software (Malvern, UK). Throughout, the nanoparticles' dispersions were diluted with water (1 mg in 1 mL) to obtain the operating range of nanoparticle concentration. The measurements were performed at 25.0 ± 0.1 °C, and at the syringe pump infusion rate set to 100 µL/min.

3.5. *Photocatalytic Studies*

The photocatalytic studies were performed using a red-light LED lamp (EcoEnergy, Gdańsk, Poland) at wavelength 665 nm, and a power adjusted to 10 mW/cm^2 with the use of an Optel radiometer. The measurements were conducted in a 10 mm quartz cuvette in *N,N'*-dimethylformamide (DMF). In the experiments with a reference standard 1,3-diphenylisobenzofurane (DPBF), 1 mL of 0.1 mM DPBF solution in DMF was mixed with 1 mL of TiO_2 dispersion in DMF (0.1 mg/mL). In the experiments with active pharmaceutical ingredients (diclofenac sodium salt and ibuprofen), DPBF was replaced by 1 mL of 0.3 mM diclofenac sodium salt solution in DMF or 1 mL of 1.5 mM ibuprofen solution in DMF. The irradiations of mixtures were performed with an LED lamp within 8 min. The UV–Vis spectra were recorded every 2 min on an Ocean Optics USB 2000+ spectrometer (Ocean Optics Inc., Largo, FL, USA).

4. Conclusions

Two novel sulfanyl magnesium(II) and zinc(II) porphyrazines with morpholinylethyl periphery were synthesized in the cyclotetramerization reaction using a dimercaptomaleonitrile derivative. The obtained macrocyclic compounds were broadly characterized by ESI MS spectrometry, 1D and 2D NMR techniques, and UV–Vis spectroscopy. Both porphyrazines were subjected to electrochemical studies. Subsequently, the obtained porphyrazines were embedded on titanium(IV) oxide nanoparticles' surface and characterized in terms of particle size and distribution. The obtained hybrid materials' applicability was assessed in photocatalytic studies with a singlet oxygen quencher (DPBF) and selected drug active pharmaceutical ingredients (diclofenac sodium salt and ibuprofen). In the UV–Vis and NMR studies, the characteristic features of porphyrazines were confirmed. The electrochemical studies revealed four irreversible redox processes for both porphyrazines. In addition, the calculated electrochemical band gap values were found to be in agreement with the optical ones. Interestingly, the obtained hybrid materials presented four times higher particle sizes compared with unmodified titanium(IV) oxide P25 nanoparticles and were monodispersive. The **4@P25** material was found to be the most active in comparative photocatalytic tests with 1,3-diphenylisobenzofurane, and it was therefore used in the photooxidation studies of diclofenac sodium salt and ibuprofen. The **4@P25** material revealed good photocatalytic potential. For this reason, it can be considered in future photocatalytic experiments with various organic compounds and active pharmaceutical ingredients as a potential hybrid material for the photodegradation of various organic pollutants.

Supplementary Materials: The following are available online: Figure S1. ^{1}H NMR spectrum of **4** in pyridine-d_5. # indicates solvent residual peaks. Figure S2. ^{13}C NMR spectrum of **4** in pyridine-d_5. # indicates solvent residual peaks. Figure S3. ^{1}H-^{1}H COSY NMR spectrum of **4** in pyridine-d_5. Figure S4. ^{1}H NMR spectrum of **5** in pyridine-d_5. * indicates solvent residual peaks and # stands for water residual. Figure S5. ^{13}C NMR spectrum of **5** in pyridine-d_5. * indicates solvent residual peaks. Figure S6. ^{1}H-^{1}H COSY NMR spectrum of **5** in pyridine-d_5.

Author Contributions: Conceptualization, T.K.; methodology, T.K.; investigation, T.K., W.S. and A.T.; writing—original draft preparation, T.K.; writing—review and editing, W.S. and T.G.; visualization, A.T.; supervision, T.G.; funding acquisition, T.G. All authors have read and agreed to the published version of the manuscript.

Funding: This research was funded by National Science Centre, Poland under Grant No. 2016/21/B/NZ9/00783.

Institutional Review Board Statement: Not applicable.

Informed Consent Statement: Not applicable.

Data Availability Statement: Not applicable.

Acknowledgments: This study was supported by the National Science Centre, Poland under Grant No. 2016/21/B/NZ9/00783 and the European Fund Regional Development Fund No. UDA-POIG.02.01.00-30-182/09. The authors thank Beata Kwiatkowska and Rita Kuba for excellent technical assistance.

Conflicts of Interest: The authors declare no conflict of interest

Sample Availability: Samples of the compounds **3**, **4**, and **5** are available from the authors.

References

1. Rodríguez-Morgade, M.S.; Stuzhin, P.A. The chemistry of porphyrazines: An overview. *J. Porphyrins Phthalocyanines* **2004**, *8*, 1129–1165. [CrossRef]
2. Fuchter, M.J.; Zhong, C.; Zong, H.; Hoffman, B.M.; Barrett, A.G.M. Porphyrazines: Designer Macrocycles by Peripheral Substituent Change. *Aust. J. Chem.* **2008**, *61*, 235–255. [CrossRef]
3. Belviso, S.; Ricciardi, G.; Lelj, F.; Scolaro, L.M.; Bencini, A.; Carbonera, C. Inducing asymmetry in free-base, MnIII, NiII and CuII (ethylsulfanyl)porphyrazines: Synthetic aspects and spectro-electrochemical implications. *J. Chem. Soc. Dalton Trans.* **2001**, 1143–1150. [CrossRef]
4. Belviso, S.; Giugliano, A.; Amati, M.; Ricciardi, G.; Lelj, F.; Scolaro, L.M. Two-electron reduction of alkyl(sulfanyl)porphyrazines: A route to free-base and peripherally metallated asymmetric porphyrazines. *Dalton Trans.* **2003**, 305–312. [CrossRef]
5. Belviso, S.; Capasso, A.; Santoro, E.; Najafi, L.; Lelj, F.; Superchi, S.; Casarini, D.; Villani, C.; Spirito, D.; Bellani, S.; et al. Thioethyl-Porphyrazine/Nanocarbon Hybrids for Photoinduced Electron Transfer. *Adv. Funct. Mater.* **2018**, *28*, 1705418. [CrossRef]
6. Lochman, L.; Machacek, M.; Miletin, M.; Uhlířová, Š.; Lang, K.; Kirakci, K.; Zimcik, P.; Novakova, V. Red-Emitting Fluorescence Sensors for Metal Cations: The Role of Counteranions and Sensing of SCN− in Biological Materials. *ACS Sens.* **2019**, *4*, 1552–1559. [CrossRef]
7. Cao, L.; Yang, C.; Zhang, B.; Lv, K.; Li, M.; Deng, K. Synergistic photocatalytic performance of cobalt tetra(2-hydroxymethyl-1,4-dithiin)porphyrazine loaded on zinc oxide nanoparticles. *J. Hazard. Mater.* **2018**, *359*, 388–395. [CrossRef]
8. Belviso, S.; Santoro, E.; Penconi, M.; Righetto, S.; Tessore, F. Thioethyl Porphyrazines: Attractive Chromophores for Second-Order Nonlinear Optics and DSSCs. *J. Phys. Chem. C* **2019**, *123*, 13074–13082. [CrossRef]
9. Yuzhakova, D.V.; Lermontova, S.A.; Grigoryev, I.S.; Muravieva, M.S.; Gavrina, A.I.; Shirmanova, M.V.; Balalaeva, I.V.; Klapshina, L.G.; Zagaynova, E.V. In vivo multimodal tumor imaging and photodynamic therapy with novel theranostic agents based on the porphyrazine framework-chelated gadolinium (III) cation. *Biochim. Biophys. Acta Gen. Subj.* **2017**, *1861*, 3120–3130. [CrossRef]
10. Kunt, H.; Gonca, E. Synthesis and characterization of novel metal-free and metallo-porphyrazines with eight 3-thiopropylpentaflu orobenzoate units. *Polyhedron* **2012**, *38*, 218–223. [CrossRef]
11. Gonca, E. Metallo-porphyrazines with eight [5-thiopentyl 3,4,5-tris(benzyloxy)benzoate] groups: Synthesis, characterization, aggregation, and solubility behavior. *J. Mol. Struct.* **2017**, *1130*, 10–18. [CrossRef]
12. Gonca, E. New soluble porphyrazine derivatives containing electron-rich substituents. *J. Coord. Chem.* **2013**, *66*, 1720–1729. [CrossRef]
13. Yoshida, T.; Furuyama, T.; Kobayashi, N. Synthesis and optical properties of tetraazaporphyrin phosphorus(V) complexes with electron-rich heteroatoms. *Tetrahedron Lett.* **2015**, *56*, 1671–1674. [CrossRef]
14. Fernandez-Ariza, J.; Urbani, M.; Rodriguez-Morgade, M.S.; Torres, T. Panchromatic Photosensitizers Based on Push-Pull, Unsymmetrically Substituted Porphyrazines. *Chem. A Eur. J.* **2017**, *24*, 2618–2625. [CrossRef]
15. Tuncer, S.; Koca, A.; Gul, A.; Avcıata, U. Synthesis, characterization, electrochemistry and spectroelectrochemistry of novel soluble porphyrazines bearing unsaturated functional groups. *Dye Pigment.* **2012**, *92*, 610–618. [CrossRef]
16. Belviso, S.; Amati, M.; Rossano, R.; Crispini, A.; Lelj, F. Non-symmetrical aryl- and arylethynyl-substituted thioalkyl-porphyrazines for optoelectronic materials: Synthesis, properties, and computational studies. *Dalton Trans.* **2014**, *44*, 2191–2207. [CrossRef] [PubMed]
17. Falkowski, M.; Rebis, T.; Kryjewski, M.; Popenda, L.; Lijewski, S.; Jurga, S.; Mielcarek, J.; Milczarek, G.; Goslinski, T. An enhanced electrochemical nanohybrid sensing platform consisting of reduced graphene oxide and sulfanyl metalloporphyrazines for sensitive determination of hydrogen peroxide and l-cysteine. *Dye Pigment.* **2017**, *138*, 190–203. [CrossRef]
18. Falkowski, M.; Rebis, T.; Piskorz, J.; Popenda, L.; Jurga, S.; Mielcarek, J.; Milczarek, G.; Goslinski, T. Multiwalled carbon nanotube/sulfanyl porphyrazine hybrids deposited on glassy carbon electrode—Effect of nitro peripheral groups on electrochemical properties. *J. Porphyrins Phthalocyanines* **2017**, *21*, 295–301. [CrossRef]
19. Falkowski, M.; Rebis, T.; Piskorz, J.; Popenda, L.; Jurga, S.; Mielcarek, J.; Milczarek, G.; Goslinski, T. Improved electrocatalytic response toward hydrogen peroxide reduction of sulfanyl porphyrazine/multiwalled carbon nanotube hybrids deposited on glassy carbon electrodes. *Dye Pigment* **2016**, *134*, 569–579. [CrossRef]

20. Koczorowski, T.; Rębiś, T.; Szczolko, W.; Antecka, P.; Teubert, A.; Milczarek, G.; Goslinski, T. Reduced graphene oxide/iron(II) porphyrazine hybrids on glassy carbon electrode for amperometric detection of NADH and L-cysteine. *J. Electroanal. Chem.* **2019**, *848*. [CrossRef]
21. Rębiś, T.; Falkowski, M.; Milczarek, G.; Goslinski, T. Electrocatalytic NADH Sensing using Electrodes Modified with 2-[2-(4-Nitrophenoxy)ethoxy]ethylthio-Substituted Porphyrazine/Single-Walled Carbon Nanotube Hybrids. *ChemElectroChem* **2020**, *7*, 2838–2850. [CrossRef]
22. Piskorz, J.; Lijewski, S.; Gierszewski, M.; Gorniak, K.; Sobotta, L.; Wicher, B.; Tykarska, E.; Düzgüneş, N.; Konopka, K.; Sikorski, M.; et al. Sulfanyl porphyrazines: Molecular barrel-like self-assembly in crystals, optical properties and in vitro photodynamic activity towards cancer cells. *Dye Pigment* **2017**, *136*, 898–908. [CrossRef]
23. Piskorz, J.; Mlynarczyk, D.T.; Szczolko, W.; Konopka, K.; Düzgüneş, N.; Mielcarek, J. Liposomal formulations of magnesium sulfanyl tribenzoporphyrazines for the photodynamic therapy of cancer. *J. Inorg. Biochem.* **2018**, *184*, 34–41. [CrossRef] [PubMed]
24. Sobotta, L.; Skupin-Mrugalska, P.; Piskorz, J.; Mielcarek, J. Porphyrinoid photosensitizers mediated photodynamic inactivation against bacteria. *Eur. J. Med. Chem.* **2019**, *175*, 72–106. [CrossRef] [PubMed]
25. Kabay, N.; Baygu, Y.; Alpoguz, H.K.; Kaya, A.; Gök, Y. Synthesis and characterization of porphyrazines as novel extractants for the removal of Ag(I) and Hg(II) from aqueous solution. *Dye Pigment* **2013**, *96*, 372–376. [CrossRef]
26. Liu, Y.; Zhou, X.; Zhang, Z.; Zhang, B.; Deng, K. Effect of carrier and axial ligand on the photocatalytic activity of cobalt thioporphyrazine. *Chin. J. Catal.* **2017**, *38*, 330–336. [CrossRef]
27. Yang, C.; Gao, L.; Zhang, B.; Zhang, Z.; Deng, K. Uniform zinc thioporphyrazine nanosphere by self-assembly and the photocatalytic performance. *J. Porphyrins Phthalocyanines* **2018**, *22*, 868–876. [CrossRef]
28. Zhang, Z.; Wen, X.; Deng, K.; Zhang, B.; Lv, K.; Sun, J. Photodegradation of rhodamine B with molecular oxygen catalyzed by a novel unsymmetrical iron porphyrazine under simulated sunlight. *Catal. Sci. Technol.* **2013**, *3*, 1415–1422. [CrossRef]
29. Yang, C.; Sun, J.; Deng, K.; Wang, D. Synthesis and photocatalytic properties of iron(II)tetramethyl- tetra(1,4-dithiin)porphyrazine. *Catal. Commun.* **2008**, *9*, 321–326. [CrossRef]
30. Su, R.; Sun, J.; Sun, Y.; Deng, K.; Cha, D.; Wang, D. Oxidative degradation of dye pollutants over a broad pH range using hydrogen peroxide catalyzed by FePz(dtnCl2)4. *Chemosphere* **2009**, *77*, 1146–1151. [CrossRef]
31. Tang, J.; Chen, L.; Sun, J.; Lv, K.; Deng, K. Synthesis and properties of iron(II) tetra(1,4-dithiin)porphyrazine bearing peripheral long-chain alkyl group of active end-bromine. *Inorg. Chem. Commun.* **2010**, *13*, 236–239. [CrossRef]
32. Wieczorek, E.; Mlynarczyk, D.T.; Kucinska, M.; Dlugaszewska, J.; Piskorz, J.; Popenda, L.; Szczolko, W.; Jurga, S.; Murias, M.; Mielcarek, J.; et al. Photophysical properties and photocytotoxicity of free and liposome-entrapped diazepinoporphyrazines on LNCaP cells under normoxic and hypoxic conditions. *Eur. J. Med. Chem.* **2018**, *150*, 64–73. [CrossRef] [PubMed]
33. Mlynarczyk, D.T.; Lijewski, S.; Falkowski, M.; Piskorz, J.; Szczolko, W.; Sobotta, L.; Stolarska, M.; Popenda, L.; Jurga, S.; Konopka, K.; et al. Dendrimeric Sulfanyl Porphyrazines: Synthesis, Physico-Chemical Characterization, and Biological Activity for Potential Applications in Photodynamic Therapy. *ChemPlusChem* **2016**, *81*, 460–470. [CrossRef] [PubMed]
34. Mlynarczyk, D.T.; Dlugaszewska, J.; Falkowski, M.; Popenda, L.; Kryjewski, M.; Szczolko, W.; Jurga, S.; Mielcarek, J.; Goslinski, T. Tribenzoporphyrazines with dendrimeric peripheral substituents and their promising photocytotoxic activity against Staphylococcus aureus. *J. Photochem. Photobiol. B Biol.* **2020**, *204*, 111803. [CrossRef] [PubMed]
35. Mlynarczyk, D.T.; Piskorz, J.; Popenda, L.; Stolarska, M.; Szczolko, W.; Konopka, K.; Jurga, S.; Sobotta, L.; Mielcarek, J.; Düzgüneş, N.; et al. S-seco-porphyrazine as a new member of the seco-porphyrazine family—Synthesis, characterization and photocytotoxicity against cancer cells. *Bioorg. Chem.* **2020**, *96*, 103634. [CrossRef]
36. Lijewski, S.; Gierszewski, M.; Sobotta, L.; Piskorz, J.; Kordas, P.; Kucinska, M.; Baranowski, D.; Gdaniec, Z.; Murias, M.; Karolczak, J.; et al. Photophysical properties and photochemistry of a sulfanyl porphyrazine bearing isophthaloxybutyl substituents. *Dye Pigment* **2015**, *113*, 702–708. [CrossRef]
37. Gierszewski, M.; Falkowski, M.; Sobotta, L.; Stolarska, M.; Popenda, L.; Lijewski, S.; Wicher, B.; Burdzinski, G.; Karolczak, J.; Jurga, S.; et al. Porphyrazines with peripheral isophthaloxyalkylsulfanyl substituents and their optical properties. *J. Photochem. Photobiol. A Chem.* **2015**, *307–308*, 54–67. [CrossRef]
38. Falkowski, M.; Kucinska, M.; Piskorz, J.; Wieczorek-Szweda, E.; Popenda, L.; Jurga, S.; Sikora, A.; Mlynarczyk, D.T.; Murias, M.; Marszall, M.P.; et al. Synthesis of sulfanyl porphyrazines with bulky peripheral substituents—Evaluation of their photochemical properties and biological activity. *J. Photochem. Photobiol. A Chem.* **2021**, *405*, 112964. [CrossRef]
39. Kucinska, M.; Skupin-Mrugalska, P.; Szczolko, W.; Sobotta, L.; Sciepura, M.; Tykarska, E.; Wierzchowski, M.; Teubert, A.; Fedoruk-Wyszomirska, A.; Wyszko, E.; et al. Phthalocyanine Derivatives Possessing 2-(Morpholin-4-yl)ethoxy Groups As Potential Agents for Photodynamic Therapy. *J. Med. Chem.* **2015**, *58*, 2240–2255. [CrossRef] [PubMed]
40. Dlugaszewska, J.; Szczolko, W.; Koczorowski, T.; Skupin-Mrugalska, P.; Teubert, A.; Konopka, K.; Kucinska, M.; Murias, M.; Düzgüneş, N.; Mielcarek, J.; et al. Antimicrobial and anticancer photodynamic activity of a phthalocyanine photosensitizer with N -methyl morpholiniumethoxy substituents in non-peripheral positions. *J. Inorg. Biochem.* **2017**, *172*, 67–79. [CrossRef]
41. Kucinska, M.; Plewinski, A.; Szczolko, W.; Kaczmarek, M.; Goslinski, T.; Murias, M. Modeling the photodynamic effect in 2D versus 3D cell culture under normoxic and hypoxic conditions. *Free Radic. Biol. Med.* **2021**, *162*, 309–326. [CrossRef] [PubMed]
42. Rak, J.; Pouckova, P.; Benes, J.; Vetvicka, D. Drug Delivery Systems for Phthalocyanines for Photodynamic Therapy. *Anticancer Res.* **2019**, *39*, 3323–3339. [CrossRef]
43. Ghosh, S.; Carter, K.A.; Lovell, J.F. Liposomal formulations of photosensitizers. *Biomaterials* **2019**, *218*, 119341. [CrossRef]

44. Chelminiak-Dudkiewicz, D.; Rybczynski, P.; Smolarkiewicz-Wyczachowski, A.; Mlynarczyk, D.T.; Wegrzynowska-Drzymalska, K.; Ilnicka, A.; Goslinski, T.; Marszałł, M.P.; Ziegler-Borowska, M. Photosensitizing potential of tailored magnetite hybrid nanoparticles functionalized with levan and zinc (II) phthalocyanine. *Appl. Surf. Sci.* **2020**, *524*, 146602. [CrossRef] [PubMed]
45. Aksenova, N.A.; Savko, M.A.; Uryupina, O.Y.; Roldugin, V.I.; Timashev, P.S.; Kuz'Min, P.G.; Shafeev, G.A.; Solov'Eva, A.B. Effect of the preparation method of silver and gold nanoparticles on the photosensitizing properties of tetraphenylporphyrin–amphiphilic polymer—nanoparticle systems. *Russ. J. Phys. Chem. A* **2017**, *91*, 124–129. [CrossRef]
46. Bera, K.; Maiti, S.; Maity, M.; Mandal, C.; Maiti, N.C. Porphyrin–Gold Nanomaterial for Efficient Drug Delivery to Cancerous Cells. *ACS Omega* **2018**, *3*, 4602–4619. [CrossRef] [PubMed]
47. Setaro, F.; Wennink, J.W.H.; Mäkinen, P.I.; Holappa, L.; Trohopoulos, P.N.; Ylä-Herttuala, S.; Van Nostrum, C.F.; De La Escosura, A.; Torres, T. Amphiphilic phthalocyanines in polymeric micelles: A supramolecular approach toward efficient third-generation photosensitizers. *J. Mater. Chem. B* **2020**, *8*, 282–289. [CrossRef]
48. Ziental, D.; Czarczynska-Goslinska, B.; Mlynarczyk, D.T.; Glowacka-Sobotta, A.; Stanisz, B.; Goslinski, T.; Sobotta, L. Titanium Dioxide Nanoparticles: Prospects and Applications in Medicine. *Nanomaterials* **2020**, *10*, 387. [CrossRef]
49. Musial, J.; Krakowiak, R.; Mlynarczyk, D.T.; Goslinski, T.; Stanisz, B.J. Titanium Dioxide Nanoparticles in Food and Personal Care Products—What Do We Know about Their Safety? *Nanomaterials* **2020**, *10*, 1110. [CrossRef]
50. Lü, X.-F.; Li, J.; Wang, C.; Duan, M.-Y.; Luo, Y.; Yao, G.-P.; Wang, J.-L. Enhanced photoactivity of CuPp-TiO2 photocatalysts under visible light irradiation. *Appl. Surf. Sci.* **2010**, *257*, 795–801. [CrossRef]
51. Forsyth, T.P.; Williams, D.B.G.; Montalban, A.G.; Stern, C.L.; Barrett, A.G.M.; Hoffman, B.M. A Facile and Regioselective Synthesis of Trans-Heterofunctionalized Porphyrazine Derivatives. *J. Org. Chem.* **1998**, *63*, 331–336. [CrossRef]
52. Linstead, R.P.; Whalley, M. 944. Conjugated macrocylces. Part XXII. Tetrazaporphin and its metallic derivatives. *J. Chem. Soc.* **1952**, *1952*, 4839–4846. [CrossRef]
53. Roncucci, G.; Dei, D.; De Filippis, M.P.; Fantetti, L.; Masini, I.; Cosimelli, B.; Jori, G. Zinc-Phthalocyanines and Corresponding Conjugates, Their Preparation and Use in Photodynamic Therapy and as Diagnostic Agents. U.S. Patent 5,965,598, 12 October 1999.
54. Koczorowski, T.; Ber, J.; Sokolnicki, T.; Teubert, A.; Szczolko, W.; Goslinski, T. Electrochemical and catalytic assessment of peripheral bromoaryl-substituted manganese and iron porphyrazines. *Dye Pigment* **2020**, *178*, 108370. [CrossRef]
55. Tuncer, S.; Koca, A.; Gül, A.; Avciata, U. 1,4-Dithiaheterocycle-fused porphyrazines: Synthesis, characterization, voltammetric and spectroelectrochemical properties. *Dye Pigment* **2009**, *81*, 144–151. [CrossRef]
56. Fuchter, M.J.; Beall, L.S.; Baum, S.M.; Montalban, A.G.; Sakellariou, E.G.; Mani, N.S.; Miller, T.; Vesper, B.J.; White, A.J.; Williams, D.J.; et al. Synthesis of porphyrazine-octaamine, hexamine and diamine derivatives. *Tetrahedron* **2005**, *61*, 6115–6130. [CrossRef]
57. Kryjewski, M.; Tykarska, E.; Rębiś, T.; Dlugaszewska, J.; Ratajczak, M.; Teubert, A.; Gapiński, J.; Patkowski, A.; Piskorz, J.; Milczarek, G.; et al. Porphyrazine with bulky 2-(1-adamantyl)-5-phenylpyrrol-1-yl periphery tuning its spectral and electrochemical properties. *Polyhedron* **2015**, *98*, 217–223. [CrossRef]
58. Rębiś, T.; Lijewski, S.; Nowicka, J.; Popenda, L.; Sobotta, L.; Jurga, S.; Mielcarek, J.; Milczarek, G.; Goslinski, T. Electrochemical properties of metallated porphyrazines possessing isophthaloxybutylsulfanyl substituents: Application in the electrocatalytic oxidation of hydrazine. *Electrochim. Acta* **2015**, *168*, 216–224. [CrossRef]
59. Hajri, A.; Touaiti, S.; Jamoussi, B. Preparation of Organic Zn-Phthalocyanine-Based Semiconducting Materials and Their Optical and Electrochemical Characterization. *Adv. Optoelectron.* **2013**, *2013*, 321563. [CrossRef]
60. Pommerehne, J.; Vestweber, H.; Guss, W.; Mahrt, R.F.; Bässler, H.; Porsch, M.; Daub, J. Efficient two layer leds on a polymer blend basis. *Adv. Mater.* **1995**, *7*, 551–554. [CrossRef]
61. Sakamoto, K.; Furuya, N.; Soga, H.; Yoshino, S. Cyclic voltammetry of non-peripheral thioaryl substituted phthalocyanines. *Dye Pigment* **2013**, *96*, 430–434. [CrossRef]
62. Sułek, A.; Pucelik, B.; Kuncewicz, J.; Dubin, G.; Dąbrowski, J.M. Sensitization of TiO2 by halogenated porphyrin derivatives for visible light biomedical and environmental photocatalysis. *Catal. Today* **2019**, *335*, 538–549. [CrossRef]
63. Oliveira, D.F.; Batista, P.S.; Muller, P.S.; Velani, V.; França, M.D.; De Souza, D.R.; Machado, A.E. Evaluating the effectiveness of photocatalysts based on titanium dioxide in the degradation of the dye Ponceau 4R. *Dye Pigment* **2012**, *92*, 563–572. [CrossRef]
64. Kozlov, N.K.; Natashina, U.A.; Tamarov, K.P.; Gongalsky, M.B.; Solovyev, V.V.; Kudryavtsev, A.A.; Sivakov, V.; Osminkina, L.A. Recycling of silicon: From industrial waste to biocompatible nanoparticles for nanomedicine. *Mater. Res. Express* **2017**, *4*, 095026. [CrossRef]
65. Sobotta, L.; Fita, P.; Szczolko, W.; Wrotynski, M.; Wierzchowski, M.; Goslinski, T.; Mielcarek, J. Functional singlet oxygen generators based on porphyrazines with peripheral 2,5-dimethylpyrrol-1-yl and dimethylamino groups. *J. Photochem. Photobiol. A Chem.* **2013**, *269*, 9–16. [CrossRef]

MDPI

Article

A Concise Synthesis of a BODIPY-Labeled Tetrasaccharide Related to the Antitumor PI-88

Juan Ventura [1], Clara Uriel [1], Ana M. Gomez [1,*], Edurne Avellanal-Zaballa [2], Jorge Bañuelos [2,*], Inmaculada García-Moreno [3] and Jose Cristobal Lopez [1,*]

1 Instituto de Química Orgánica General, IQOG-CSIC, Juan de la Cierva 3, 28006 Madrid, Spain; juan.ventura@csic.es (J.V.); clara.uriel@csic.es (C.U.)
2 Departamento de Química Física, Universidad del Pais Vasco-EHU, Apartado 644, 48080 Bilbao, Spain; eavellanal002@ehu.eus
3 Instituto de Química-Física "Rocasolano", IQFR-CSIC, Serrano 119, 28006 Madrid, Spain; i.garcia-moreno@iqfr.csic.es
* Correspondence: ana.gomez@csic.es (A.M.G.); jorge.banuelos@ehu.es (J.B.); jc.lopez@csic.es (J.C.L.)

Abstract: A convergent synthetic route to a tetrasaccharide related to PI-88, which allows the incorporation of a fluorescent BODIPY-label at the reducing-end, has been developed. The strategy, which features the use of 1,2-methyl orthoesters (MeOEs) as glycosyl donors, illustrates the usefulness of suitably-designed BODIPY dyes as glycosyl labels in synthetic strategies towards fluorescently-tagged oligosaccharides.

Keywords: PI-88; glycosylation; 1,2-methyl orthoesters; BODIPY; fluorescent labeling

Citation: Ventura, J.; Uriel, C.; Gomez, A.M.; Avellanal-Zaballa, E.; Bañuelos, J.; García-Moreno, I.; Lopez, J.C. A Concise Synthesis of a BODIPY-Labeled Tetrasaccharide Related to the Antitumor PI-88. *Molecules* **2021**, *26*, 2909. https://doi.org/10.3390/molecules26102909

Academic Editor: Ugo Caruso

Received: 14 April 2021
Accepted: 12 May 2021
Published: 14 May 2021

1. Introduction

The recognition of the crucial role of heparanase enzyme, an endo-β-D-glucuronidase able to degrade heparan sulfate (HS) in the extracellular matrix and basement membranes, in a number of pathological processes, such as metastasis and angiogenesis, has triggered the development of heparanase inhibitors [1–6]. Among these inhibitors PI-88, a mixture of functionalized mannose oligosaccharides, i.e., **1**, **2**, (Figure 1), that potently inhibited heparanase and in vitro angiogenesis, has been considered as a promising candidate and has received considerable attention [7]. On the other hand, fluorescent labeling of biomolecules has been recognized as a research topic of great significance, since such labeling facilitates the investigation of glycoconjugates and their interaction in biological systems at high sensitivity [8–12]. Among the types of fluorescent dyes commonly employed as tags, borondipyrromethene (BODIPY, 4, 4-difluoro-4-bora-3a, 4a-diaza-s-indacene) dyes, e.g., **3** (Figure 1) have excelled. Several reasons can be cited for this preference: their high fluorescent quantum yields (∅), excellent photochemical and chemical stabilities, and, arguably, the relatively facile modulation of their photophysical and/or chemical properties by means of synthetic postfunctionalization of their indacene core [13–18].

Based on these precedents, we thought it would be of interest to investigate the feasibility of a synthetic approach to BODIPY-labeled PI-88 saccharide components, where the fluorescent dye is incorporated from the beginning of the synthetic sequence. Additionally, we envisioned that the incorporation of the lipophilic BODIPY moiety at the reducing end of a PI-88 saccharide analogue would bring one additional advantage by facilitating the visualization and detection of the synthetic intermediates along the saccharide synthesis [19]. Additionally, in light of some reported literature precedents [20], the incorporation of the lipophilic BODIPY core to the saccharidic ensemble could lead to ameliorated biological activity in the ensuing saccharides. Thus, in this *Article*, we report a synthetic approach to a PI-88 tetrasaccharide analogue [21] featuring the use of 1,2-methyl orthoester (MeOE) glycosyl donors, i.e., **4** (Figure 1), in which a BODIPY-type fluorescent probe could be attached at the reducing end of the saccharides from the beginning of the synthesis.

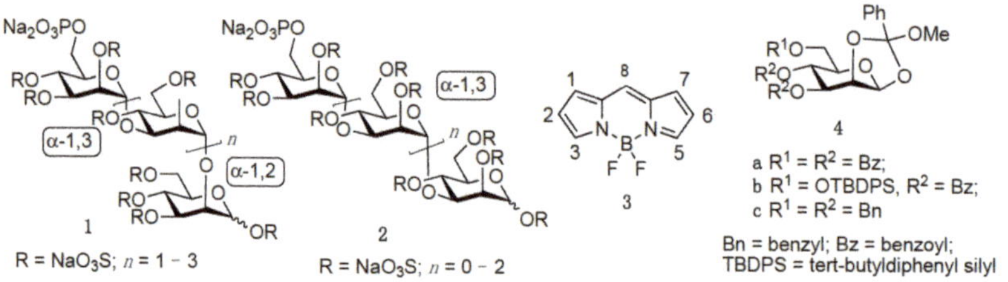

Figure 1. PI-88 (**1**,**2**), BODIPY (**3**), IUPAC numbering) and 1,2-methyl orthoester glycosyl donors (**4**) used in this study.

2. Results

Several synthetic approaches to analogues of tetra- and pentasaccharides **1** have already been reported [22–28]. In those approaches, different types glycosyl donors have been employed. We have been interested in the use of 1,2-methyl orthoesters (MeOEs) [29–33] as an inexpensive alternative to Fraser–Reid's *n*-pentenyl orthoester glycosyl donors (NPOEs) [34–36]. The former derivatives were shown to display good regioselectivity, similar to that of NPOEs [31]. Accordingly, our convergent 1 + 1 + 2 approach to BODIPY-tagged tetramannan, **10** (Scheme 1) was based on our previous studies on the selective mono-glycosylation at position *O*-3 in mannopyranose substrates possessing a 2,3,4 triol moiety, i.e., **6**, **7** (Scheme 1), with MeOE glycosyl donors, **4a**, **4b**, and **9** [31]. Thus, our strategy involved two glycosidic disconnections **A** and **B** (Scheme 1). Disconnection **A** (Scheme 1), leading to BODIPY-disaccharide **6**, was envisioned by glycosylation of a benzylated mannopyranoside **5**, exposing only one hydroxyl group (*O*-2) for the glycosyl coupling, with MeOE donor **4a**. On the other hand, disconnection **B** (Scheme 1) was imagined by regioselective mono-glycosylation of triol **6**, at position *O*-3′, according to precedents from our research group [31], by a disaccharide MeOE donor, **9**.

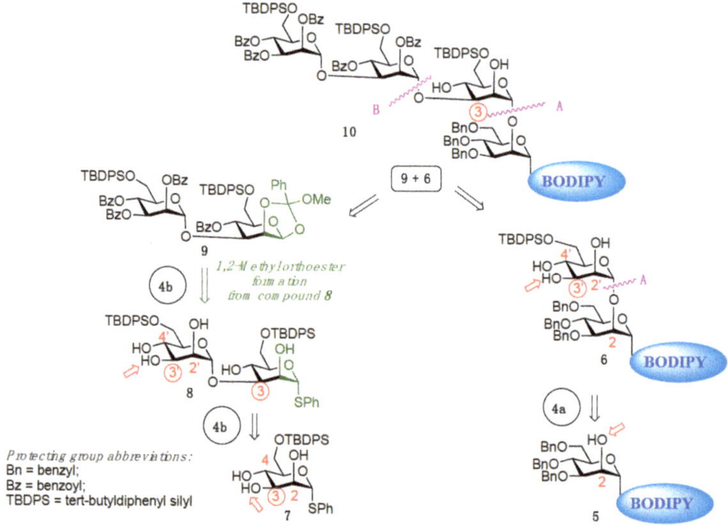

Scheme 1. Retrosynthesis of PI-88 tetrasaccharide analogue **10**, involving the sole use of 1,2-methyl orthoester glycosyl donors, **4** and **9**.

The synthetic route started with the glycosylation of BODIPY **11** [37], readily available through a one-pot transformation using phthalide and pyrrole as the starting materials, followed by in situ coordination with $BF_3.OEt_2$ [38], with methyl orthoester **4c**, to yield BODIPY glycoside **12** (80% yield, Scheme 2a). However, attempted saponification of the

2-*O*-Bz substituent in **12**, by treatment with Et_3N or NaOMe in methanol, resulted in the production of the undesired $B(OMe)_2$-BODIPY derivatives **13** and **14**, where the fluorine atoms were replaced by methoxy groups (Scheme 2a). These derivatives, although also fluorescent [39], proved to be labile under the acidic conditions required in the next glycosylation events. We then turned our attention to borondiphenyl BODIPY **15**, a more chemically robust yet fluorescent BODIPY analogue [40]. Access to **15** was also affected by slightly modifying our one-pot procedure [38], from phthalide and pyrrole, by simply replacing $BF_3.OEt_2$ by $B(Ph)_3$ in the borondipyrromethene ring closing reaction (Scheme 2b) [41].

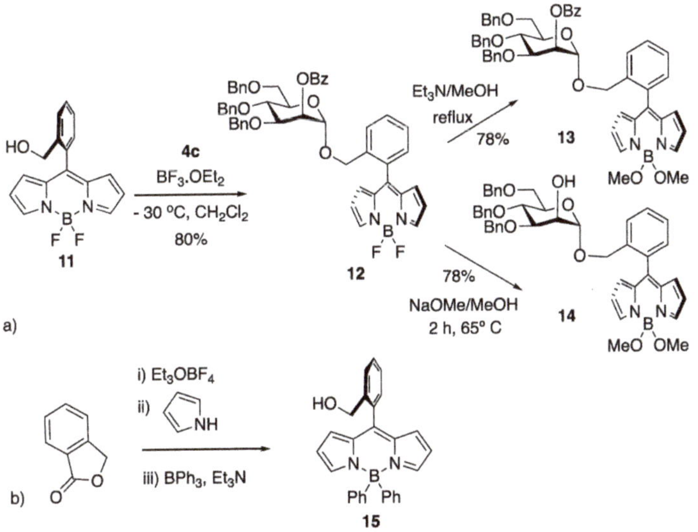

Scheme 2. (**a**) Glycosylation of BODIPY **11** with MeOE **4c** and saponification attempts; (**b**) one-pot synthesis of 4,4-diphenyl BODIPY **15**.

The synthetic routes to the two fragments in the convergent approach to PI-88 tetrasaccharide analogue **10**, glycosyl acceptor **6** and glycosyl donor **9**, are depicted in Scheme 3a,b. Thus, glycosylation of hydroxymethyl BODIPY **15** with benzylated MeOE **4c**, followed by saponification (NaOMe/MeOH), led to BODIPY-mannopyranoside **5** (Scheme 3a). The latter was then glycosylated with tri-*O*-benzoyl MeOE **4a**, to yield BODIPY disaccharide **16a** in moderate yield (52%). Next, protecting group manipulations in compound **16a**, including de-*O*-benzoylation leading to tetraol **16b**, and selective monosilylation at the primary hydroxyl group at O-6′ in the latter [terbutyldiphenylsilyl chloride, 4-dimethylaminopyridine (DMAP), in dimethyl formamide (DMF)] produced 2′,3′,4′-triol **6**, (Scheme 3a).

On the other hand, the route to MeOE-disaccharide donor **9** started from phenyl thiomannopyranoside **7** (Scheme 3b). Accordingly, regioselective mono-glycosylation at *O*-3 of 2,3,4-triol **7** with MeOE **4b**, based on precedents from our research group [31,42], yielded phenyl 1-thiomannopyranoside **8** in good yield (Scheme 3b). To conclude, a three-step sequence from **8**, including perbenzoylation, anomeric bromination, and orthoester formation from an intermediate glycosyl bromide according to Wei et al. [43], allowed its conversion to MeOE disaccharide donor **9** (Scheme 3b).

Finally, glycosyl acceptor **6**, containing a mannopyranoside triol unit, was regioselectively mono-glycosylated at *O*-3′with MeOE disaccharide **9** [44], to yield PI-88 tetrasaccharide precursor analogue **10**, in 53% yield (Scheme 4). An alternative glycosylation of **6** with a MeOE monosaccharide, e.g., **4a** or **4b**, rather than with a disaccharide, i.e., **9**, in our hands consistently led to low yields of the trisaccharide analogue.

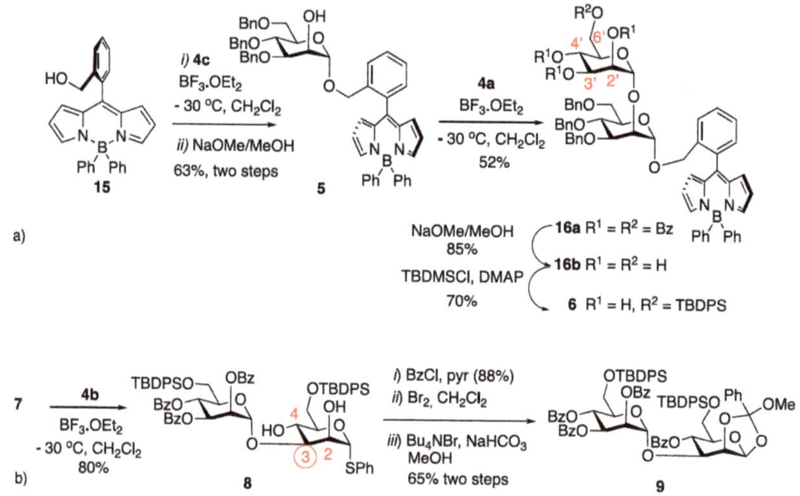

Scheme 3. (**a**) Synthesis of BODIPY disaccharide acceptor **6**; (**b**) access to MeOE-disaccharide donor **9**, from thioglycoside **7**, by regioselective glycosylation with MeOE **4b** and subsequent orthoester formation on the ensuing disaccharidic thioglycoside intermediate **8**.

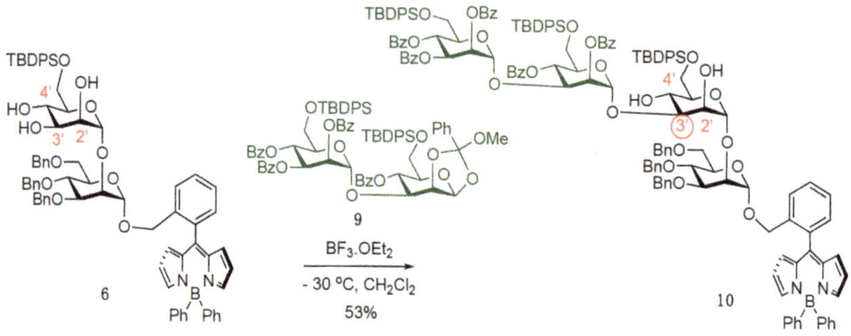

Scheme 4. Regioselective glycosylation of triol **6** with MeOE-disaccharide **9** leading to tetrasaccharide **10**.

To impel the advanced applications of the new glycoprobes, we analyzed the photonic behavior of BODIPY **15** and its saccharide derivatives **16a** and **10**, under low (photophysical properties) and high (laser properties) irradiation regimes. The replacement of the fluorine atoms at the boron bridge, as in **11** [38], by phenyl groups, as in **15** [45], had low impact on the spectral properties of BODIPY (Figure 2) but induced both a decrease of the emission efficiency (Table 1) and a biexponential character of the fluorescent lifetime (Table S1), regardless of the environmental properties. The free motion of these phenyl rings chelating the boron atom reduced the planarity of the dipyrrin core (computed bending angles in the dipyrrin core up to 16° in the excited state, Figure 2), increasing the internal conversion processes with a deleterious effect on the fluorescence signal. The labelling of a disaccharide or tetrasaccharide with BODIPY **15**, as in compound **16a** and **10**, respectively, widened the absorption spectrum, while the spectral profile of fluorescence matched that of its precursor (Figure 2). It is noteworthy that BODIPYs **10** and **16a** displayed a brighter fluorescence with longer lifetimes than their non-glycosylated counterpart **15** in all tested media and regardless of the number of saccharide units appended (Table 1 and Table S1). Thus, glycosylation of the *ortho*-hydroxymethyl group of the C-8-aryl residue led to a more rigid and compact molecular structure, e.g., **16a** and **10**, owing to the higher steric hindrance imposed by the bulky disaccharide. In fact, the structural arrangement of the C-8-benzyl residue in **16a** was nearly orthogonal (twisting dihedral angle computed in the ground state of 75° in **15** vs. 85° in **16a**), reducing the internal conversion pathways associated to conformational freedom. Consequently, BODIPY-saccharides **16a** and **10** behaved as efficient and stable fluorescent glycoprobes even under laser irradiation conditions, exhibiting a lasing efficiency up to 38% in the green spectral region (540 nm, Table 1) with high photostability, since their laser emission remained at the initial level even after 70,000 pump pulses. This good tolerance to intense and prolonged irradiation is a highly desirable property for fluorescent labels to provide long-lasting bioimages. Therefore, from a photonic point of view, the *ortho-* position of 8-phenyl BODIPYs is highlighted as a suitable grafting position to tag (oligo)saccharides, even resulting in an amelioration of the photonic performance of the original labeling dye.

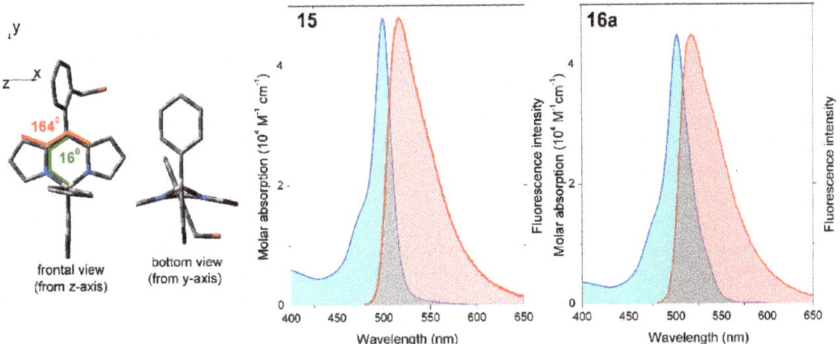

Figure 2. Absorption (in blue) and normalized fluorescence (in red) spectra of BODIPY **15** and its glycosylated derivative **16a** in diluted solution of ethyl acetate. The spectral profiles of **10** fully resemble those of **16a**. The excited state optimized geometry of **15** in two different views is also enclosed with key dihedral angles to show the bending of the chromophore.

Table 1. Photophysical [1] and laser [2] properties of BODIPY **15**, and glycosylated derivatives **16a** and **10**, in ethyl acetate. For the sake of comparison, the corresponding photophysical data of the F-BODIPY counterpart (in ethanol) have been added. For additional photophysical data, see Table S1.

Compound	λ_{ab} (nm)	ε_{max}	λ_{fl} (nm)	$\varnothing$	$<\tau>$ (ns)	λ_{la} (nm)	Eff (%)
11	499.5	6.4	513.0	0.74	6.45	540	60
15	497.0	4.8	511.5	0.22	2.83	533	18
16a	500.0	4.5	516.0	0.31	3.19	545	38
10	500.5	4.1	515.0	0.31	3.57	547	36

[1] Registered under a soft irradiation regime; dye concentration: 2 μM. Absorption (λ_{ab}) and fluorescence (λ_{fl}) wavelength, molar absorption (ε_{max}) (10^4 M^{-1} cm^{-1}), fluorescence quantum yield ($\varnothing$), and amplitude-average lifetime ($<\tau>$). [2] Recorded under a hard irradiation regime; dye concentration 2 mM. Peak wavelength for the laser emission (λ_{la}) and efficiency (Eff (%)) defined as the ratio between the energy of the laser output and the pump energy incident on the cell surface.

3. Conclusions

In summary, we developed a convergent, efficient, synthetic strategy to BODIPY-labeled PI-88 tetrasaccharide components (**10**) [46], which serves to illustrate the scope and usefulness of MeOEs as glycosyl donors. The inclusion of the BODIPY-tag from the beginning of the synthesis facilitates the visual recognition (thin-layer chromatography, TLC) of the labeled-saccharide acceptor and the glycosylated products therefrom, among the rest of the non-fluorescent side-products arising from side-reactions of the MeOE glycosyl donor [36]. This feature becomes particularly appealing when excess amounts of glycosyl donors are required to lead the glycosylation to completion. On the other hand, the chemically stable 4, 4′-diphenyl BODIPY derivative (**15**), used as a tag, displayed good fluorescent properties and photostability under strong and prolonged irradiation, and was also able to withstand all reaction conditions employed in the synthetic sequence leading to **10** [20]. Our results also indicate that the incorporation of carbohydrate subunits at the *ortho*-hydroxymethyl group of the C-8-aryl substituent has a beneficial effect on the, already, good photophysical features of the BODIPY dye.

4. Materials and Methods

4.1. General Information

The solvents and reagents used in the transformations included in the manuscript were obtained from commercial sources. In the glycosylation experiments, the adventitious water content was removed by repeated evaporation of the sample with toluene. The temperature at which the reactions were carried out will be mentioned unless room temperature was used. The glycosylation reactions were carried out in dried flasks fitted with rubber septa under an argon atmosphere.

A 5.0 M stock solution of triethyloxonium tetrafluoroborate, employed in the preparation of the BODIPYs, was prepared by dissolving 25 g (0.131 mmol) of the salt in 26.3 mL of anhydrous methylene chloride.

Anhydrous $MgSO_4$ was used to dry organic solutions during workup. Evaporation of the solvents was performed using a rotary evaporator (Buchi, Flawil, Switzerland). Flash column chromatography was used to purify or separate the samples. Thin-layer chromatography (TLC) was conducted on Kieselgel 60 F254. Spots corresponding to BODIPY-containing molecules were spotted under visible light. TLCs were then inspected under UV irradiation (254 nm) followed by charring with a solution of 20% aqueous H_2SO_4 (200 mL) in AcOH (800 mL). ^{1}H and ^{13}C-NMR spectra were recorded in $CDCl_3$ at 300, 400, or 500 MHz and 75, 101, or 126 MHz, respectively. Chemical shifts are expressed in parts per million (δ scale) downfield from tetramethylsilane and are referenced to residual protium in the NMR solvent ($CHCl_3$: δ 7.25 ppm, CD_3OD: δ 4.870 ppm). Coupling constants (*J*) are given in Hz. All presented ^{13}C-NMR spectra are proton decoupled. Mass spectra were recorded by direct injection with an Accurate Mass Q-TOF LC/MS spectrometer (Agilent Technologies, Santa Clara, CA, USA) equipped with an electrospray ion source in positive mode.

4.2. General Procedures

4.2.1. General Procedure for Glycosylation. Procedure A

A previously dried mixture of a glycosyl donor and glycosyl acceptor was dissolved in anhydrous dichloromethane (≈3 mL/0.1 mmol). Previously dried (200 °C, one night) 4Å molecular sieves were added to the mixture. The reaction was cooled to −30 °C, and then $BF_3.OEt_2$ (3.0 equiv) was added. After 5–10 min, the reaction mixture was diluted with dichloromethane and the ensuing solution washed with saturated aqueous $NaHCO_3$ solution. The organic layer was dried over Na_2SO_4, filtered, and evaporated under vacuum. The resulting crude mixture was purified by chromatography on silica gel (eluent: hexane-ethyl acetate mixtures).

4.2.2. General Procedure for Debenzoylation. Procedure B

The corresponding compound was dissolved in methanol (25 mL/ mmol) and triethylamine (6 mL/ mmol) was added to the resulting solution. The reaction mixture was refluxed overnight, the solvents evaporated, and the ensuing residue concentrated. Purification by flash chromatography was carried out using hexane-ethyl acetate mixtures, as eluent.

4.2.3. General Procedure for Silylation. Procedure C

The corresponding compound was dissolved in dry DMF (20 mL/mmol), and to this solution imidazole (4 equiv.) was added. After stirring for 5–10 min in an ice bath, under argon, terbutyldiphenylsilyl chloride (1.2 equiv.) and a small amount of dimethylaminopiridine DMAP were added. The ice bath was removed, and the reaction mixture was left with stirring at room temperature for 24 h. The reaction mixture was then diluted with ethyl acetate and extracted with a saturated aqueous $NaHCO_3$ solution and brine. The combined organic solutions were dried over anhydrous $MgSO_4$ and evaporated under vacuum. The resulting crude mixture was purified by chromatography on silica gel (eluent: hexane-ethyl acetate mixtures).

4.2.4. General Procedure for Characterization of Polyol Derivatives as Peracetates. Procedure D

The corresponding polyol was dissolved in pyridine (1 mL/0.1 mmol substrate) and acetic anhydride (0.5 mL/mmol substrate) was then added. The reaction mixture was stirred at room temperature (normally 24 h). After completion of the reaction (t.l.c.), the solvent was evaporated, and the resulting crude mixture was purified by chromatography on silica gel (eluent: hexane-ethyl acetate mixtures).

4.3. Synthesis

4.3.1. BF_2-Bodipy **11**

This compound was prepared according to our previously described method [38], from phthalide and pyrrole.

4.3.2. Mannopyranosyl BODIPY **12**

This compound was prepared by glycosylation of BODIPY **11** (26 mg, 0.088 mmol) with orthoester **4c** (100 mg, 0.178 mmol), according to the general procedure for glycosylation, procedure A. Flash chromatography (hexane- ethyl acetate; 85:15) yielded BODIPY-mannoside **12** (64 mg, 80%). ^{1}H-NMR (500 MHz, $CDCl_3$) δ 8.05–7.16 (m, 24H), 6.68 (dd, J = 12.4, 4.2 Hz, 2H), 6.47 (dd, J = 4.2, 1.8 Hz, 1H), 6.33 (dd, J = 4.3, 1.8 Hz, 1H), 5.40 (dd, J = 3.1, 1.9 Hz, 1H), 4.86 (d, J = 1.9 Hz, 1H), 4.79 (d, J = 10.7 Hz, 1H), 4.72–4.65 (m, 3H), 4.52–4.42 (m, 4H), 4.37 (d, J = 11.8 Hz, 1H), 4.01 (t, J = 9.6 Hz, 1H), 3.88–3.75 (m, 3H), 3.58 (dd, J = 10.9, 1.9 Hz, 1H), 3.50–3.44 (m, 1H) ^{13}C-NMR (125 MHz, $CDCl_3$) δ 165.7, 145.3, 145.2, 144.8, 138.7, 138.6, 138.2, 135.7, 135.4, 133.2, 132.9, 132.6, 131.2, 131.0 (×2), 130.2, 130.1 (×2), 130.0, 128.9, 128.5, 128.4 (×5), 128.2 (×2), 127.9, 127.7 (×2), 127.6 (×2), 119.2, 118.8, 97.5, 78.1, 75.3, 74.1, 73.5, 72.0, 71.6, 68.8 (×3), 68.3, 67.3, 51.1, 29.8; ^{19}F-NMR (376 MHz, $CDCl_3$) δ −145.4 (q, $J_{B–F}$ = 28.9 Hz), −145.4 (q, $J_{B–F}$ = 28.9 Hz); HRMS (ESI-TOF): calc for $C_{50}H_{49}BF_2N_3O_7$ $[M + NH_4]^+$; 852.37407 found 852.36969.

4.3.3. Mannopyranosyl BODIPY **13**

This compound was obtained when applying the general procedure for debenzoylation, procedure B, to BODIPY-mannoside **12** (60 mg, 0.087 mmol) followed by flash chromatography (hexane- ethyl acetate; 6:4). Compound **13** (56 mg, 78%): ^{1}H-NMR (300 MHz, $CDCl_3$) δ 8.04–7.12 (m, 24H), 6.66 (ddd, J = 10.4, 4.2, 1.1 Hz, 2H), 6.48 (ddd, J = 4.2, 1.8, 0.7 Hz, 1H), 6.43 (ddd, J = 4.2, 1.8, 0.7 Hz, 1H), 5.42 (dd, J = 3.1, 2.0 Hz, 1H), 4.87–4.77 (m, 2H), 4.74–4.60 (m, 3H), 4.51–4.36 (m, 4H), 4.04 (t, J = 9.5 Hz, 1H), 3.98–3.88 (m, 1H), 3.78 (dd, J = 10.6, 3.5 Hz, 1H), 3.66–3.53 (m, 2H) 3.12 (s, 3H), 3.03 (s, 3H); ^{19}F-NMR (376 MHz, $CDCl_3$): showed not peaks at all; HRMS (ESI-TOF): calc for $C_{52}H_{51}BN_2NaO_9$ $[M + Na]^+$; 881.35682 found 881.35690.

4.3.4. Mannopyranosyl BODIPY **14**

Treatment of compound **12** (80 mg, 0.096 mmoles) with sodium methoxide in MeOH, at 65 °C for 1 h resulted in the consumption of the starting material. Then, addition of solid NH_4Cl and stirring for 15 min followed by filtration and solvent evaporation provided a residue that was purified by chromatography on silica gel (hexane-ethyl acetate; 3:7) to give dimethoxyboron-derivative **14** (56 mg, 78%). ^{1}H-NMR (300 MHz, $CDCl_3$) δ 7.88–7.03 (19H), 6.68 (dd, J = 4.2, 1.3 Hz, 1H), 6.60 (dd, J = 4.2, 1.3 Hz, 1H), 6.42 (dt, J = 4.3, 1.8 Hz, 1H), 4.74 (d, J = 10.8 Hz, 1H), 4.59–4.35 (m, 9H), 3.78–3.50 (m, 6H), 3.38–3.33 (m, 1H), 3.20 (s, 3H), 3.06 (s, 3H); ^{13}C-NMR (125 MHz, $CDCl_3$) δ 152.8, 144.9, 144.5, 136.7, 136.6, 135.7, 133.7, 133.6, 130.4, 130.0, 129.8, 129.7, 129.4, 128.5 (×2), 128.4 (×2), 128.3 (×2), 128.1 (×2),

128.0 (×2), 127.8 (×2), 127.7 (×2), 118.5, 118.1, 100.6, 79.9, 75.2, 74.1, 73.6, 71.7, 71.4, 69.1, 68.5, 68.3, 67.3, 50.3, 49.8 HRMS (ESI-TOF): calc for $C_{45}H_{47}BN_2NaO_8$ $[M + Na]^+$; 777.33254 found 777.33546.

4.3.5. $B(Ph)_2$-BODIPY **15**

This compound was prepared by treatment of a toluene solution of a dipyrromethane intermediate (obtained by reaction of phthalide with pyrrole) (44 mg, 0.198 mmol), with triphenylborane (48 mg, 0.198 mmol). The mixture, under argon, was refluxed for 24 h. Then, the solvent was evaporated, and the residue dissolved in dichloromethane (10 mL), to which 10 mL of a 1M solution of NaOH were added. The ensuing mixture was kept with stirring for one night. The organic layer was separated and dried over MgSO4, filtered, and evaporated. The residue was purified by for chromatography on silica gel (hexane-ethyl acetate; 8:2) to afford compound **15** (38 mg, 46%). ^{1}H-NMR (400 MHz, $CDCl_3$) δ 7.64–6.97 (m, 14H), 6.75 (d, *J* = 4.2 Hz, 2H), 6.55–6.39 (m, 2H), 4.35 (s, 2H) ^{13}C-NMR (100 MHz, $CDCl_3$) δ 145.5 (×2), 145.1, 139.7, 135.5, 133.8 (×2), 132.5, 131.7 (×2), 129.8, 129.7, 128.7 (×2), 128.2, 127.6 (×3), 127.1, 126.7, 126.2, 118.0 (×2), 62.9 HRMS (ESI-TOF): calc for $C_{28}H_{24}BN_2O_8$ $[M + H]^+$; 415.19812 found 415.19815.

4.3.6. Mannopyranosyl BODIPY **5**

This compound was prepared by glycosylation of BODIPY **15** (73 mg, 0.176 mmol) with MeOE **4c** (150 mg, 0.264 mmol), according to the general procedure for glycosylation, procedure A. After standard work-up, purification by flash chromatography (hexane- ethyl acetate; 9:1) furnished a mixture of the 2-*O*-benzoyl derivative of compound **5** (102 mg, 72%) and "rearranged" methyl mannopyranoside (^{1}HNMR). This mixture was dissolved in MeOH, under argon, and treated with NaOMe/MeOH at room temperature for 24 h. When t.l.c. showed disappearance of the fluorescent starting material, solid NH4Cl was added. The resulting solution was kept with stirring for 15 min, filtered, and the solvent was evaporated. Purification by flash chromatography (hexane-ethyl acetate; 7:3) allowed the isolation of BODIPY-mannoside **5** (92 mg, 87%). ^{1}H-NMR (500 MHz, $CDCl_3$) δ 7.61 (s, 1H), 7.52 (s, 1H) 7.47–7.09 (m, 29H), 6.70 (dd, *J* = 26.9, 4.3 Hz, 2H), 6.43–6.34 (m, 2H), 4.74 (d, *J* = 10.7 Hz, 1H), 4.60 (d, *J* = 12.2 Hz, 1H), 4.53–4.39 (m, 4H), 4.37 (bs, 2H), 4.28 (d, *J* = 11.6 Hz, 1H), 3.73 (t, *J* = 9.5 Hz, 1H), 3.64–3.46 (m, 4H), 3.39–3.35 (m, 1H); ^{13}C-NMR (125 MHz, $CDCl_3$) δ 145.3 (×2), 145.2, 138.4, 138.1, 136.0, 135.6, 134.9, 133.8, 133.5 (×2), 132.2 (×3), 130.2, 129.6, 129.5, 129.2, 129.1, 128.6 (×2), 128.5 (×2), 128.4 (×3), 128.1 (×2), 128.0 (×2), 127.9, 127.8 (×2), 127.7 (×2), 127.6 (×2), 126.6, 126.3, 117.8, 117.6, 98.6, 80.2, 75.3, 74.1, 73.5 (×2), 71.7, 71.3, 68.8, 67.7, 67.3 HRMS (ESI-TOF): calc for C_{55}H55BN_3O_6 $[M + NH_4]^+$; 846.41876 found 846.41891.

4.3.7. BODIPY Disaccharide **16a**

This compound was prepared by glycosylation of **5** (70 mg, 0.083 mmol) with MeOE **4a** (57 mg, 0.100 mmol), according to the general procedure for glycosylation, procedure A. Purification by flash chromatography (hexane-ethyl acetate; 7:3) afforded compound **16a** (62 mg, 52%). ^{1}H-NMR (400 MHz, $CDCl_3$) δ 8.08–7.05 (m, 49 H), 6.69 (ddd, *J* = 12.0, 4.3, 1.2 Hz, 2H), 6.39 (ddd, *J* = 19.1, 4.3, 1.8 Hz, 2H), 6.11 (t, *J* = 10.1 Hz, 1H), 5.93 (dd, *J* = 10.1, 3.2 Hz, 1H), 5.87 (dd, *J* = 3.2, 1.9 Hz, 1H), 5.17 (d, *J* = 2.0 Hz, 1H), 4.84 (d, *J* = 10.9 Hz, 1H), 4.80 (d, *J* = 1.9 Hz, 1H), 4.66 (d, *J* = 12.3 Hz, 1H), 4.60–4.39 (m, 7H), 4.42 (d, *J* = 12.1 Hz, 1H), 4.33 (dd, *J* = 12.3, 3.6 Hz, 1H), 4.24 (d, *J* = 12.0 Hz, 1H), 4.00 (t, *J* = 9.6 Hz, 1H), 3.79 (d, *J* = 2.4 Hz, 1H), 3.75–3.63 (m, 2H), 3.61–3.52 (m, 2H); ^{13}C-NMR (125 MHz, $CDCl_3$) δ 166.2, 165.6, 165.3, 165.1, 145.5, 145.4, 144.9, 138.5, 138.5, 138.3, 136.0, 135.2 (×2), 133.5 (×2), 133.2 (×3), 133.1 (×2), 132.5 (×2), 132.4, 130.3, 130.2, 130.0 (×6),129.9 (×2), 129.6 (×2), 129.4 (×2), 129.1, 128.9, 128.7 (×2), 128.6 (×2), 128.5 (×4), 128.4 (×5), 128.3 (×2), 127.9 (×3), 127.8, 127.7 (×2), 127.6 (×2), 127.5 (×3), 126.4 (×2) (65 signals for aromatic carbons) 117.8, 117.7, 99.4, 97.8, 79.5, 75.9, 75.2, 74.4, 73.3 (×2), 72.2, 72.0, 70.2, 70.0, 69.1, 68.7, 66.8, 66.7, 62.6 HRMS (ESI-TOF): calc for $C_{89}H_{77}BKN_2O_{15}$ $[M + K]^+$; 1447.51131 found 1447.51484 $(M + K)^+$.

4.3.8. BODIPY-Disaccharide Tetraol **16b**

This compound was prepared according to the general procedure for debenzoylation, procedure B, from compound **16a** (60 mg, 0.087 mmol). Purification by flash chromatography (hexane-ethyl acetate; 2:8) afforded fluorescent disaccharide **16b** (40 mg, quantitative yield). ^{1}H-NMR (500 MHz, $CDCl_3$) δ 7.56–7.11 (m, 29H), 6.69–6.66 (m,2H), 6.37 (ddd, *J* = 12.6, 4.3, 1.8 Hz, 2H), 4.93 (bs, 1H), 4.74 (d, *J* = 10.8 Hz, 1H), 4.67 (bs, 1H) 4.56–4.51 (m, 2H), 4.42–4.36 (m, 4H), 4.25 (d, *J* = 12.3 Hz, 1H), 3.98 (bs, 1H), 3.87–3.77 (m, 5H), 3.69–3.64 (m,3H), 3.59–3.52 (m, 4H), 3.44–3.37 (m, 4H) ^{13}C-NMR (125 MHz, $CDCl_3$) δ 145.5, 145.3, 145.0, 138.5 (×2), 138.4, 136.0, 135.2, 135.1, 133.7, 133.4, 133.2 (×2), 133.1, 132.9, 132.5 (×2), 130.3, 130.2, 129.6, 129.1, 128.8 (22), 128.5 (×5), 128.4 (×2), 128.0 (×2), 127.8 (×3), 127.7 (×3), 127.6 (×4), 126.5, 126.4, 117.9 (×2), 101.5, 97.6, 79.5, 75.3, 75.2, 74.4, 73.4, 72.5, 72.1, 72.0, 71.6, 70.9, 68.8, 66.6, 61.6; API-ES positive mode: $[M + Na]^+$ = 1031.3.

4.3.9. BODIPY-Disaccharide Silylated Triol **6**

The general procedure for silylation, procedure C, was applied to tetraol **16b** (56 mg, 0.056 mmol), although this time pyridine, rather than DMF, was used as solvent. Purification by flash chromatography (hexane-ethyl acetate; 6:4) yielded triol **6** (48 mg, 69%). ^{1}H-NMR (400 MHz, $CDCl_3$) δ 7.72–7.10 (m, 39H), 6.64 (ddd, *J* = 5.6, 4.3, 1.2 Hz, 2H), 6.34 (ddd, *J* = 12.6, 4.3, 1.8 Hz, 2H), 4.97 (d, *J* = 1.7 Hz, 1H), 4.76 (d, *J* = 10.8 Hz, 1H), 4.65–4.59 (m, 2H), 4.54–4.38 (m, 4H), 4.36 (bs 2H), 4.18 (d, *J* = 12.2 Hz, 1H), 3.99 (bs, 1H), 3.87–3.57 (m, 13H), 3.45–3.40 (m, 2H),1.04 (s, 9H) ^{13}C-NMR (100 MHz, $CDCl_3$) δ 145.5, 145.4, 145.1, 138.8, 138.7, 138.5, 136.1, 135.8, 135.2, 133.4, 133.2, 133.1, 133.0, 132.7, 130.2, 129.5, 129.3, 128.6, 128.5, 128.1, 128.0, 127.8, 127.7, 127.6, 127.5, 126.5 (48 aromatic carbons), 118.0, 117.9, 101.2, 98.1, 79.9, 75.3, 74.6, 73.5 (×2), 72.3, 72.1 (×2), 71.5 (×2), 71.4, 70.5 (×2), 70.1, 69.1, 66.6, 65.1, 53.7, 27.1 (×3), 19.5; API-ES, positive mode: $[M + Na]^+$ = 1269.30; HRMS (ESI-TOF): calc for $C_{156}H_{155}BN_2O_{26}Si_3$ $[M + Na]^+$; 2591.02041 found 2591.01683.

4.3.10. Thioglycosyl Disaccharide **8**

According to the general procedure for glycosylation, procedure A, thioglycoside **7** (50 mg, 0.098 mmol) was glycosylated with orthoester **4b** (146 mg, 0.196 mmol) in CH_2Cl_2 (3 mL). After 5 min, the reaction was quenched, the organic extract concentrated, and the resulting residue chromatographed over silica gel flash column (hexane-ethyl acetate; 9:1) to give diol **8** (65 mg, 58%). For the sake of characterization, and according to the general procedure for acetylation, procedure D, the corresponding peracetyl derivative **8-Acet** was prepared (68 mg, quantitative yield). $[\alpha]_D^{21}$: −39.3°, (*c* 0.8, $CHCl_3$) ^{1}H-NMR (400 MHz, $CDCl_3$) δ 8.18–7.10 (m, 40H), 6.35 (t, *J* = 10.2 Hz, 1H), 5.71 (dd, *J* = 10.3, 3.3 Hz, 1H), 5.53–5.45 (m, 4H), 5.32 (d, *J* = 1.8 Hz, 1H), 4.38–4.34 (m, 1H), 4.25–4.20 (m, 2H), 3.92–3.80 (m, 2H), 3.78–3.61 (m, 2H), 2.20 (s, 3H), 2.12 (s, 3H), 1.07 (s, 9H), 1.06 (s, 9H) ^{13}C-NMR (100 MHz, $CDCl_3$) δ 170.5, 170.1, 165.8, 165.6, 165.5, 136.0 (×4), 135.9 (×2), 135.7 (×2), 133.9, 133.7, 133.5, 133.4, 133.3, 133.2, 133.0, 131.9 (×2), 130.2 (×2), 130.0 (×4), 129.9 (×2), 129.8, 129.7 (×2), 129.5 (×2), 129.3 (×2), 128.8 (×2), 128.6 (×2), 128.5 (×2), 127.9 (×4), 127.8 (×2), 99.4, 86.3, 75.6, 73.3, 72.9, 72.2, 71.2, 70.3, 68.2, 66.2, 63.2, 61.9, 26.95 (×3), 26.86 (×3), 21.7, 21.0, 19.5 (×2); API-ES, positive mode:1324.5 $[M + NH_4]^+$.

4.3.11. 1,2-Methyl Orthoester Disaccharide **9**

Benzoylation (BzCl, pyridine) of compound **8** (498 mg, 0.41 mmol), followed by purification by flash chromatography (hexane-ethyl acetate; 85:15) provided the corresponding benzoylated disaccharide intermediate **8-Bzl** (450 mg, 88%). A portion of this compound (298 mg, 0.208 mmol) was dissolved in CH_2Cl_2 (3 mL), and cooled to 0 °C, in the darkness, then bromine (15 µL, 0.314 mmol) was added. When the starting material had disappeared (t.l.c.) the reaction mixture was washed with 10% aqueous sodium thiosulfate solution containing sodium bicarbonate saturated aqueous solution, and water. The organic extract was dried over Na_2SO_4, filtered, and concentrated. Without further purification, the residue was dissolved in acetonitrile (1 mL) to which solution, methanol (84 µL), tetrabutylammonium bromide (Bu_4NBr) (47 mg), sodium bicarbonate (35 mg), and molecular sieves 4Å (previously dried) were added. The resulting reaction mixture was stirred for one night, the molecular sieves were then filtered, and the solvent concentrated. The ensuing residue was purified by flash chromatography (hexane-ethyl acetate; 9:1, containing 1% of triethylamine) to afford the 1,2-methyl orthoester disaccharide **9** (182.6 mg, 65% (two steps)) $[\alpha]_D^{21}$: −89.2 (*c* 1.2, $CHCl_3$) ^{1}H-NMR (400 MHz, $CDCl_3$) δ 8.05–7.09 (m, 45H), 6.20 (t, *J* = 10.1 Hz, 1H), 5.84 (dd, *J* = 10.2, 3.3 Hz, 1H), 5.62 (dd, *J* = 9.8, 8.7 Hz, 1H), 5.50 (d, *J* = 3.0 Hz, 1H), 5.44 (dd, *J* = 3.4, 1.8 Hz, 1H), 5.27 (d, *J* = 1.8 Hz, 1H), 4.84 (dd, *J* = 4.1, 3.0 Hz, 1H), 4.50 (d, *J* = 10.1 Hz, 1H), 4.23 (d, *J* = 9.8 Hz, 1H), 3.95–3.88 (m, 2H), 3.77–3.62 (m, 3H), 2.96 (s, 3H), 1.06 (s, 9H), 0.91 (s, 9H).^{13}C-NMR (100 MHz, $CDCl_3$) δ 165.5, 165.3 (×2), 165.2, 136.2, 136.0 (×2), 135.8 (×6), 133.5, 133.4, 133.3, 133.2, 133.1, 130.1 (×4), 130.0 (×2), 129.9 (×3), 129.8 (×2), 129.7 (×3), 129.6, 129.5, 129.4, 128.7 (×2), 128.6 (×2), 128.5 (×2), 128.4 (×4), 127.8 (×3), 127.7 (×4), 126.9 (×2), 122.9, 100.1, 98.1, 78.6, 77.9, 75.5, 72.2, 71.0, 70.5, 68.2, 66.6, 63.7, 62.4, 51.5, 26.9 (×3), 26.9 (×3), 19.5, 19.3 HRMS (ESI-TOF): calc for $C_{80}H_{84}NO_{16}Si_2$ $[M + NH_4]^+$ 1370.5323; found 1370.5323 $[M + NH_4]^+$.

4.3.12. BODIPY-Tetrasaccharide **10**

BODIP disaccharide **6** (21 mg, 0.017 mmol) was glycosylated with glycosyl donor **9** (vide infra) (45 mg, 0.034 mmol), according to the general procedure for glycosylation, procedure A. Purification by flash chromatography (hexane-ethyl acetate; 8:2) yielded tetrasaccharide **10** (23 mg, 53%). ^{1}H-NMR (500 MHz, $CDCl_3$) δ 8.05–6.98 (m, 84H), 6.62 (td, *J* = 4.3, 1.3 Hz, 2H), 6.31 (dd, *J* = 4.3, 1.8 Hz, 1H), 6.28 (dd, *J* = 4.3, 1.8 Hz, 1H), 6.21 (t, *J* = 10.1 Hz, 1H), 5.66 (d, *J* = 8.7 Hz, 1H), 5.59–5.52 (m, 2H), 5.40 (bs, 1H) 5.37 (m, 1H), 5.31 (bs, 1H), 5.20 (bs, 1H), 4.71–4.69 (m, 2H), 4.62–4.45 (m, 6H), 4.39–4.28 (m, 5H), 4.21–4.03 (m, 4H), 3.99–3.68 (m, 12H), 3.64–3.56 (m, 2H), 3.49–3.35 (m, 4H), 1.02 (s, 9H), 0.99 (s, 9H), 0.95 (s, 9H) ^{13}C-NMR (125 MHz, $CDCl_3$) δ 165.7, 165.5, 165.2, 165.0 (×2), 145.4, 145.3, 144.9, 138.6, 138.6, 138.4, 138.0, 136.2, 135.9 (×4), 135.8 (×3), 135.7 (×3), 135.6 (×4), 135.1, 133.6, 133.5, 133.4, 133.3, 133.2, 133.1(×3), 133.0, 132.9, 132.7 (×2), 132.6 (×2), 132.3, 130.2 (×2), 130.1 (×2), 130.0 (×2), 129.9, 129.8 (×6), 129.7 (×2), 129.5 (×2), 129.4 (×2), 129.2 (×3), 128.8, 128.7, 128.5 (×4), 128.4 (×11), 128.2, 128.0, 127.9, 127.8, 127.7 (×5), 127.6, 127.4 (×4), 127.2, 126.5, 126.4, 125.4, 117.9 (×2), 101.1, 99.1, 98.2, 98.0, 81.5, 80.1, 76.2, 75.1, 74.6, 73.4 (×2), 72.9, 72.3 (×2), 72.2, 72.0, 71.9 (×2), 71.8, 70.6, 70.4, 69.5, 69.1, 68.3, 66.6, 66.2, 65.8, 64.4, 63.8, 61.5, 53.6, 27.0 (×3), 26.8 (×3), 26.7 (×3), 19.5, 19.3, 19.0; HRMS (ESI-TOF): calc for $C_{156}H_{155}BN_2O_{26}Si_3$ $[M + Na]^+$; 2591.02041 found 2591.01683.

Supplementary Materials: The following are available online, Phtophysical data (Table S1). Experimental conditions for photophysical properties, and quantum mechanical calculations. ^{1}H- and ^{13}C-NMR of all compounds.

Author Contributions: Development of the synthetic scheme, structural analysis of the saccharide derivatives, conceptualization of the results, J.V.; Design of the saccharide synthesis, conceptualization of the results, C.U.; Design of the saccharide and BODIPY syntheses, preparation of the manuscript, A.M.G.; Photophysical experiments, E.A.-Z.; Photophysical experiments, conceptualization of the results, preparation of the manuscript, J.B.; Laser experiments, supervision, I.G.-M.; Conceptualization of the results, supervision, manuscript preparation, J.C.L.; Funding acquisition, A.M.G., J.B., I.G.-M., J.C.L. All authors have read and agreed to the published version of the manuscript.

Funding: This research was funded by Spanish MINISTERIO DE ECONOMIA Y COMPETITIVIDAD, GOBIERNO DE ESPAÑA (projects MAT2017–83856-C3-1-P and 3-P, PiD2020-1147555GB-C33), the MINISTERIO DE CIENCIA INNOVACION Y UNIVERSIDADES (project RTI2018-094862-B-I00), and the GOBIERNO VASCO (project IT912-16).

Data Availability Statement: Data sharing is not applicable to this article.

Acknowledgments: E.A.-Z. Thanks the MINISTERIO DE ECONOMIA Y COMPETITIVIDAD for a postdoctoral contract. The authors thank SGIker of UPV/EHU for technical support with the computational calculations, which were carried out in the "arina" informatic cluster.

Conflicts of Interest: The authors declare no conflict of interest. The funders had no role in the design of the study; in the collection, analyses, or interpretation of data; in the writing of the manuscript, or in the decision to publish the results.

Sample Availability: No samples of the compounds are available from the authors.

References

1. Ferro, V.; Hammond, E.; Fairweather, J.K. The development of inhibitors of heparanase, a key enzyme involved in tumour metastasis, angiogenesis and inflammation. *Rev. Med. Chem.* **2004**, *4*, 693–702. [CrossRef] [PubMed]
2. Hammond, E.; Bytheway, I.; Ferro, V. Heparanase as a target for anticancer therapeutics: New developments and future prospects. In *New Developments in Therapeutic Glycomics*; Delehedde, M., Lortat-Jacob, H., Eds.; Research Signpost: Trivandrum, India, 2006; pp. 251–282.
3. Miao, H.Q.; Liu, H.; Navarro, E.; Kussie, P.; Zhu, Z. Development of heparanase inhibitors for anti-cancer therapy. *Curr. Med. Chem.* **2006**, *13*, 2101–2111.
4. McKenzie, E.A. Heparanase: A target for drug discovery in cancer and inflammation. *Br. J. Pharmacol.* **2007**, *151*, 1–14. [CrossRef]
5. Jia, L.; Ma, S. Recent advances in the discovery of heparanase inhibitors as anticancer agents. *Eur. J. Med. Chem.* **2016**, *121*, 209–220. [CrossRef]
6. Chhabra, M.; Ferro, V. The development of assays for heparanase enzymatic activity: Towards a gold standard. *Molecules* **2018**, *23*, 2971. [CrossRef]
7. Chhabra, M.; Ferro, V. PI-88 and related heparan sulfate mimetics. In *Heparanase: From Basic Research to Clinical Applications*, 1st ed.; Vlodavsky, I., Sanderson, R.D., Ilan, N., Eds.; Springer-Nature: Cham, Switzerland, 2020; Chapter 19; pp. 473–492.
8. Mechref, Y.; Novotny, M.V. Structural investigations of glycoconjugates at high sensitivity. *Chem. Rev.* **2002**, *102*, 321–369. [CrossRef]
9. Xia, B.; Kawar, Z.S.; Ju, T.; Alvarez, R.A.; Sachdev, G.P.; Cummings, R.D. Versatile fluorescent derivatization of glycans for glycomic analysis. *Nat. Methods* **2005**, *2*, 845–850. [CrossRef] [PubMed]
10. Kobayashi, H.; Ogawa, M.; Alford, R.; Choyke, P.L.; Urano, Y. New strategies for fluorescent probe design in medical diagnostic imaging. *Chem. Rev.* **2010**, *110*, 2620–2640. [CrossRef] [PubMed]
11. Ueno, T.; Nagano, T. Fluorescent probes for sensing and imaging. *Nat. Methods* **2011**, *8*, 642–645. [CrossRef] [PubMed]
12. Keithley, R.B.; Rosenthal, A.S.; Essaka, D.C.; Tanaka, H.; Yoshimura, Y.; Palcic, M.M.; Hindsgaul, O.; Dovichi, N.J. Capillary electrophoresis with three-color fluorescence detection for the analysis of glycosphingolipid metabolism. *Analyst* **2013**, *138*, 164–170. [CrossRef]
13. Louret, A.; Burgess, K. BODIPY Dyes and their derivatives: Syntheses and spectroscopic properties. *Chem. Rev.* **2007**, *107*, 4891–4932.
14. Ziessel, R.; Ulrich, G.; Harriman, A. The chemistry of BODIPY: A new El Dorado for fluorescence tools. *New. J. Chem.* **2007**, *31*, 496–501. [CrossRef]
15. Ulrich, G.; Ziessel, R.; Harriman, A. The chemistry of fluorescent BODIPY dyes: Versatility unsurpassed. *Angew. Chem. Int. Ed.* **2008**, *47*, 1184–1201. [CrossRef] [PubMed]
16. Boens, N.; Verbelen, B.; Dehaen, W. Postfunctionalization of the BODIPY core: Synthesis and spectroscopy. *Eur. J. Org. Chem.* **2015**, *30*, 6577–6595. [CrossRef]
17. Bañuelos, J. BODIPY Dye, the most versatile fluorophore ever? *Chem. Rec.* **2016**, *16*, 335–348. [CrossRef]
18. Boens, N.; Verbelen, B.; Ortiz, M.J.; Jiao, L.; Dehaen, W. Synthesis of BODIPY dyes through postfunctionalization of the boron dipyrromethene core. *Coord. Chem. Rev.* **2019**, *399*, 213024. [CrossRef]
19. Uriel, C.; Permingeat, C.; Ventura, J.; Avellanal-Zaballa, E.; Bañuelos, J.; Garcia-Moreno, I.; Gómez, A.M.; López, J.C. BODIPYs as chemically stable fluorescent tags for synthetic glycosylation strategies towards fluorescently labeled saccharides. *Chem. Eur. J.* **2020**, *26*, 5388–5399. [CrossRef]
20. Ekblad, M.; Adamiak, B.; Bergstrom, T.; Johnstone, K.D.; Karoli, T.; Liu, L.; Ferro, V.; Trybala, E. A highly lipophilic sulfated tetrasaccharide glycoside related to muparfostat (PI-88) exhibits virucidal activity against herpes simplex virus. *Antivir. Res.* **2010**, *86*, 196–203. [CrossRef]
21. Handley, P.N.; Carroll, A.; Ferro, V. New structural insights into the oligosaccharide phosphate fraction of *Pichia* (Hansenula) *holstii* NRRL Y2448 phosphomannan. *Carbohydr. Res.* **2017**, *446–447*, 68–75. [CrossRef]
22. Gu, G.; Wei, G.; Du, Y. Synthesis of a 6^{V}-sulfated mannopentasaccharide analogue related to PI-88. *Carbohydr. Res.* **2004**, *339*, 1155–1162. [CrossRef]
23. Namme, R.; Mitsugi, T.; Takahashi, H.; Ikegami, S. Synthesis of PI-88 analogue using novel *O*-glycosidation of *exo*-methylenesugars. *Tetrahedron Lett.* **2005**, *46*, 3033–3036. [CrossRef]
24. Fairweather, J.K.; Hammond, E.; Johnstone, K.D.; Ferro, V. Synthesis and heparanase inhibitory activity of sulfated mannooligosaccharides related to the antiangiogenic agent PI-88. *Bioorg. Med. Chem.* **2008**, *16*, 699–709. [CrossRef]
25. Valerio, S.; Pastore, A.; Adinolfi, M.; Iadonisi, A. Sequential one-pot glycosidations catalytically promoted: Unprecedented strategy in oligosaccharide synthesis for the straightforward assemblage of the antitumor PI-88 pentasaccharide. *J. Org. Chem.* **2008**, *73*, 4496–4503. [CrossRef] [PubMed]

26. Liu, L.; Johnstone, K.D.; Fairweather, J.K.; Dredge, K.; Ferro, V. An improved synthetic route to the potent angiogenesis inhibitor benzyl Manα(1→3)-Manα(1→3)-Manα(1→3)-Manα(1→2)-Man hexadecasulfate. *Aust. J. Chem.* **2009**, *62*, 546–552. [CrossRef]
27. Pastore, A.; Adinolfi, M.; Iadonisi, A.; Valerio, S. One-pot catalytic glycosidation/Fmoc removal—An iterable sequence for straightforward assembly of oligosaccharides related to HIC gp120. *Eur. J. Org. Chem.* **2010**, *4*, 711–718. [CrossRef]
28. Mong, K.-K.T.; Shiau, K.-S.; Lin, Y.H.; Cheng, K.-C.; Lin, C.-H. A concise synthesis of single components of partially sulfated oligomannans. *Org. Biomol. Chem.* **2015**, *13*, 11550–11560. [CrossRef]
29. Kochetkov, N.K.; Khorlin, A.J.; Bochkov, A.F. New synthesis of glycosides. *Tetrahedron Lett.* **1964**, *5*, 289–293. [CrossRef]
30. Kochetkov, N.K.; Khorlin, A.J.; Bochkov, A.F. A new method of glycosylation. *Tetrahedron* **1967**, *23*, 693–707. [CrossRef]
31. Uriel, C.; Ventura, J.; Gómez, A.M.; López, J.C.; Fraser-Reid, B. Methyl 1,2-orthoesters as useful glycosyl donors in glycosylation reactions: A comparison with *n*-pent-4-enyl 1,2 orthoesters. *Eur. J. Org. Chem.* **2012**, *16*, 3122–3131. [CrossRef]
32. Uriel, C.; Ventura, J.; Gómez, A.M.; López, J.C.; Fraser-Reid, B. Unexpected stereocontrolled access to 1α,1′β-disaccharides from methyl 1,2-ortho esters. *J. Org. Chem.* **2012**, *77*, 795–800. [CrossRef] [PubMed]
33. Uriel, C.; Rijo, P.; Fernandes, A.S.; Gómez, A.M.; Fraser-Reid, B.; López, J.C. Methyl 1,2-orthoesters in acid-washed molecular sieves mediated glycosylations. *Chem. Select* **2016**, *1*, 6011–6015. [CrossRef]
34. Lu, J.; Fraser-Reid, B. The antituberculosis, antitumor, multibranched dodecafuranoarabinan of *Mycobacterium* species has been assembled from a single *n*-pentenylfuranoside source. *Chem. Commun.* **2005**, *7*, 862–864. [CrossRef] [PubMed]
35. Jayaprakash, K.N.; Lu, J.; Fraser-Reid, B. Synthesis of a lipomannan component of the cell-wall complex of *Mycobacterium tuberculosis* is based on Paulsen's concept of donor/acceptor "match". *Angew. Chem. Int. Ed.* **2005**, *44*, 5894–5898. [CrossRef]
36. Fraser-Reid, B.; Loópez, J.C. Orthoesters and Related Derivatives. In *Handbook of Chemical Glycosylation: Advances in Stereoselectivity and Therapeutic Relevance*; Demchenko, A.V., Ed.; Wiley-VCH: New York, NY, USA, 2008; Chapter 5.1.
37. Roacho, R.I.; Metta-Magaña, A.J.; Peña-Cabrera, E.; Pannell, K.H. Synthesis, structural characterization, and spectroscopic properties of the *ortho*, *meta* and *para* isomers of 8-($HOCH_2$-C_6H_4)-BODIPY and 8-($MeOC_6H_4$)-BODIPY. *J. Phys. Org. Chem.* **2013**, *26*, 345–351. [CrossRef]
38. Del Rio, M.; Lobo, F.; Lopez, J.C.; Oliden, A.; Bañuelos, J.; Lopez-Arbeloa, I.; Garcia-Moreno, I.; Gomez, A.M. One-pot synthesis of rotationally restricted, conjugatable, BODIPY derivatives from phthalides. *J. Org. Chem.* **2017**, *82*, 1240–1247. [CrossRef]
39. Bodio, E.; Goze, C. Investigation of B-F substitution of BODIPY and aza-BODIPY dyes: Development of B-O and B-C BODIPYs. *Dyes Pigment.* **2019**, *160*, 700–710. [CrossRef]
40. Yang, L.; Simionescu, R.; Lough, A.; Yan, H. Some observations relating to the stability of the BODIPY fluorophore under acidic and basic conditions. *Dyes Pigment.* **2011**, *91*, 264–267. [CrossRef]
41. Nguyen, A.L.; Fronczek, F.R.; Smith, K.M.; Vicente, M.G.H. Synthesis of 4,4′-functionalized BODIPYs from dipyrrins. *Tetrahedron Lett.* **2015**, *56*, 6348–6351. [CrossRef]
42. Lopez, J.C.; Agocs, A.; Uriel, C.; Gomez, A.M.; Fraser-Reid, B. Iterative, orthogonal strategy for oligosaccharide synthesis based on the regioselective glycosylation of polyol acceptors with partially unprotected *n*-pentenyl-orthoesters: Further evidence for reciprocal donor acceptor selectivity. *Chem. Commun.* **2005**, *40*, 5088–5090. [CrossRef] [PubMed]
43. Wei, S.; Zhao, J.; Shao, H. A facile method for the preparation of sugar orthoesters promoted by anhydrous sodium bicarbonate. *Can. J. Chem.* **2009**, *87*, 1733–1737. [CrossRef]
44. Vidadala, S.R.; Thadke, S.A.; Hotha, S. Orthogonal activation of propargyl and *n*-pentenyl glycosides and 1,2-orthoesters. *J. Org. Chem.* **2009**, *74*, 9233–9236. [CrossRef] [PubMed]
45. Duran-Sampedro, G.; Esnal, I.; Agarrabeitia, A.R.; Bañuelos-Prieto, J.; Cerdan, L.; Garcia-Moreno, I.; Costela, A.; Lopez-Arbeloa, I.; Ortiz, M.-J. First highly efficient and photostable *E* and *C* derivatives of 4,4,-difluoro-4-bora-3a,4a-diaza-s-indacene (BODIPY) as dye lasers in the liquid phase, thin films, and solid-state rods. *Chem. Eur. J.* **2014**, *20*, 2646–2653. [CrossRef] [PubMed]
46. Patel, I.; Woodcock, J.; Beams, R.; Stranick, S.J.; Niewendaal, R.; Gilman, J.W.; Mulenos, M.R.; Sayes, C.M.; Salari, M.; DeLoid, G.; et al. Fluorescently labeled cellulose nanofibers for environmental health and safety studies. *Nanomaterials* **2021**, *11*, 1015. [CrossRef] [PubMed]

 MDPI

Article

Novel Orange Color Pigments Based on La_3LiMnO_7

Ryohei Oka [1], Jun-ichi Koyama [2], Takuro Morimoto [3] and Toshiyuki Masui [2,*]

[1] Field of Advanced Ceramics, Department of Life Science and Applied Chemistry, Graduate School of Engineering, Nagoya Institute of Technology, Gokiso, Showa, Nagoya 466-8555, Aichi, Japan; oka.ryohei@nitech.ac.jp
[2] Center for Research on Green Sustainable Chemistry, Department of Chemistry and Biotechnology, Faculty of Engineering, Tottori University, 4-101, Koyama-cho Minami, Tottori 680-8552, Japan; b15t3045h@gmail.com
[3] Department of Engineering, Graduate School of Sustainability Science, Tottori University, 4-101, Koyama-cho Minami, Tottori 680-8552, Japan; M21J5037H@edu.tottori-u.ac.jp
* Correspondence: masui@tottori-u.ac.jp; Tel.: +81-857-31-5264

Abstract: $La_3LiMn_{1-x}Ti_xO_7$ ($0 \leq x \leq 0.05$) samples were synthesized by a solid-state reaction method, and a single-phase form was observed for the samples in the range of $x \leq 0.03$. Crystal structure, optical properties, and color of the $La_3LiMn_{1-x}Ti_xO_7$ ($0 \leq x \leq 0.03$) samples were characterized. Strong optical absorption was observed at a wavelength between 400 and 550 nm, and a shoulder absorption peak also appeared around 690 nm in all samples; orange colors were also exhibited. Among the samples synthesized, the most brilliant orange color was obtained at $La_3LiMn_{0.97}Ti_{0.03}O_7$. The redness ($a^*$) and yellowness ($b^*$) values of this pigment were higher than those of the commercially available orange pigments. Therefore, the orange color of this pigment is brighter than those of the commercial products. Since the $La_3LiMn_{0.97}Ti_{0.03}O$ pigment is composed of non-toxic elements, it could be a new environmentally friendly inorganic orange pigment.

Keywords: inorganic pigments; orange color; environment-friendly; Mn^{4+} ion; d–d transition

Citation: Oka, R.; Koyama, J.-i.; Morimoto, T.; Masui, T. Novel Orange Color Pigments Based on La_3LiMnO_7. *Molecules* **2021**, *26*, 6243. https://doi.org/10.3390/molecules26206243

Academic Editors: Jorge Bañuelos Prieto and Ugo Caruso

Received: 13 September 2021
Accepted: 13 October 2021
Published: 15 October 2021

1. Introduction

Inorganic pigments have been widely applied to paints, glasses, ceramics, etc., because of their high hiding power and thermal stability compared to organic pigments [1]. Several orange pigments such as cadmium orange (CdS·CdSe), molybdate orange ($PbCrO_4 \cdot PbMoO_4 \cdot PbSO_4$), and bayferrox orange ($Fe_2O_3 \cdot FeOOH$) are conventionally used. However, the use of the cadmium and molybdate orange pigments has been forbidden or restricted because they contain toxic elements which have harmful effects on the human body and the environment. Although the bayferrox orange pigment is environmentally friendly, the vividness of this pigment is not sufficient. Therefore, development of environmentally friendly inorganic orange pigments is required, and several studies have been reported [2–17]. Some compounds, such as $La_2Ce_{2-x}W_{0.5x}Fe_{0.5x}O_{7+\delta}$ [14], $Sr_4Mn_2Cu_{0.5}Zn_{0.5}O_9$ [15], and $SrBaCe_{0.6}Tb_{0.4}O_4$ [3], for example, have been proposed and exhibit an orange color. Unfortunately, the colors of $La_2Ce_{2-x}W_{0.5x}Fe_{0.5x}O_{7+\delta}$ and $Sr_4Mn_2Cu_{0.5}Zn_{0.5}O_9$ are pale or dark, and $SrBaCe_{0.6}Tb_{0.4}O_4$ contains Tb, for which raw materials are expensive. Hence, environment-friendly and low-cost orange pigments are desirable.

Because of this situation, we focused on Mn^{4+} as an orange coloring source. Mn^{4+} has been investigated as an activator for the red-light emitting phosphors [18–25]. These Mn^{4+}-activated phosphors absorb/emit visible light, due to the d-d transition. In general, the absorption intensity and wavelength of the optical absorption band corresponding to the d-d transition are strongly influenced by the content of Mn^{4+} and the coordination environment around the Mn^{4+} ions, respectively. In the case of phosphors, the concentration of Mn^{4+} is limited by $\leq$1% mol, but the absorption becomes stronger as the Mn^{4+} concentration increases. Recently, the materials based on Li_2MnO_3 containing a large amount of Mn^{4+} have been reported as environment-friendly red pigments [26–28]. The pure Li_2MnO_3 pigment shows orange color, while the color becomes reddish by doping with other cations. In other words, Mn^{4+} ion is a promising coloring source for not only red but also orange.

In this study, we selected La_3LiMnO_7 as a host material for environmentally friendly orange pigment. This compound could have quite low toxicity as compared with the conventional pigments containing toxic elements, because toxicity of the constituent elements is quite lower than that of toxic ones, such as Cd and Pb. In addition, raw materials of this are cost-effective. The components of this material are similar to those of Li_2MnO_3. Therefore, it is expected that the host La_3LiMnO_7 material exhibit orange color due to the d–d transition of Mn^{4+}. As mentioned above, the optical absorption of the d–d transition is influenced by the geometric structure around Mn^{4+} and the concentration of Mn^{4+}. If other cations such as Ti^{4+} (ionic radius: 0.0605 nm [29]), the ionic radius of which is close to that of Mn^{4+} (ionic radius: 0.053 nm [29]), are doped into the Mn^{4+} site, it is possible to tune the color by controlling the content and geometric structure for Mn^{4+}. For these reasons, the $La_3LiMn_{1-x}Ti_xO_7$ ($0 \leq x \leq 0.05$) samples were synthesized by a solid-state reaction method, and their optical and color properties were characterized as environmentally friendly inorganic orange pigments.

2. Results and Discussion

2.1. X-ray Powder Diffraction (XRD) and Scanning Electron Microscopy (SEM) Image

Figure 1 shows the XRD patterns of the $La_3LiMn_{1-x}Ti_xO_7$ ($0 \leq x \leq 0.05$) samples. The positions, full width at half maximum (FWHM), and relative integrated intensities (RII) of diffraction peaks for La_3LiMnO_7 phase are also listed in Table 1. A single-phase form was observed for the samples in the range of $x \leq 0.03$, and all diffraction peaks were assigned to the La_3LiMnO_7 phase. On the other hand, an impurity phase (TiO_2) was detected in the sample with $x = 0.05$.

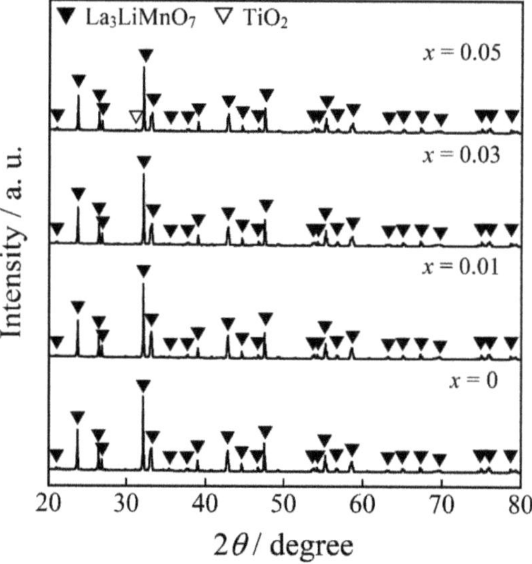

Figure 1. XRD patterns of the $La_3LiMn_{1-x}Ti_xO_7$ ($0 \leq x \leq 0.05$) samples.

Table 1. Peak positions, full width at half maximum (FWHM), and relative integrated intensities (RII) of diffraction peaks for La_3LiMnO_7 phase.

x = 0			x = 0.01			x = 0.03			x = 0.05		
2θ/deg.	**FWHM**	**RII/%**	**2θ/deg.**	**FWHM**	**RII/%**	**2θ/deg.**	**FWHM**	**RII/%**	**2θ/deg.**	**FWHM**	**RII/%**
21.00	0.12	4	20.99	0.48	6	20.98	0.40	11	20.99	0.08	4
23.67	0.06	51	23.67	0.07	50	23.67	0.07	51	23.67	0.07	50
26.36	0.07	28	26.36	0.07	25	26.37	0.07	27	26.36	0.07	25
26.78	0.07	15	26.77	0.07	15	26.77	0.07	15	26.77	0.07	14
32.16	0.07	100	32.16	0.07	100	32.16	0.07	100	32.16	0.07	100
33.07	0.19	47	33.11	0.22	69	33.04	0.17	39	33.22	0.23	54
35.41	0.05	2	35.38	0.11	3	35.41	0.08	2	35.39	0.06	2
37.70	0.24	7	37.66	0.20	8	37.69	0.27	7	37.61	0.19	6
38.99	0.07	15	38.99	0.08	15	38.99	0.08	15	38.99	0.08	16
42.76	0.14	30	42.82	0.18	55	42.73	0.13	27	42.73	0.13	30
44.66	0.07	11	44.66	0.08	10	44.67	0.08	10	44.66	0.08	10
46.77	0.08	5	46.77	0.09	5	46.77	0.09	5	46.77	0.08	5
47.54	0.08	42	47.53	0.09	42	47.52	0.08	41	47.52	0.08	42
53.72	0.29	10	53.76	0.30	13	53.67	0.27	11	53.67	0.31	10
54.25	0.08	5	54.24	0.08	5	54.25	0.08	6	54.25	0.08	5
55.17	0.08	17	55.16	0.09	16	55.15	0.09	19	55.15	0.10	19
55.29	0.09	21	55.29	0.09	21	55.29	0.09	20	55.29	0.08	18
56.76	0.21	10	56.71	0.17	9	56.77	0.21	9	56.65	0.13	5
58.49	0.22	21	58.60	0.27	36	58.40	0.26	27	58.41	0.27	26
63.21	0.32	7	63.14	0.29	8	63.15	0.39	9	63.22	0.25	5
65.07	0.20	9	65.05	0.17	9	65.01	0.21	10	65.03	0.22	9
67.27	0.11	9	67.26	0.10	10	67.25	0.11	10	67.26	0.11	10
69.44	0.72	11	69.40	0.29	6	69.27	0.51	8	69.28	0.63	9
74.99	0.13	9	74.97	0.13	9	74.94	0.13	10	74.94	0.12	9
75.96	0.32	24	75.87	0.31	23	75.74	0.37	22	75.68	0.21	10
78.64	0.12	7	78.62	0.12	7	78.62	0.14	7	78.60	0.13	7
79.07	0.37	7	79.10	0.39	13	79.16	0.48	10	78.97	0.32	6

La_3LiMnO_7 belongs to the layered perovskite type structure, and it crystallizes into a tetragonal structure with space group of $P4_2/mnm$ (No. 136) [30]. The Li^+ and Mn^{4+} ions occupy one 8j site according to the Wyckoff notation. These cations form the [Li/MnO_6] octahedra surrounded by six O^{2-} ions. Figure 2 shows the composition dependence of the lattice volume for the $La_3LiMn_{1-x}Ti_xO_7$ ($0 \leq x \leq 0.05$) samples. Cell volume increased as the Ti^{4+} content increased in the range of $x \leq 0.03$. This result indicates that some Mn^{4+} (ionic radius: 0.053 nm [29]) ions in the host lattice were substituted with larger Ti^{4+} (ionic radius: 0.0605 nm [29]) ones. The lattice volumes of the samples for x = 0.03 and 0.05 were even equal. Therefore, the solubility limit of Ti^{4+} into the La_3LiMnO_7 lattice was x = 0.03.

The SEM images of the $La_3LiMn_{1-x}Ti_xO_7$ (x = 0 and 0.03) samples are shown in Figure 3. The particle size was about 1 μm in both samples, and particle aggregation was observed. Colors of materials are affected by various factors such as crystal structure, chemical composition, and particle size. In both present samples, there was no significant change in particle size and morphology as seen in Figure 3. These results indicate that the change in color properties was attributed to the changes in crystal structure and optical absorption caused by doping Ti^{4+}.

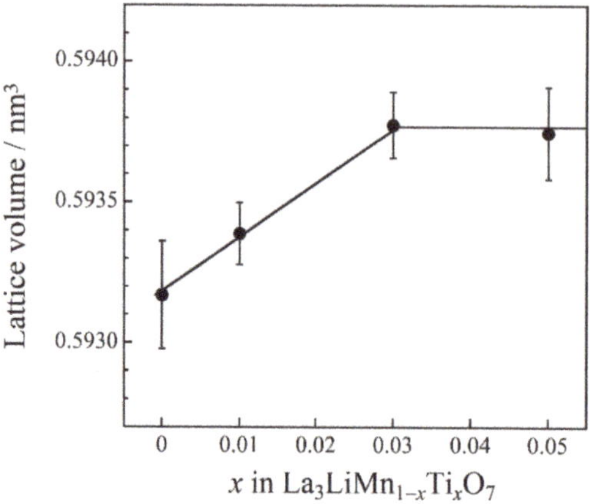

Figure 2. Compositional dependence of the lattice volume for $La_3LiMn_{1-x}Ti_xO_7$ ($0 \leq x \leq 0.05$).

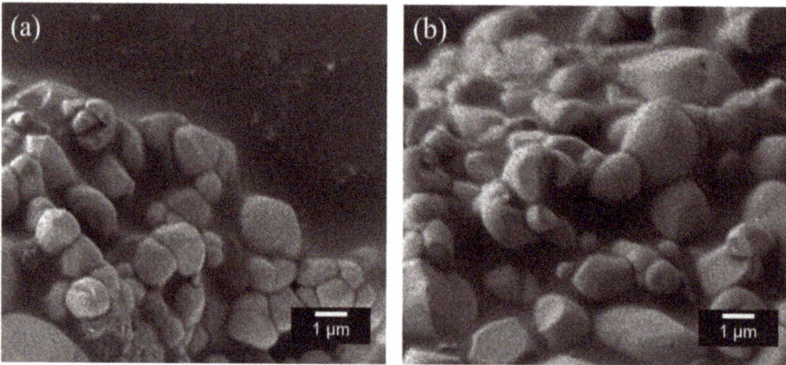

Figure 3. SEM images of (**a**) La_3LiMnO_7 and (**b**) $La_3LiMn_{0.97}Ti_{0.03}O_7$.

2.2. X-ray Fluorescence Analysis (XRF)

The element ratios of La, Mn, and Ti for the $La_3LiMn_{1-x}Ti_xO_7$ ($0 \leq x \leq 0.03$) samples, which were obtained in a single-phase form, were analyzed by XRF, and the results are summarized in Table 2. They were in approximate agreement with the stoichiometric ratios of the starting mixtures.

Table 2. Elemental ratios of La, Mn, and Ti for the $La_3LiMn_{1-x}Ti_xO_7$ ($0 \leq x \leq 0.03$) samples.

Samples	Stoichiometry (La:Mn:Ti)	Analyzed Ratio (La:Mn:Ti)
La_3LiMnO_7	3:1	3.03:0.97
$La_3LiMn_{0.99}Ti_{0.01}O_7$	3:0.99:0.01	3.01:0.96:0.03
$La_3LiMn_{0.97}Ti_{0.03}O_7$	3:0.97:0.03	3.01:0.94:0.05

2.3. Optical Reflectance Spectra

The UV–Vis reflectance spectra of the $La_3LiMn_{1-x}Ti_xO_7$ ($0 \leq x \leq 0.03$) samples, which were obtained in a single-phase form, are depicted in Figure 4. The optical absorption below 400 nm in the UV light region corresponded to the O_{2p}-Mn_{3d} charge transfer transition, while those around 500 and 690 nm in the visible region were attributed to the d-d transition of Mn^{4+}. The former was spin-allowed $^4A_{2g}$–$^4T_{2g}$ transition and the latter was spin-forbidden $^4A_{2g}2212^2E_g$ and $^4A_{2g}$–$^2T_{1g}$ transitions, respectively [22,26–28].

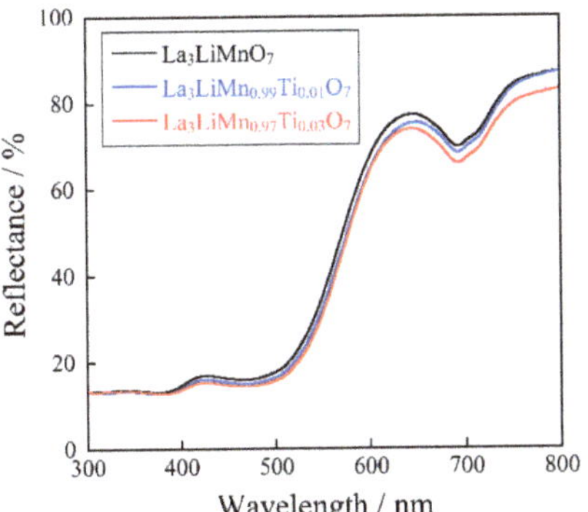

Figure 4. UV–Vis reflectance spectra for the $La_3LiMn_{1-x}Ti_xO_7$ ($0 \leq x \leq 0.03$) samples.

The absorption intensity of the d-d transition bands was increased by doping with Ti^{4+}. This behavior was due to the increased distortion of the $[MnO_6]$ octahedra, which was caused by the Ti^{4+} substitution, because Ti^{4+} (0.0605 nm [29]) was larger than Mn^{4+} (0.053 nm [29]). Although the d-d transitions were essentially forbidden, they were partially allowed due to the loss of symmetry. As a result, the absorption intensity of the d-d transitions was increased by the Ti^{4+} doping.

The UV–Vis reflectance spectrum of $La_3LiMn_{0.97}Ti_{0.03}O_7$ was compared to those of the commercial orange pigments such as Bayferrox® 960 and Bayferrox® 4960 (Fe_2O_3-FeOOH, Ozeki), as shown in Figure 5. The reflectance values in the green-blue light region (480–490 nm), corresponding to the complementary color of orange, of the current $La_3LiMn_{0.97}Ti_{0.03}O_7$ and commercial pigments were almost the same. On the other hand, the $La_3LiMn_{0.97}Ti_{0.03}O_7$ pigment showed higher reflectance in the yellow-red light region (580–750 nm) than those of the conventional ones. Accordingly, the $La_3LiMn_{0.97}Ti_{0.03}O_7$ pigment exhibited more vibrant orange color than the commercially available orange pigments.

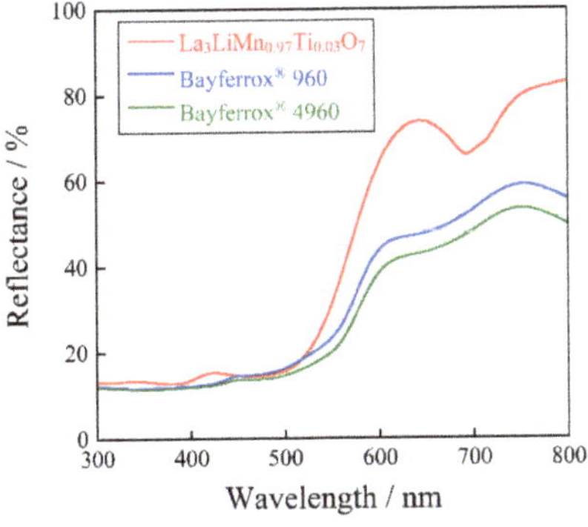

Figure 5. UV–Vis reflectance spectra for $La_3LiMn_{0.97}Ti_{0.03}O_7$, Bayferrox® 960 and Bayferrox® 4960.

2.4. Color Properties

The $L^*a^*b^*Ch°$ color coordinate data of the $La_3LiMn_{1-x}Ti_xO_7$ ($0 \leq x \leq 0.03$) and commercially available orange Bayferrox® 960 and Bayferrox® 4960 pigments are summarized in Table 3 (See 3.2. Characterization section described in later for the detail of these parameters). The photographs of these orange pigments are also shown in Figure 6. The hue angle ($h°$) values of the samples synthesized in this study have fallen within the orange color region (35°–70°). The redness (a^*), yellowness (b^*), and chroma (C) values were slightly increased by doping with Ti^{4+} into the Mn^{4+} site. As already discussed above on the results in Figure 4, the optical absorption from 480 to 490 nm (green-blue) became stronger when the Ti^{4+} ions were introduced into the Mn^{4+} site in the host lattice. Therefore, the sample color became more vivid orange.

Table 3. $L^*a^*b^*Ch°$ color coordinate data of the $La_3LiMn_{1-x}Ti_xO_7$ ($0 \le x \le 0.03$) and commercially available orange (Bayferrox® 960 and Bayferrox® 4960) pigments.

Pigment	L^*	a^*	b^*	C	$h°$
La_3LiMnO_7	69.5	+25.7	+63.8	68.8	68.1
$La_3LiMn_{0.99}Ti_{0.01}O_7$	68.7	+26.1	+64.4	69.5	67.9
$La_3LiMn_{0.97}Ti_{0.03}O_7$	67.2	+27.3	+65.4	70.9	67.3
Bayferrox® 960	59.0	+21.0	+47.5	51.9	66.1
Bayferrox® 4960	55.9	+23.5	+47.3	52.8	63.6

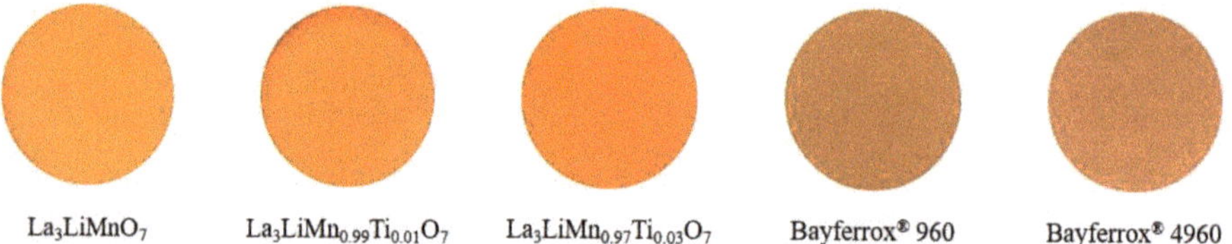

Figure 6. Photographs of the $La_3LiMn_{1-x}Ti_xO_7$ ($0 \le x \le 0.03$) and commercially available orange Bayferrox® 960 and Bayferrox® 4960 pigments.

As recognized in Table 3, the C value of the present $La_3LiMn_{0.97}Ti_{0.03}O_7$ pigment was much higher than those of the commercial orange pigments. In addition, the $h°$ value of these pigments were almost equivalent. These results elucidate that the $La_3LiMn_{0.97}Ti_{0.03}O_7$ pigment exhibited a bright orange color with high color purity, compared to the commercial orange pigments.

2.5. Chemical Stability Test

The chemical stability of the $La_3LiMn_{0.97}Ti_{0.03}O_7$ pigment was tested in the acid/base solutions. The pigment was dispersed into the 4% acetic acid and 4% ammonium bicarbonate solutions. These sample solutions for the acid and base conditions were left at room temperature for 1 and 24 h. After that, the samples were washed with deionized water and ethanol, and then dried at room temperature. The color of the pigment after the chemical stability test was evaluated by using the colorimeter. The $L^*a^*b^*Ch°$ color coordinate data of the $La_3LiMn_{0.97}Ti_{0.03}O_7$ pigment before and after the soaking test are tabulated in Table 4, and the photographs are also displayed in Figure 7. After the soaking test, the b^* value decreased slightly, but the $h°$ value was almost constant and no significant color degradation was observed.

Table 4. $L^*a^*b^*Ch°$ color coordinate data of the $La_3LiMn_{0.97}Ti_{0.03}O_7$ pigment before and after the chemical stability test.

Treatment	L^*	a^*	b^*	C	$h°$
As synthesized	67.2	+27.3	+65.4	70.9	67.3
4% CH_3COOH	65.5	+27.0	+58.7	64.6	65.3
4% NH_4HCO_3	62.0	+26.9	+58.2	64.1	65.2

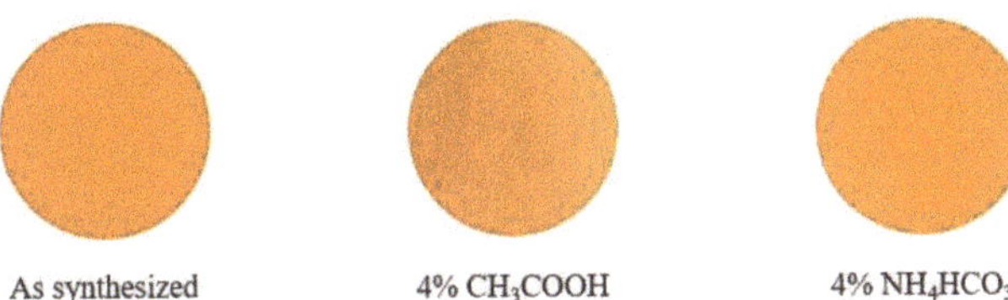

Figure 7. Photographs of the $La_3LiMn_{0.97}Ti_{0.03}O_7$ pigment before and after the chemical stability test.

3. Materials and Methods

3.1. Synthesis

The $La_3LiMn_{1-x}Ti_xO_7$ ($0 \leq x \leq 0.05$) samples were synthesized by a conventional solid state reaction method. La_2O_3 (99.99%), Li_2CO_3 (99.0%), Mn_2O_3 (99.0%), and TiO_2 (99.0%) powders were used as starting reagents. Stoichiometric amounts of metal oxides and three times the stoichiometric amount of Li_2CO_3 were mixed in an agate mortar. The homogeneous mixtures were calcined in an alumina crucible at 900 °C for 6 h in air. Before characterization, the samples were ground in an agate mortar.

3.2. Characterization

The element ratios of La, Mn, and Ti for the samples were confirmed by using X-ray fluorescence spectroscopy (XRF; Rigaku, ZSX Primus). The crystal structure of the samples was identified by X-ray powder diffraction (XRD; Rigaku, Ultima IV) with Cu-Kα radiation, operating with voltage and current settings of 40 kV and 40 mA. The lattice parameters and volumes were calculated from the XRD peak angles refined, using α-Al_2O_3 as a standard and using the CellCalc Ver. 2.20 software. The morphologies and particle sizes were observed by using field-emission-type scanning electron microscopy (FE-SEM; JEOL, JSM-6701F).

The optical reflectance spectra were measured with a UV-Vis-NIR spectrometer (JASCO, V-770), using a standard white plate as a reference. The color properties were evaluated in terms of the Commission Internationale de l'Éclairage (CIE) $L^*a^*b^*Ch°$ system, using a colorimeter (Konica-Minolta, CR-300). The L^* parameter represents the brightness or darkness in a neutral grayscale. The a^* and b^* values indicate the red–green and yellow-blue axes, respectively. The chroma parameter (C) expresses the color saturation and is calculated with the formula, $C = [(a^*)^2 + (b^*)^2]^{1/2}$. The hue angle ($h°$) ranges from 0 to 360° and is estimated according to the following formula: $h° = \tan^{-1}(b^*/a^*)$.

4. Conclusions

$La_3LiMn_{1-x}Ti_xO_7$ ($0 \leq x \leq 0.05$) samples were synthesized using a solid-state reaction technique as environmentally friendly inorganic orange pigments. The $La_3LiMn_{1-x}Ti_xO_7$ (x = 0, 0.01, and 0.03) samples were obtained in a single-phase form, but an impurity phase was observed for x = 0.05. In the visible light region, the optical absorption band at a wavelength below 550 nm and the shoulder absorption peak around 690 nm were attributed to the spin-allowed and spin-forbidden d–d transitions of Mn^{4+}, respectively. These absorption intensities were increased by the Ti^{4+} doping, because the $[MnO_6]$ octahedra were more distorted. Accordingly, the sample color became more vivid orange. Among the samples synthesized in this study, the $La_3LiMn_{0.97}Ti_{0.03}O_7$ pigment exhibited the most brilliant orange color. In addition, the orange color of the present pigment was brighter than those of the commercially available orange pigments, because the a^* and b^* values of this pigment were higher than those of the commercial ones. Since the orange color of the $La_3LiMn_{0.97}Ti_{0.03}O_7$ pigment has chemical stability, it has a potential to be a novel environmentally friendly inorganic orange pigment.

Author Contributions: The following are the author contributions to this study: Conceptualization, R.O. and T.M. (Toshiyuki Masui); methodology, R.O., J.-i.K., T.M. (Takuro Morimoto) and T.M. (Toshiyuki Masui); data curation, J.-i.K. and T.M. (Takuro Morimoto); writing—original draft preparation, R.O.; writing—review and editing, T.M. (Toshiyuki Masui); supervision, T.M. (Toshiyuki Masui); funding acquisition, T.M. (Toshiyuki Masui) All authors have read and agreed to the published version of the manuscript.

Funding: This research was partially funded by the JSPS KAKENHI, grant number JP19K05668 and JP20H02439.

Institutional Review Board Statement: Not applicable.

Informed Consent Statement: Not applicable.

Data Availability Statement: Data is contained within the article.

Conflicts of Interest: The authors declare no conflict of interest.

Sample Availability: Samples of the compounds are not available from the authors.

References

1. Faulkner, E.B.; Schwartz, R.J. *High Performance Pigments*, 2nd ed.; Wiley-VCH: Weinhein, Germany, 2009.
2. Fortuño-Morte, M.; Beltrán-Mir, H.; Cordoncillo, E. Study of the role of praseodymium and iron in an environment-friendly reddish orange pigment based on Fe doped $Pr_2Zr_2O_7$: A multifunctional material. *J. Alloys Compd.* **2020**, *845*, 155841. [CrossRef]
3. Raj, A.K.V.; Rao, P.P.; Sreena, T.S. Color tunable pigments with high NIR reflectance in terbium-doped cerate systems for sustainable energy saving applications. *ACS Sustain. Chem. Eng.* **2019**, *7*, 8804–8815. [CrossRef]
4. Šulcová, P.; Trojan, M. Thermal synthesis of the $(Bi_2O_3)_{1-x}(Er_2O_3)_x$ pigments. *J. Therm. Anal. Calorim.* **2007**, *88*, 111–113. [CrossRef]
5. Šulcová, P.; Trojan, M. Thermal synthesis and properties of the $(Bi_2O_3)_{1-x}(Ho_2O_3)_x$ pigments. *J. Therm. Anal. Calorim.* **2006**, *83*, 557–559. [CrossRef]
6. Schildhammer, D.; Fuhrmann, G.; Petschnig, L.; Schottenberger, H.; Huppertz, H. Synthesis and optical properties of new highly NIR reflective inorganic pigments $RE_6Mo_2O_{15}$ (RE = Tb, Dy, Ho, Er). *Dyes Pigm.* **2017**, *140*, 22–28. [CrossRef]
7. Oka, R.; Kosaya, T.; Masui, T. Novel environmentally friendly inorganic orange pigments based on $Ca_{14}Al_{10}Zn_6O_{35}$. *Chem. Lett.* **2018**, *47*, 1522–1525. [CrossRef]
8. Oka, R.; Shobu, Y.; Masui, T. Synthesis and color evaluation of Ta^{5+}-doped Bi_2O_3. *ACS Omega* **2019**, *4*, 7581–7585. [CrossRef] [PubMed]
9. Jiang, P.; Li, J.; Sleight, A.W.; Subramanian, M.A. New oxides showing an intense orange color based on Fe^{3+} trigonal-bipyramidal coordination. *Inorg. Chem.* **2011**, *50*, 5858–5860. [CrossRef] [PubMed]
10. Kusumoto, K. Synthesis of Bi_2O_3–ZrO_2 solid solutions for environment-friendly orange pigments. *J. Ceram. Soc. Jpn.* **2017**, *125*, 396–398. [CrossRef]
11. Rao, R.G.; Divya, D. Synthesis and characterization of orange pigments from rare earth metal ions. *Asian J. Chem.* **2017**, *29*, 1673–1676. [CrossRef]
12. Giampaoli, G.; Li, J.; Hermann, R.P.; Stalick, J.K.; Subramanian, M.A. Tuning color through sulfur and fluorine substitutions in the defect tin (II, IV) niobate pyrochlores. *Solid State Sci.* **2018**, *81*, 32–42. [CrossRef]
13. Kusumoto, K. Synthesis of Bi_2O_3–Nb_2O_5 solid solutions for environmental-friendly reddish yellow pigments. *J. Ceram. Soc. Jpn.* **2016**, *124*, 926–928. [CrossRef]
14. Chen, J.; Xiao, Y.; Huang, B.; Sun, X. Sustainable cool pigments based on iron and tungsten co-doped lanthanum cerium oxide with high NIR reflectance for energy saving. *Dyes Pigm.* **2018**, *154*, 1–7. [CrossRef]
15. Bae, B.; Takeuchi, N.; Tamura, S.; Imanaka, N. Environmentally friendly orange pigments based on hexagonal perovskite-type compounds and their high NIR reflectivity. *Dyes Pigm.* **2017**, *147*, 523–528. [CrossRef]
16. Schildhammer, D.; Fuhrmann, G.; Petschnig, L.; Weinberger, N.; Schottenberger, H.; Huppertz, H. Synthesis and characterization of a new high NIR reflective ytterbium molybdenum oxide and related doped pigments. *Dyes Pigm.* **2017**, *138*, 90–99. [CrossRef]
17. Llusar, M.; García, E.; García, M.T.; Gargori, C.; Badenes, J.A.; Monrós, G. Synthesis, stability and coloring properties of yellow–orange pigments based on Ni-doped karrooite (Ni,Mg)Ti_2O_5. *J. Eur. Ceram. Soc.* **2015**, *35*, 357–376. [CrossRef]
18. Yang, F.; Qiao, L.; Ren, H.; Yan, F. Luminescence analysis of Mn^{4+},Zn^{2+}: Li_2TiO_3 red phosphors. *J. Lumin.* **2018**, *194*, 179–184. [CrossRef]
19. Takeda, Y.; Kato, H.; Kobayashi, M.; Kobayashi, H.; Kakihana, M. Photoluminescence properties of Mn^{4+}-activated perovskite-type titanates, La_2MTiO_6:Mn^{4+} (M = Mg and Zn). *Chem. Lett.* **2015**, *44*, 1541–1543. [CrossRef]
20. Yang, Z.; Yang, L.; Ji, C.; Xu, D.; Zhang, C.; Bu, H.; Tan, X.; Yun, X.; Sun, J. Studies on luminescence properties of double perovskite deep red phosphor La_2ZnTiO_6:Mn^{4+} for indoor plant growth LED applications. *J. Alloys Compd.* **2019**, *802*, 628–635. [CrossRef]
21. Cao, R.; Ye, Y.; Peng, Q.; Zheng, G.; Ao, H.; Fu, J.; Guo, Y.; Guo, B. Synthesis and luminescence characteristics of novel red-emitting $Ba_2TiGe_2O_8$:Mn^{4+} phosphor. *Dyes Pigm.* **2017**, *146*, 14–19. [CrossRef]

22. Hasegawa, T.; Nishiwaki, Y.; Fujishiro, F.; Kamei, S.; Ueda, T. Quantitative determination of the effective Mn^{4+} concentration in a Li_2TiO_3:Mn^{4+} phosphor and its effect on the photoluminescence efficiency of deep red emission. *ACS Omega* **2019**, *4*, 19856–19862. [CrossRef]
23. Liu, Z.; Yuwen, M.; Liu, J.; Yu, C.; Xuan, T.; Li, H. Electrospinning, optical properties and white LED applications of one-dimensional $CaAl_{12}O_{19}$:Mn^{4+} nanofiber phosphors. *Ceram. Int.* **2017**, *43*, 5674–5679. [CrossRef]
24. Adachi, S. Review—Mn^{4+}-activated red and deep red-emitting phosphors. *ECS J. Solid State Sci. Technol.* **2020**, *9*, 016001. [CrossRef]
25. Chen, Y.; Chen, J.; Liang, J.; He, J.; Liu, Z.-Q.; Yin, Y. Localized charge accumulation driven by Li^+ incorporation for efficient LED phosphors with tunable photoluminescence. *Chem. Mater.* **2020**, *32*, 9551–9559. [CrossRef]
26. Oka, R.; Kusukami, K.; Masui, T. Effect of [MnO_6] octahedra to the coloring mechanism of $(Li_{1-x}Na_x)_2MnO_3$. *ACS Omega* **2019**, *4*, 19856–19862. [CrossRef]
27. Tamilarasan, S.; Laha, S.; Natarajan, S.; Gopalakrishnan, J. Li_2MnO_3: A rare red-coloured manganese (IV) oxide exhibiting tunable red–yellow–green emission. *J. Mater. Chem. C* **2015**, *3*, 4794–4800. [CrossRef]
28. Gramm, G.; Fuhrmann, G.; Zimmerhofer, F.; Wieser, M.; Huppertz, H. Development of high NIR-reflective red Li_2MnO_3 pigments. *Z. Anorg. Allg. Chem.* **2020**, *646*, 1722–1729. [CrossRef]
29. Shannon, R.D. Revised effective ionic radii and systematic studies of interatomic distances in halides and chalcogenides. *Acta Crystallogr. Sect. A Found. Adv.* **1976**, *32*, 751–767. [CrossRef]
30. Battle, P.D.; Burley, J.C.; Gallon, D.J.; Grey, C.P.; Sloan, J. Magnetism and structural chemistry of the $n = 2$ Ruddlesden–Popper phase La_3LiMnO_7. *J. Solid State Chem.* **2004**, *177*, 119–125. [CrossRef]

Review

The Role of Zinc(II) Ion in Fluorescence Tuning of Tridentate Pincers: A Review

Rosita Diana and Barbara Panunzi *

Department of Agriculture, University of Napoli Federico II, via Università 100, 80055 Portici NA, Italy; rosita.diana@unina.it

* Correspondence: barbara.panunzi@unina.it

Academic Editors: Jorge Bañuelos Prieto and Ugo Caruso
Received: 6 October 2020; Accepted: 25 October 2020; Published: 28 October 2020

Abstract: Tridentate ligands are simple low-cost pincers, easy to synthetize, and able to guarantee stability to the derived complexes. On the other hand, due to its unique mix of structural and optical properties, zinc(II) ion is an excellent candidate to modulate the emission pattern as desired. The present work is an overview of selected articles about zinc(II) complexes showing a tuned fluorescence response with respect to their tridentate ligands. A classification of the tridentate pincers was carried out according to the binding donor atom groups, specifically nitrogen, oxygen, and sulfur donor atoms, and depending on the structure obtained upon coordination. Fluorescence properties of the ligands and the related complexes were compared and discussed both in solution and in the solid state, keeping an eye on possible applications.

Keywords: zinc ion; fluorescence; tridentate ligand

1. Introduction

Over the past 20 years, fluorescence-responsive compounds are increasingly required for many technological applications, from lighting and switch devices to bio-imaging and analytical probes. Materials based on transition metal complexes were advantageously utilized. In this area, interest is growing in the abundant, less expensive, and environmentally "green" zinc(II) metal cation. Today, science is in great demand to address the challenge of sustainability. Scientific innovations and advances must play a major role in technological breakthroughs thanks to the choice of sustainable green matter as a substitute for highly toxic, expensive, and difficult to dispose products. So, green chemistry as the design of less hazardous chemical products and processes is a hot topic today. The replacement of heavy metal atoms in aromatic macrostructures with the small eco-friendly zinc cation, able to modulate the properties of coordination environments, easier to synthetize, and low cost, can be a way to meet the challenge.

From a research point of view, to the advantage of a large variety of coordination geometries and elaborate molecular architectures, zinc(II) complexes add the versatility of the luminescent levels, both in solution and in the solid state. Real breakthroughs for the novel luminescent technologies were obtained by employing highly efficient, stable, and cheap emitters. Among them, several zinc(II) complexes have to be included [1–6].

In the coordination complexes, electronic charge can be transferred between different molecular entities in the complex, from the electron donor zone to the receiving electron acceptor zone. In most complexes, charge-transfer electronic bands involve electron transfer between metal atoms and ligands. The charge-transfer bands in transition metal complexes are due to the shift of charge density between orbitals predominantly metal in character and those predominantly ligand in character. Most transitions are ligand-to-metal charge-transfer (LMCT) if the transfer occurs from the ligand orbitals to the metal, or metal-to-ligand charge-transfer (MLCT) in the reverse case. Due to its d^{10} closed shell configuration, zinc(II) ion has no optical signature. The d–d electronic transitions are not expected in zinc emissive complexes and the lowest energy excited states are mainly of a ligand-centered charge transfer (LCT) nature (as intramolecular charge transfer, ICT, and intraligand charge transfer, ILCT) and/or ligand-to-ligand charge transfer (LLCT) nature, rarely due to ligand-to-metal charge transfer (LMCT) states involving s or p empty orbitals of the metal [7]. The absence of ligand field stabilization energy leads to the formation of complexes with various coordination numbers, such as 4, 5, or 6, with tetrahedral, square pyramidal/trigonal bipyramidal, and octahedral geometries.

In dependence of the type of ligands and the coordination pattern imposed by the ligands, zinc(II) complexes exhibit fluorescence tuning both in intensity and/or emission maximum. Specifically, fluorescence enhancement (chelation enhanced fluorescence, CHEF mechanism [8–11]) or fluorescence reduction (metal-binding-induced fluorescence quenching [12]) can occur upon coordination. In addition, due to the lowering of the excited state of the bonded ligand upon coordination, a qualitative fluorescence tuning (blue or red shift of the emission maximum from ligand to metal) can be observed. The two different mechanisms can activate in solution and/or in the solid state, with drastic variations in fluorescence with respect to the ligand. Recently [13], the ability of the fluorophore ligand to form a π-contact with the metal cation was correlated with the fluorescence quenching or enhancement ability. Therefore, the information about the excited state geometry and the frontier orbital arrangement of the excited states is essential to understanding and foresight of the phenomenon.

The fluorescence enhancement effect is often the result of a stabilization of the excited state in poorly emissive ligands upon coordination [14]. Zinc(II) cation often causes a CHEF effect. Typically, in emissive zinc(II) complexes, the fluorescence emission results from a π-π^* LCT and the role of the zinc ion is to freeze the favorable re-emissive conformation. In this way, it is possible to obtain strongly emissive materials for lighting devices by increasing the emission of organic molecules upon zinc coordination. Ligands with flexible spacers and appropriate aromatic moieties able to fold over the metal in a locked conformation show good potential in analytical and bio-chemistry. Many sensing systems for zinc cation detection exploit the CHEF effect [15–17]. At the opposite, fluorescence quenching is quite unusual in zinc complexes. Nevertheless, ICT or ILCT transitions can be responsible of zinc binding-induced fluorescence quenching. Fluorescence quenching is observed in rigid pyridine-based ligands upon coordination, due to a decreased HOMO–LUMO energy gap, or in zinc-induced quenching of the protein intrinsic fluorescence due to conformational perturbations [18–21].

Finally, the aggregate nature of materials consisting of fluorophores frozen into polymeric chains or networks give rise to noticeable changes in the energetic levels of the ligands. The assembly of emissive pincers by zinc coordination produces the most varied polymeric structures: coordination polymers (CPs) obtained by zinc bridges [22–25], metallated polymers obtained by coordination with pre-formed chains [26,27], and polymeric networks obtained by interlacing of flexible zinc-crossed fluorophores [28–30]. Owing to both the restrictions imposed to fluorophore and the efficient electron hopping in the tight structure, relevant emission tuning with respect to the free ligands and to mononuclear structures is envisaged.

As modifications of the ligand are expected to change energy levels and structural features, the number of binding sites in the pincer ligand plays a decisive role on the spectroscopic properties. Tridentate ligands are quite simple pincers, often low cost and easy to synthetize. At the same time, tridentate ligands guarantee good stability to the complex thanks to a relevant chelate effect. Ligands that can bind zinc(II) through three donor atom groups, such as *O*, *N*, and *S* donor atom groups, represent an interesting class due to the variety of behavior and applications. Tridentate zinc complexes are known to produce simple and mononuclear or intricate even polymeric structures, by themselves or with auxiliary ligands. By doing so, they can cause relevant fluorescence tuning with respect to the free ligand, in solution and/or in the solid state. In addition, depending on the charge of the pincer ligand, the zinc-binding reaction leaves unoccupied coordination sites for additional ligands, which in turn can modulate the structure and properties of the derived complexes.

Some representative functional groups involved in the build of fluorescence-responsive tridentate pincers can be identified. The Schiff base moiety, obtained by reaction of amines and carbonyl-containing compounds, is a good candidate for the synthesis of *N*, *O*, *S* donors containing ligands. In particular, in half-*salen*-type ligands, one nitrogen atom and one oxygen atom group chelate the metal, while the C=N functional group constitutes a versatile bridge between the *N*,*O* pincer and the branch of the ligand containing the third binding site. This role could be played by a third nitrogen, oxygen, or sulfur atom, as an example the carbonyl oxygen, the thione sulfur, or the donor atom of an heteroaromatic ring. The formation of a coordination core consisting of five- and/or six-membered rings between the pincer and zinc cation produces stable and sometimes highly emissive coordination complexes. Finally, the coordination core can be all made up of aromatic rings with *N*, *O*, *S* donor heteroatoms fixed in a rigid pincer or able to fold up as a flexible pincer.

This is an overview of selected cutting-edge examples of tridentate zinc(II) complexes causing fluorescence tuning with respect to the ligand. In this research area, many articles have been produced in the last 15 years. Starting from the synthetic and purely phenomenological approach, to the complete and detailed analysis of the energy levels, until the synergistic approach to the structure–property relationship, there is much to tell on this subject.

In the following sections, ligands will be classified according to the binding donor atom groups, and fluorescence properties of ligands and related complexes will be comparatively discussed. By use of an easy "cartoon" representation (as in Figure 1), we will propose a quick intuitive overview of groups of structures, examined according to the photoluminescence (PL) response and the application area.

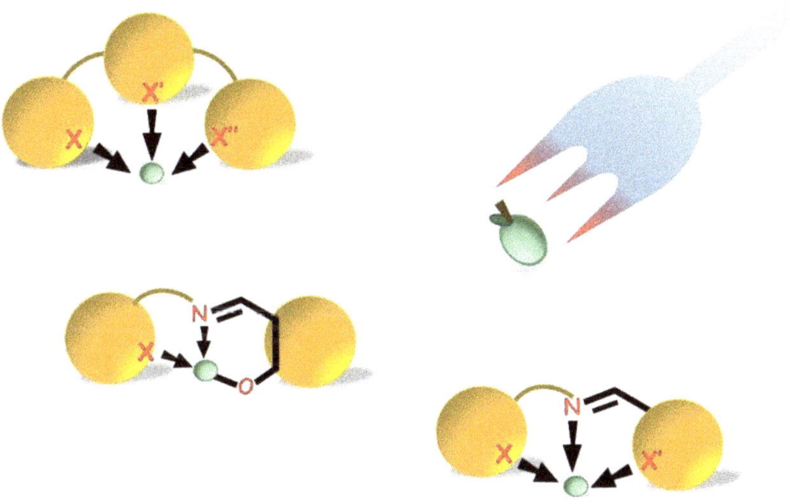

Figure 1. Schematic representation of most of the coordination cores achievable from tridentate pincers binding zinc(II) cation (in green). X, X′, and X″ can be *N*, *O*, and *S* atom groups.

2. Nitrogen Binding Sites

The study of polydentate ligands with available *N* donor sites is a prolific research area in coordination chemistry [31–43]. Typically, nitrogen binding sites are neutral sites where the lone pair of nitrogen atom is available for donation. Nitrogen aromatic heterocycles are excellent building blocks for the synthesis of *N*-donor polydentate ligands. Single or fused five- and/or six-term rings produce stable and soluble pincers able to direct the chemical and chemo-physical properties. A substantial amount of literature articles on pyridine-containing tridentate ligands has emerged. Related to the enhanced π-electron delocalization upon zinc coordination, in some cases, the zinc-binding-induced fluorescence quenching phenomenon was detected. At the opposite, pyridine and other nitrogen heterocycles assembled in flexible architectures often produce a CHEF effect.

2.1. Terpyridine-Type Ligands

Rigid pyridine-based tridentate ligands as terpyridine (Tpy) are well-known chelate ligands for transition metals [44–46]. Many Tpy ligands were found to be emissive in the solid state, with photoluminescence quantum yields (PLQYs) strongly depending on their own molecular structure and on substituents, with a relevant emission tuning due to metal coordination. Zinc-binding-induced fluorescence reduction/quenching was often detected, as well strong color emission tuning. In 2009, a series of zinc(II) bis(4′- phenyl-terpyridine) complexes with substituted Tpy were studied by J. Popp and coworkers [47] (Figure 2). The zinc ion is coordinated with the three nitrogen atoms from each of two terpyridine ligands and with two PF_6^- as auxiliary ligands. Tuning of the color emission from violet to cyan (425–487 nm) in dependence of the extension of the π-conjugated system of the ligand was observed, and from low to high PLQYs (from 6 to 64%) were recorded. Thin solid films obtained by low-concentration dye-doped poly(methylmethacrylate) (PMMA) matrix show bright emission, with PLQYs up to 0.30. By DFT study, the HOMO energy level was also significantly influenced by the ligand structure, whereas the LUMO energy appeared to be independent of the electronic pattern of the Tpy ligand, with the enhanced π-electron delocalization leading to a decreased HOMO–LUMO energy gap.

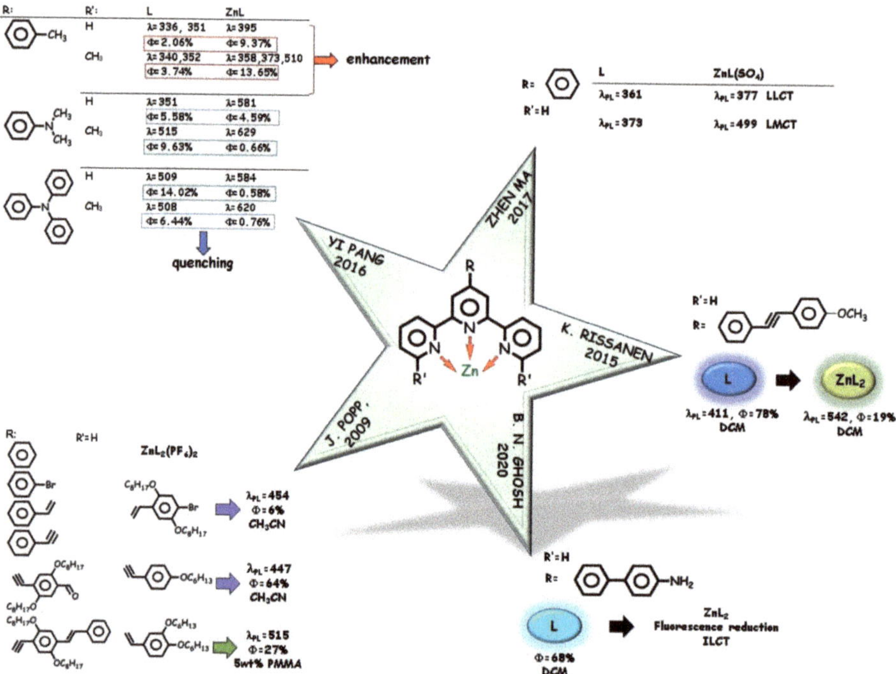

Figure 2. Tpy-type tridentate pincers causing zinc-binding fluorescence reduction/quenching.

The crystalline structure of a Tpy-type ligand upon coordination with different transition metals was examined in 2015 by K. Rissanen and coworkers [48] (Figure 2). The ligand displayed bright blue emission and high PLQYs dissolved in several organic solvents. A distorted octahedral arrangement with two tridentate terpyridine ligands was detected in the zinc complex and a significant greenish-yellow tuned emission attributed to ILCT states, still appreciable but lowered in intensity with respect to the ligand. DFT analysis rationalized the ICT-type electronic transitions involved from the diphenylacetylene moiety to the terpyridine group. Except for the other d^{10} closed shell cadmium (II) ion, a complete metal-binding-induced fluorescence quenching was observed in the presence of the other divalent metal ions.

Tpy-type ligands with different substituents were examined for their binding ability specifically toward zinc(II) cation in 2016 [20] and in 2017 [21] (Figure 2). In [21], Zhen Ma and coworkers examined the compound obtained by reacting 4′-phenyl-terpyridine and $ZnSO_4{\cdot}7H_2O$ by X-ray crystallographic analysis. A neutral 1:1 (Zn:Tpy) complex with a coordinated sulfate group, qualitatively emissive in the solid state, was obtained. The emission spectrum displayed two bands at ca. 377 and 499 nm red-shifted with respect to 361 and 373 nm for the ligand in the solid state. The former higher energy band of the complex was assigned to LLCT whereas the latter band to LMCT [49,50]. In 2016, the influence of the coordination stoichiometry was explored by Yi Pang and coworkers [20], by reacting substituted Tpy-type ligands (R in Figure 2 is a *p*-substituted phenyl ring bearing a donor group and R′=H or R′=CH_3 on the lateral pyridine groups) with $ZnCl_2$. The authors pointed out that the zinc complex forms in a 2:1 (ligand: metal) ratio with a low zinc(II) concentration and in a 1:1 ratio with a high concentration of metal. In ethanol, fluorescence quenching occurs in 2:1 complexes, whereas, turning into 1:1 complexes, fluorescence increases in the opposite direction, marking the role of the coordination pattern in the emission intensity. In 1:1 complexes, temperature-dependent fluorescence spectroscopy elucidated the role of the ICT mechanism between donor and acceptor groups. Due to the occurrence of a relevant ICT process, the strong donor substituents induce zinc-binding fluorescence quenching and red-shift of the emission maximum. Conversely, the weak donor substituent *p*-CH_3-phenyl group causes an increase in the fluorescence emission of the 1:1 complex.

Very recently [51], B.N. Gosh and coworkers examined the X-ray crystal structure of 4′-functionalized terpyridine complexes (Figure 2). Zinc cation did show a trans-arrangement of the terminal pyridine nitrogen atoms with respect to the central pyridine ring in the 2:1 ligand zinc tridentate complex. The terpyridine ligand exhibits a bright blue emission in dichloromethane with 68% PLQY and produces a different coordination pattern in dependence of the coordinated metal. Upon zinc complexation, ligand emission shows a significant reduction while other transition metal ions completely quench the fluorescence of the ligand. The weak fluorescence of the zinc complex can be imputed as an ILCT transition from the amine moiety to the metal coordinated terpyridine fragment [48,52,53]. The non-fluorescence nature of the complexes with other metal cations can be imputed as MLCT-type transitions.

Recently, fluorogens exhibiting aggregation-induced emissive properties (AIEgens) have grabbed scholars' attention in various scientific areas. In contrast to conventional aggregation-caused quenching (ACQ) molecules, AIEgens are weakly fluorescent/non-emissive in diluted solution and emit intensely in their aggregate form (concentrated solution and solid state), owing to the restriction of intramolecular motions [54]. Tpy-based zinc complexes AIEgens have been proved to be attractive and versatile tools for biological imaging and chemical sensing. Under physiological conditions, Tpy-type ligands are employed in bio-medical applications, such as for living cell imaging. As an example, in 2005 [55], Valery N. Kozhevnikov and Burkhard König presented 2,2′-bi- and 2,2′:6′,2″-terpyridines with aminomethyl and aryl substituents employable as luminescent probes coordinating in a pentadentate coordination core zinc cation in physiological media.

In 2019, a thiophene bridged Tpy tridentate Zn(II) complex (Figure 3) was designed for RNA-specific targeting thanks to its AIE bright yellow-green fluorescence emission under physiological conditions by Ju Yupeng Tian, Dandan Li, and coworkers [56]. On the other hand, in 2016, a new fluorescence sensor for citrate (acting itself as a tridentate ligand) detection was developed by integrating an AIE Tpy ligand with zinc(II) by Ju Mei, Jianli Hua, and coworkers [57]. The enhanced electron-withdrawing ability of the complex gave rise to a red-shifted fluorescence compared with the organic ligand. The zinc-based probe was not an AIEgen but showed significant fluorescence enhancement (fluorescence turn-on mechanism) under substitution of the auxiliary ligands with citrate. The bathochromic shift of the absorption maximum and the decrease of fluorescence intensity may be ascribed to the stronger ICT in the complex as compared to the ligand (Figure 3).

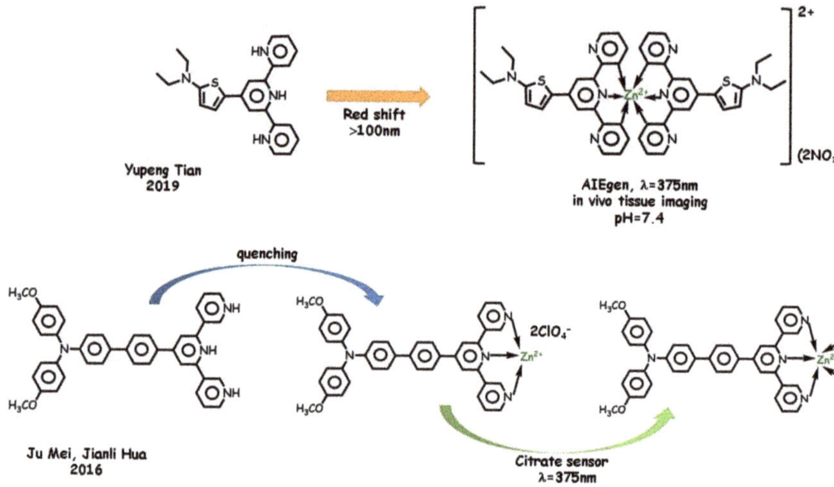

Figure 3. AIE behavior of zinc complexes from Tpy-type tridentate pincers.

2.2. *N,N,N Schiff Base-Type Ligands*

Schiff base ligands containing an additional nitrogen binding site, usually derived from a nitrogen heterocycle, have been widely reported. The versatile CH=N bridge allows the employment of a large variety of substituents and to build the most varied architectures. Unlike their complexes, this kind of ligand is often non-emissive, and a general behavior due to the CHEF effect was mostly found. In 2011, Kaushik Ghosh and coworkers [58] explored the crystal structure and photophysical properties of four zinc(II) complexes derived from the tridentate ligand Pyimpy (see Figure 4) with different auxiliary ligands. In the complexes, two pyridinic nitrogen groups are involved in the equatorial plane binding along with the iminic nitrogen group. The complexes show various fluorescence emission in toluene solution upon excitation of the charge transfer band near 350 nm (ascribable to π,π^* LC of the ligand [59]) while the free ligand Pyimpy displays no fluorescence emission in the same experimental conditions. The phenomenon was ascribed to the loss of vibrational energy decay due to ligand stiffening under coordination. In addition to the zinc-binding fluorescence intensity enhancement, a shift of the maximum of emission was detected, in dependence of the auxiliary ligands.

Other nitrogen aromatic heterocycles have recently emerged as electron donor-containing moieties. Among them, pyrimidine groups attract attention for their role in biological systems [60,61]. Susanta Kumar Kar and coworkers in 2012 [62] prepared two tridentate *N,N,N* donor Schiff base ligands [63] using pyrimidine- and pyridine-containing carbonyl compounds (see Figure 4). Whereas the ligand with a methyl substituent was fluorescent silent, its zinc complex showed a strong CHEF effect in methylcyclohexane. Probably due to photoinduced electron transfer processes in the presence of several nonbonding electron pairs on the nitrogen donor atom groups, the ligand π,π^* transitions are not allowed and the flexible bonds of the ligands cause the activation of the non-radiative channel. CHEF activates by the increase in conformational rigidity of the ligands upon strong zinc binding, which prevents non-radiative channels [64]. Pyrimidine-containing Schiff base ligands were studied in 2015 by Saugata Konar [65] for the zinc-binding ability into 1-D coordination polymers held together by $\mu_{1,5}$-bridged dicyanamide ions. The difference due to the activation of a *N,N,O* bonding site (ligand L1, achieved by the presence of a half-*salen* group) with respect to the *N,N,N* bonding site (ligand L2, see Figure 4) was pointed out. Both the ligands display low fluorescence intensity in methanol. Conversely, both their polymeric complexes show red-shifted enhanced fluorescence emission. The low emission intensity of the ligands was ascribed to photo-induced electron transfer processes, while the CHEF effect to an increase in conformational rigidity of the ligands upon complexation. The complex derived from the *N,N,O* ligand shows the highest CHEF effect compared to the complex derived from the *N,N,N* ligand. This has been attributed to the strong binding of the *salen*-type ligand L1 thanks to the *N,O* donor pincer compared to the *N,N* pincer of ligand L2. In the first case, the conformation of the coordination core is strongly trapped in a planar conjugated *habitus*.

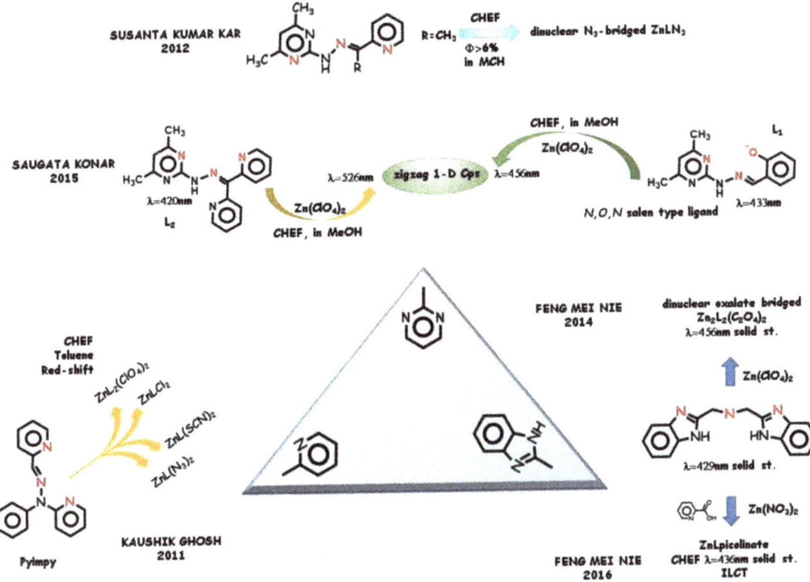

Figure 4. *N,N,N* Schiff base-type ligands with different nitrogen aromatic rings and their CHEF active complexes.

Polydentate flexible ligands containing two or more benzimidazole donor units have long been used in coordination and supramolecular chemistry [66–68]. The fused benzimidazole ring contains the *N*-binding site of the imidazole moiety. Feng-Mei Nie and coworkers [69,70] synthetized solid-state emissive zinc(II) complexes derived from tridentate and polydentate benzimidazole (Figure 4), solid-state emissive by themselves. Different architectures were obtained in dependence from the number of chelating site and on the auxiliary ligands, as analyzed by the X-ray diffraction technique. In presence of oxalate, a dinuclear oxalate-bridges Zn_2L_2 coordination core was obtained (in 2014, [69]), whereas the same ligand produced a five-coordinate ZnL coordination core in the presence of *N,O* chelating picolinate (in 2016, [70]). In both cases, the fluorescence emission was assigned to π-π^* ILCT bands. The red shift of the emission maxima from ligand to complexes is influenced by the coordination pattern. The remarkable fluorescence enhancement compared to the free ligand was ascribed to extensive π-conjugated structure formation.

2.3. Ligands for Sensing Analysis and for Supramolecular Architecture Building

Polydentate structures with flexible moieties are useful tools for the sensing of metal analytes by the fluorescence technique [71,72]. Many novel fluorescence chemosensors for zinc cations were recently explored. From a purely theoretical point of view, Hee-Seung Lee and coworkers in 2013 [14] explored the role of fluorophore–metal interaction in photoinduced electron transfer (PET) sensors and the large CHEF effect promoted by zinc(II) coordination by time-dependent density functional theory (TDDFT) study, pointing out how DFT study is the logical complement of the synthetical work about novel sensing molecules [17,30,59,73].

A significant example of a bendable *N,N,N* ligand useful as a sensor was synthetized and employed in 2018 by Ugo Caruso, Rosita Diana, and coworkers [25,59,74–76]. Specifically, the pyridine/phenol/benzoxazole-based ligand (Figure 5) able to bind various transition metals acts as *N,N,N* tridentate selective fluorogenic ligand toward zinc(II) by a sensing CHEF mechanism, in water or water/mixed solvents. DFT calculations for the free ligand and the complex were used to calculate frontier molecular orbitals. The frontier molecular orbitals undergo strong changes when the sensor folds back onto the metal cation (see Figure 5). HOMO of the ligand and of the complex are π orbitals with contributions from 2p orbitals of the carbon atoms in the benzothiazole ring. LUMO of the free ligand is a π^* orbital with contributions mainly from the benzothiazole ring, while the LUMO of the complex is a π^* orbital localized on the pyridine group. Not unusual for multidentate ligands, the same tripodal multidentate sensor acts as a tetradentate ligand toward zinc ion at pH = 8.0 [17]. In basic media, the sensor activates the phenate oxygen-binding site in addition to the *N,N,N* chelate site. The ligand results a pH-dependent sensor [77], able to detect zinc(II) ion in a neutral/slightly acidic and in a slightly basic aqueous environment with different emission responses.

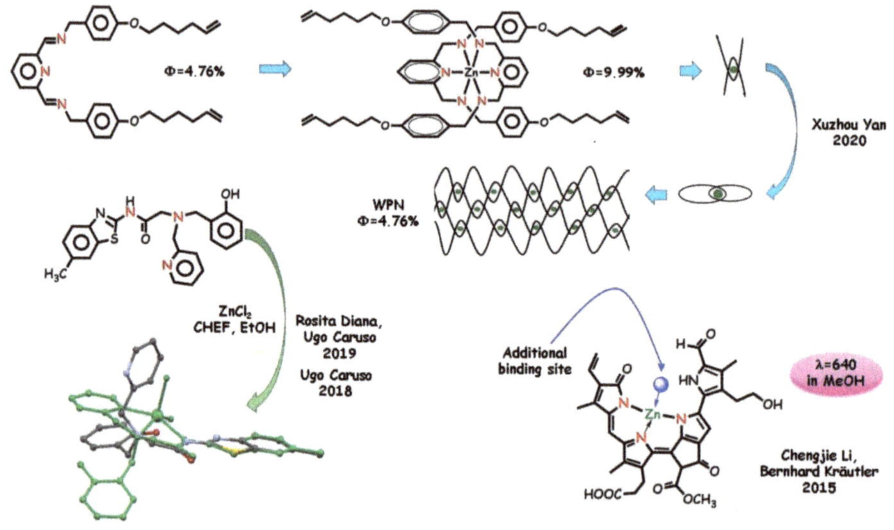

Figure 5. Supramolecular architectures produced from zinc-interlocked chains. Tridentate pincers for sensing analysis of zinc cations.

Chlorophyll-catabolite named phyllobilins may display a capacity to complex metal ions. In 2015, in a mighty article [78], Chengjie Li and Bernhard Kräutler explored pink-colored phyllobiladienes as effective tridentate ligands, leaving one unoccupied coordination site that may be used for coordination by an external additional ligand, such as proteins or nucleobases (Figure 5). Coordination of the zinc cation to the scarcely luminescent pink chlorophyll catabolites induces bright fluorescence in the complex. The zinc(II) adduct ZnL shows strong red emission in solution (band picked around 650 nm, almost two orders of magnitude more intense than the free ligand) so it can be potentially used as in vivo sensors. Analysis of the fluorescence of MeOH solutions leads to quantitative detection of the cation thanks to the linear correlation between fluorescence intensity and zinc(II) concentrations.

N,N,N tridentate complexes have a part as novel polymeric materials with intriguing structural and mechanical features for the construction of smart supramolecular architectures. The formation of polymeric architectures through zinc cation linkers can be the way to increase and/or tune the fluorescence properties of the organic ligands, and to transfer the desired emission properties to macrostructures.

Mechanically interlocked molecules, such as catenanes [79–81], are topological structures held by mechanical bonds, with intriguing potential in several fields from synthetic chemistry to materials science and nanotechnology [82–84]. In 2020, Xuzhou Yan and coworkers [28] obtained a mononuclear ZnL_2 complex by reacting a zinc salt with *N,N,N* chelating ring-like [2] catenane ligands. The synthesis of a "woven" polymer network (WPN) via ring-opening metathesis polymerization of the catenane produced a 3-D coordination polymer consisting of rigid metal-coordinated crossing points and flexible alkyl chain. The flexible and firm network obtained by interlaced fluorophore units exhibit different emission properties in the solid state with respect to the reagents. The mononuclear ZnL_2 complex is an AIEgen, relatively flexible and less restricted. It can aggregate tightly in the solid state, resulting in a strong emission. After the formation of the more interlocked network structure, the restrictions imposed to fluorophore aggregation lower the emission. The quantum yields of the three structures (9.99% for ZnL_2, 4.76% for the [2] catenane, and 8.97% for the WPN) measured in the solid state showed similar variation trends along with different topological structural transformations.

3. Nitrogen and Oxygen Binding Sites

N,O chelating Schiff bases ligands, often half-*salen*-type ligands, can be obtained by condensation of salicylaldehyde and its derivatives with a variety of primary amines. Applications of Schiff base complexes in various fields, such as molecular electronics, optical, catalysis, analytical, pharmaceutical, and biomedical [85–100], are known. The *salen* moiety owes attention to its versatility and coordination ability toward several metals as a mononegative ligand. Schiff bases ligands can form homo- and hetero-metallic complexes and 1-D, 2-D, and 3-D polymers. The emission behavior of many zinc(II) half-*salen* complexes has attracted interest due to their potential as light-emitting layers [101] and fluorescent sensors [102,103]. Photoluminescence properties of *N,O* Schiff base complexes can be changed/improved by the introduction of a third binding site at the ligand backbone. In this case, locking the metal in a strong *N,O* clamp, properties can be modulated by insertion of the third donor atom group in a suitable site of the binding architecture. Tuning of fluorescence emission is expected by varying the third donor atom and its position, by addition of substituents on the coordination core and by the auxiliary ligands. The most recent and intriguing advances in the design of *N,O,N* and *O,N,O* tridentate ligands for zinc(II) complexes are presented below.

3.1. *N,N,O Ligands*

Many *N,N,O* ligands have a relatively simple structure. By addition to the mononegative *N,O* half-*salen* block of an aromatic or non-aromatic nitrogen-containing fragment, a wide variety of structures can be obtained. In many articles, X-ray diffraction analysis constitutes the starting point to correlate structural data and theoretical analysis. On the other hand, more elaborate *N,N,O* chelating structures have been designed for specific functions. Because zinc is essential to life as part of enzymes, the detection of zinc cation in complex biological systems is a desirable goal. Moreover, zinc complexes have recently been employed as probes in fluorescence bio-imaging techniques.

3.1.1. Half-*Salen*-Type Ligands

In 2016, Shyamapada Shit and coworkers [104] presented two tridentate *salen* Schiff base ligands with two different halogen substituents in combination with azide as auxiliary ligands (Figure 6). The structure–property relationships of the coordination complexes were examined. Single-crystal X-ray diffraction studies revealed a similar dinuclear pattern for the two complexes. Spectroscopic characterizations revealed that the fluorescence pattern of the complexes is scarcely affected by the halogen substituent, in both case distant from the coordination core and with no relevant electronic effect.

As expected, the emission recorded in methanol, ascribable to ILCT of the complexes, is significantly higher than that of the corresponding ligands.

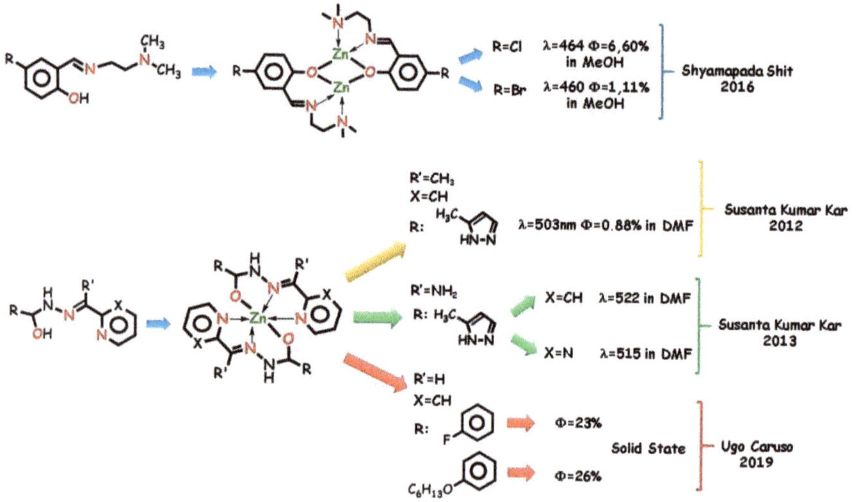

Figure 6. *N,N,O* pincers containing half-*salen* moiety and related complexes.

Variously substituted polydentate hydrazones are important scaffolds in coordination chemistry REFF. Tridentate *N,O,N* ligands can explain their chelating ability utilizing the pyridine/pyrazine *N* atom, one azomethine *N* atom, and one carbohydrazide *O* atom, and can mold itself according to the required coordination of the metal ion. Pyrazole-based flexible *N,N,O* hydrazones and their zinc(II) complexes were studied by Susanta Kumar Kar and coworkers in 2012 [63]. By the X-ray technique, the flexible pyridyl–pyrazolyl-ended ligand was found to be able to produce different coordination structures with different metal ions, and relevant changes in the luminescent pattern (Figure 6). In DMF, d^{10} ions, such as Zn(II) and Cd(II) cations, show a high CHEF effect, unlike the Ni(II) ion, which in turn causes fluorescence quenching with respect to the free ligand. In 2013 Kumer Kar and coworkers [105] observed no relevant fluorescence in the ligands, while its cadmium(II) and zinc(II) complexes were emissive in DMF, due to intraligand p,p* and n,p* transitions and also to the weak MLCT band [64]. Chelation-induced rigidity also plays an important role impeding the nonradiative channels due to the flexible bonds.

Recently, the interest in this class of complexes was promoted by the photoluminescence activity in the solid phase, as required for emitting layers of LEDs and solar cells. In 2019, Ugo Caruso and coworkers [75] obtained two complexes by reaction of zinc(II) acetate and *N,N,O* tridentate pyridinyl-hydrazone ligands (Figure 6). Both ligands have a pyridinyl-hydrazone moiety acting as mono-negative tridentate ligands toward the zinc ion in a 2:1 stoichiometric ratio, producing an octahedral environment. Ligands and complexes are scarcely emissive in diluted solution. The crystalline ligands show poor emission in the solid state while the *push-pull* more efficient pattern of the complexes guarantee intense solid-state blue fluorescence due to the AIE (aggregation-induced emission) effect [106]. In this case, the fluorescence pattern of the complexes is largely affected by the substituent. Quantum yields above 20% and large Stoke's shifts were recorded. Because large Stoke's shifts eliminate spectral overlap between absorption and emission phenomena, the detection of the fluorescence improves both in the intensity and in color purity, making the complexes promising for actual applications.

3.1.2. Ligands for Sensing Analysis and for Biological Applications

In order to achieve the goal of zinc(II) detection in complex biological systems, or to design zinc-based architectures with biological activity, the C=N moiety included in flexible ligands can be an easy synthetic solution. A half-*salen*-type *N,O,N* ligand [107] was prepared in situ by Shyamal Kumar Chattopadhyay in 2019 (Figure 7) and employed to produce a mononuclear zinc(II) complex whose structure was determined by single crystal X-ray diffraction. In an aqueous methanol solution at the physiological pH, the complex exhibits an intense greenish-blue fluorescence whose maximum does not differ substantially with respect to the free ligand while the intensity is about 17-fold stronger. The ability to give fluorescence in aqueous solutions makes the probe promising for DNA binding activity and fluorescence bio-imaging. By DFT calculations, the nature of the electronic transitions was assigned to the π,π^* transitions of the imine and heterocyclic moiety and to a n,π^* transition for the free ligand, and to π,π^* LCT transition in the complex.

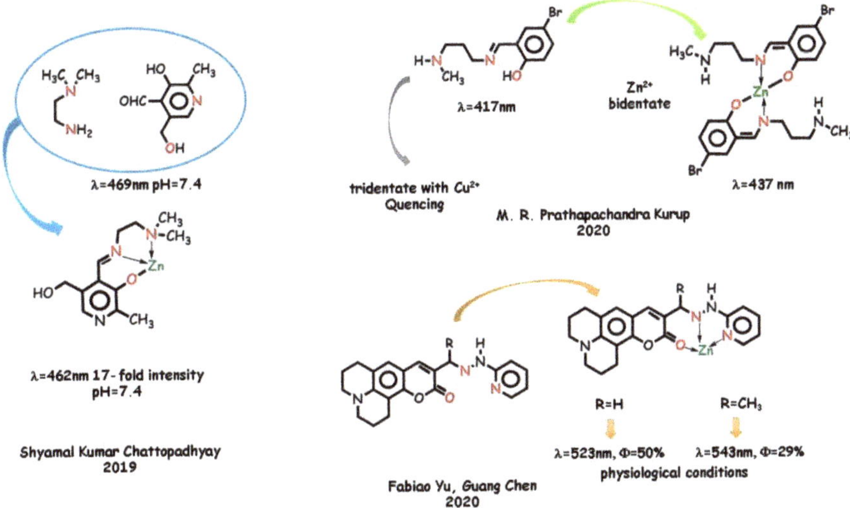

Figure 7. *N,N,O* pincers for sensing analysis and biological applications.

Designed for cytotoxic and antibacterial activity by M.R. Prathapachandra Kurup and coworkers in 2020 [108], a potentially *N,N,O* tridentate ligand worked with copper and zinc salts with unexpected different results. The two complexes (Figure 7) show a very different structural and spectroscopic pattern. In solution, the fluorescence emission intensity of the ligand decreased on complexation with Cu(II) ion, which can be attributed to the decrease in electron density on the ligand due to d orbital being involved. Contrarily, the ligand acts as a bidentate pincer toward zinc(II), coordinating to the metal cation through phenoxo oxygen and imine nitrogen. The bis-chelate metal complex produces an enhancement in the fluorescent intensity in DMF, due to the prevention of the photoinduced electron transfer process preserving ILCT bands.

Very recently [109], two *N,O,N* tridentate ligands were used by Fabiao Yu, Guang Chen, and coworkers as fluorescent sensors to monitor intracellular zinc(II) in living cells by fluorescent bioimaging. The elaborate fused-rings aromatic part guarantees a rich π-conjugated system able to give a photoluminescent response. The oxygen atom group of the C=O fragment is the third neutral jaw of the tridentate ligand. Due to the restriction of the isomerization and rotation of C=N upon coordination, the probes show fluorescence enhancement (from 4 to 7-fold at 523 and at 543 nm, respectively) and large Stokes shifts of the emission spectra. A real-time two-photon excitation wavelength apt to biological experiments and deep penetration in tissues was detected.

3.2. *O,N,O Ligands*

The ubiquitous Schiff base moiety is the useful fragment also in this case. Due to the versatile synthesis and the coordination ability of the CH=N functional group, many binegative ligands were built starting from *O,N* chelating *salen*-derivatives and a third oxygen or sulphur-containing moiety.

Tridentate furan-containing half-*salen*-type ligands were published by Debashis Ray and coworkers in 2014 [22] (Figure 8). Participation of the furan oxygen group in coordination is scarcely reported. The phenoxido-O group can be involved in coordination as a neutral donor site. The coordination abilities of the furan ring and the effect of several auxiliary triatomic bridging groups were checked by reacting zinc perchlorate salt in the absence and in the presence of auxiliary thiocyanato or azido anions. The ligand coordinates as a tridentate ligand producing a mononuclear specie, a dinuclear specie in the presence of thiocyanato, and a polymeric azido-bridged chain with azido anion. In MeOH solution, the emission bands of ligand and complexes are very similar. The PET process due to the presence of an electron lone pair of the donor atoms in the ligand produces a low PL quantum yield. Zinc-binding-induced emission greatly depends on the coordination pattern. The coordination-driven enhancement of fluorescence intensity is explainable with an increased rigidity upon complexation, so that the emission intensity in the dinuclear-bridged complex is higher than in the mono and polynuclear.

Coumarin-based molecules were recently employed as laser dyes and fluorescent probes [110–112]. Zinc-selective coumarin-based chemosensors were used in biological systems. Vinay K. Singh and coworkers in 2019 [113] produced two mononegative *O,N,O* tridentate Schiff base ligands employed in the coordination of various metal cations (Figure 8). The structural information obtained by the X-ray technique was used in the structure–activity correlation. In contrast to the fluorescence quenching upon cobalt, nickel, and copper complexation, zinc complexes show a from medium to strong emission, due to the locally excited π^*,n transition state, the nature of substituents, and the conformational rigidity of the fluorophore greatly affecting the photo-induced electron transfer processes. Another coumarine-containing tridentate ligand with a hydrazonic flexible skeleton was studied by Nader Noshiranzadeh and Mirabdullah Seyed Sadjadi in 2019 [114], focusing its catalytic activity in azide-nitrile cycloaddition reactions (Figure 8). The combination of the coumarin moiety and hydrazone functional group did show interesting optical properties. The ligand acts as a mononegative *O,N,O* tridentate trough the azomethine nitrogen and the esteric oxygen atom groups to the metal ion. Methanol and a chloride ion complete the coordination sphere. The ligand itself exhibits an intense fluorescent emission in methanol at 475 nm, which can be assigned to the p,p* transfers. Interestingly, as the ligand is encumbered, the nonradiative channels due to the flexible bonds are impeded and the fluorescence intensity is scarcely affected by zinc coordination. The higher emission band of the complex is very similar both in intensity and in the maximum wavelength, still related to intraligand emissions.

Very recently [115], a series of mononuclear acetate-containing zinc complexes derived from acylhydrazones demonstrated efficient photoluminescence in the solid state, with emission maxima from 414 to 536 nm and quantum yields from 9.5 to 64.2% depending on the nature of the acyl fragment and of the auxiliary ligand (water or pyridine). A.N. Gusev and coworkers in 2020 synthetized several hydrazones containing a phenylpyrazole fragment acting as mononegative ligands toward the cation by deprotonation of the pyrazole fragment. The ligands themselves are poor emitters in the solid state. The PL efficiency of the hydrate complexes is lower with respect to the pyridinium analogs in the case of the aromatic acid derivatives, whereas an inverse dependence was observed for the phenylalkyl derivatives.

A systematic approach based on zinc-binding aroyl- and acylhydrazones ligands with different substituents and pyridine rings as auxiliary ligands was adopted in a series of articles by B. Panunzi and coworkers (Figure 9). This approach, based on the study of a homogeneous set of the same skeleton ligands, which differ in one relevant substituent, led to highly stable mononuclear and polynuclear structures and to metallated zinc polymers emissive in the solid state.

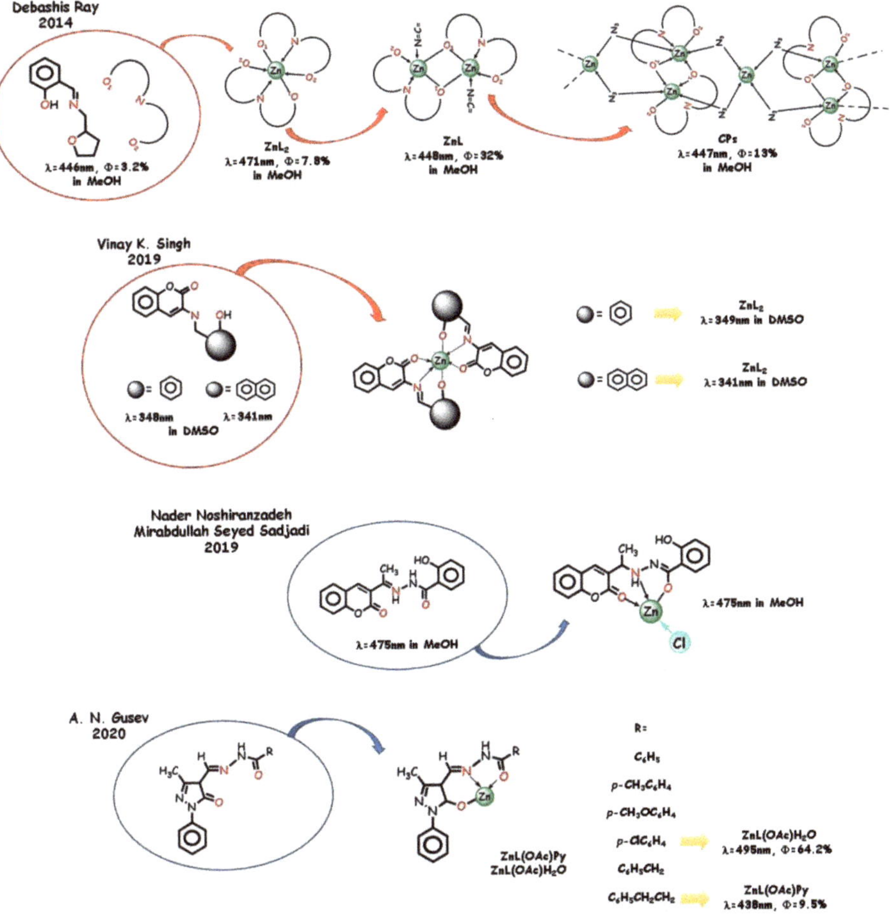

Figure 8. Selected examples of *O,N,O* ligands and zinc complexes containing the C=N moiety.

In 2014, B. Panunzi and coworkers reported the synthesis and characterization of four *O,N,O* acylhydrazono [23] and four analogous aroylhydrazono-type [116] ligands. The difference within the first mononuclear complex group and within the second (mono and dinuclear) groups is the electron-acceptor substituent R on the same tridentate chelating core. The difference between the two groups is an additional benzyloxy bulky group (Figure 9). In both cases, the enhanced fluorescence in the solid state is due to the increased rigidity upon coordination, which leads to a decreased probability of electronic nonradiative transitions from the excited states [117,118]. Tuning of the emission wavelength was achievable by varying the electron-acceptor group R with significant analogies in the two series (see Figure 9). DFT analysis produced a first rationalization of the red shift in the chromophore series. The ability of pyridine molecules to complete the coordination sphere of zinc(II) was explored and its dominant contribution to LUMO involved in the electronic transitions was pointed up. Due to self-quenching decreasing, the photoluminescence intensity enhances by increasing the distance between the emitting species in the crystalline complexes. Therefore, the second series of bulky complexes show higher PLQYs with respect to the first series. An unprecedented 64% PLQY for the R=CN bulky zinc complex was recorded, with relevant tuning in the wavelength and emission intensity with respect to the ligand; this value is suitable for lighting applications.

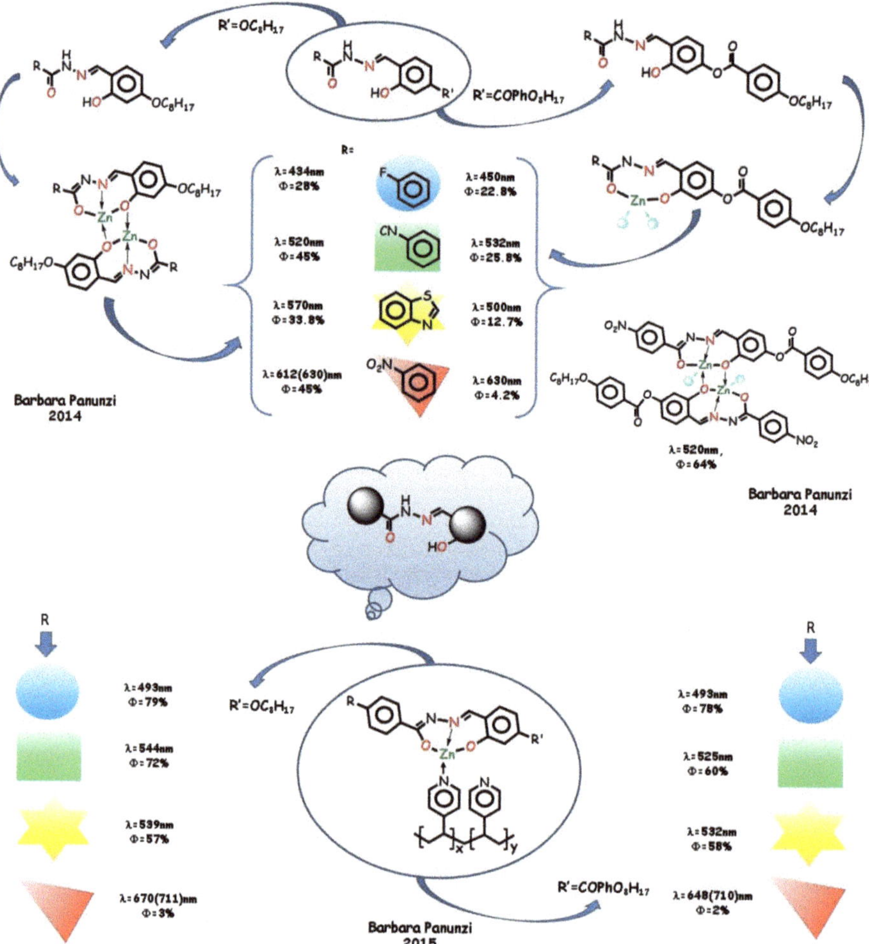

Figure 9. Aroyl- and acylhydrazones *N,O,N* tridentate pincers with different substituents, the derived complexes, and the related zinc polymers.

In order to transfer the optimal fluorescence performance of the two groups of complexes into polymeric materials, the same tridentate ligands were employed by B. Panunzi and coworkers in 2015 to prepare metallopolymers by chemical grafting of Zn(II) coordinating cores onto preformed poly(4-vinylpyridine) (PVPy) chains [119] (Figure 9). As an alternative approach to the dye-doped materials, this practice showed advantages, such as stability of the materials, synthetic easiness, and reproducibility. In the 10 wt.% grafted polymeric materials, effective emission color tuning was achieved depending on the strength of the electron acceptor substituent and high solid-state PLQYs.

As a part of the same research, other groups of aroyl- and acylhydrazones were studied for their ability to form stable zinc(II) complexes with a varied coordination environment and tunable photophysical properties. In 2019, U. Caruso and coworkers reported [74,120] on three *O,N,O* tridentate aryl-hydrazone ligands with a cationic-ended side chain and a different electron-withdrawing substituent (Figure 10). The charged chain makes both ligands and complexes very soluble in common organic solvents and aqueous mixed solvents and emissive in solution, as required in soft-matter solar cells, such as light-emitting electrochemical cells (LECs). RGB (red-green-blue) emission color tuning in ethanol was obtained by increasing the withdrawing strength of the substituent. PLQYs of the complexes are higher with respect to similar zinc coordinated systems [29,74,121–123], due to the electrostatic repulsions between the cationic chains and implemented respect to the free ligands, due to the CHEF effect.

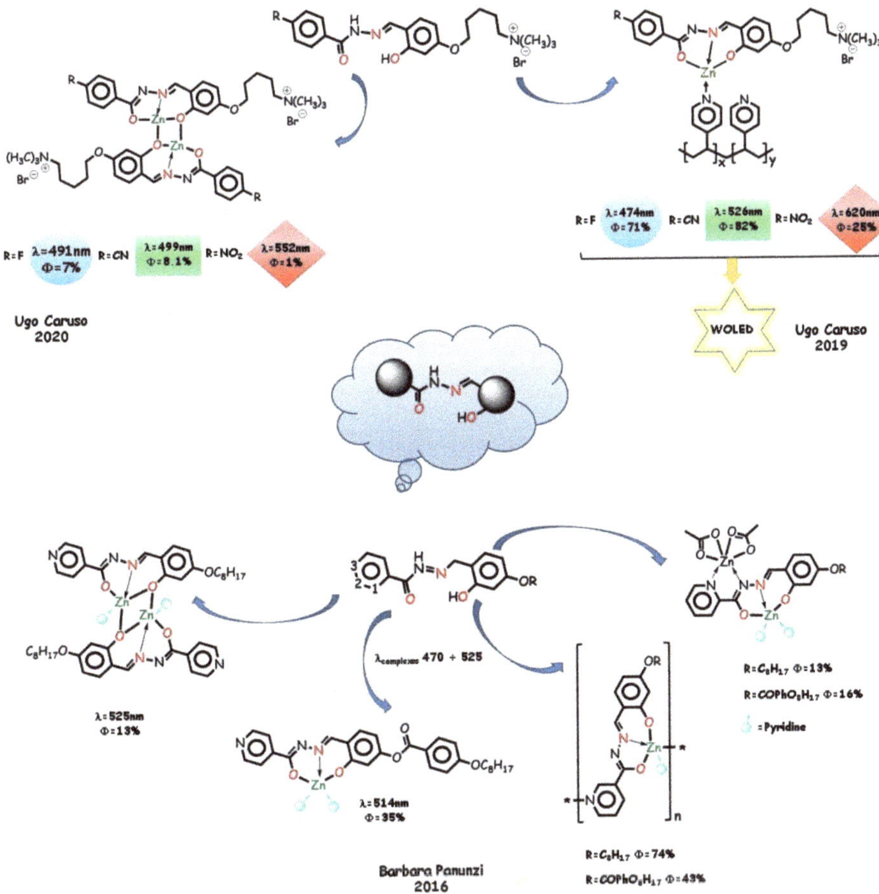

Figure 10. *O,N,O* aryl-hydrazone ligands with a cationic chain and their zinc polymers. Aroylhydrazone ligands with *orto, meta, para* pyridinoyl moiety.

The same fluoro, cyano, and nitro substituents and the charged chain guaranteeing solubility were employed by grafting the coordination moieties to a preformed PVPy (Figure 10). The resulting materials show RGB emission tuning in the solid state, with medium to excellent (more than 80% for the green-emissive polymer) PLQYs. By modulating the contents of various emissive pendants into a single polymer chain, in 2020, U. Caruso and coworkers reported a single-component highly performing white emissive material employable in the construction of white OLED devices (WOLED) with CIE coordinates (0.30, 0.31) [74].

In 2016, B. Panunzi and coworkers pointed out the exclusive role of auxiliary pyridine ligands in determining the molecular photophysical properties of the tridentate hydrazine complexes [24] and studied the effect of a pyridine moiety into the main structure of *O,N,O* aroylhydrazone ligands (Figure 10). Direct involvement of the pyridinoyl moiety in the coordination to the metal was observed when the nitrogen was in the *ortho* or *meta* position. 1-D coordination polymers were obtained with the *meta* derivatives, with 74% PLQY in the solid state. This result suggests that crystalline packed polymeric structures could provide emission enhancement for their continuous rather than discrete structure in the solid state. The tight crystal structure permits an efficient electron hopping.

4. Nitrogen, Oxygen, and Sulfur Binding Sites

In several tridentate structures, an oxygen atom was replaced by a, *S* donor binding site. Sulfur–nitrogen chelating agents are employed for their marked biological activities both as ligand and in their transition metal complexes. In many cases, the versatile C=N bond and aromatic heterocycle rings were employed in the ligand construction. Many *N,N,S* tridentate zinc complexes were explored by paying attention to both structural and spectroscopic behavior. More rarely, *S,N,S* tridentate pincer ligands with zinc salts [124] were explored, mainly screened for their reactivity and/or catalytic activity rather than for the PL properties. On the other hand, a few significative examples of mixed *N,S,O* binding sites were recently proposed. In most cases, interest was focused on the X-ray structural exploration of the coordination core and in their basic chemo-physical properties. In some cases, the observation of specific spectroscopic properties promoted the investigation of the emission properties and even moved an applicative interest.

4.1. *N,N,S Ligands*

In 2012, Jing Yang Niu and coworkers synthetized two *N,N,S* tridentate dithiocarbazate-type Schiff base ligands [125] (Figure 11). In the solid state, the ligands are in the thione tautomeric form and the derived mono or dinuclear zinc complexes show different stoichiometry and coordination core. Biological studies showed that the zinc(II) complexes are able to distinguish a leukemia cell line from a normal hepatocyte cell line by a selective fluorescence response. Still, due to their biological interest, thiosemicarbazones and 1,3,4-thiadiazole were employed to build *N,N,S* ligands by M.K. Bharty and coworkers in 2016 [126] with different metal cations. Zinc acetate was reacted with the fluorescent silent thiosemicarbazide-type ligand and with the derived fluorescent thiadiazole-type ligand producing two zinc complexes with ZnL_2 stoichiometry, where two negative nitrogen bind the metal. Interestingly, after cyclization, the same ligand acts as an *N,N* neutral bidentate ligand toward the zinc cation. The *N,N,S* tridentate complex is emissive in solution, a phenomenon ascribed to the CHEF effect by formation of four five-membered chelate rings around the cation. By DFT study, the electron density of HOMO in the thiosemicarbazide-type ligand was found on the pyridine ring nitrogen, hydrazinic nitrogen, and thione sulfur. LUMO is localized on the pyridine ring and less on hydrazinic nitrogen and sulfur. The electronic transition from HOMO to LUMO levels are associated with the π,π^* transition of ligand.

Thanks to its intrinsic fluorescence properties, triapine ligand (Figure 11) can be used to monitor the uptake and intracellular distribution in cancer cells by fluorescence microscopy. In 2010, Bernhard K. Keppler and coworkers [127] studied the triapine ligand and its tridentate zinc complex. While the compounds show similar emission spectra with a maximum at 457 nm and similar quantum yields in water, distinctly different cellular distributions of the free ligand and its complex were found. In particular, the zinc complex binds with strong affinity to a substructure within the nucleus, providing opportunities in labelling techniques. Very recently, a series of complexes from different transition metal cations were explored by V.G. Vlasenko and coworkers in 2019 [128] (Figure 11). A tridentate *N,N,S* thioxo-pyrazole Schiff base ligand was employed toward zinc cation with 1,10-phenathroline as auxiliary ligand. In the trigonal bipyramid mononuclear complexes, the amidic and iminic nitrogen atom groups and sulfur thiolate atom group constitute the tridentate site, the coordination sphere being completed by N atoms of phenanthroline. Interestingly, the related zinc complex does not display fluorescence. Computational analysis assigned the experimentally observed bands to π,π^* of the tridentate and to π(L), π^*(Phen) electronic LLCT transitions.

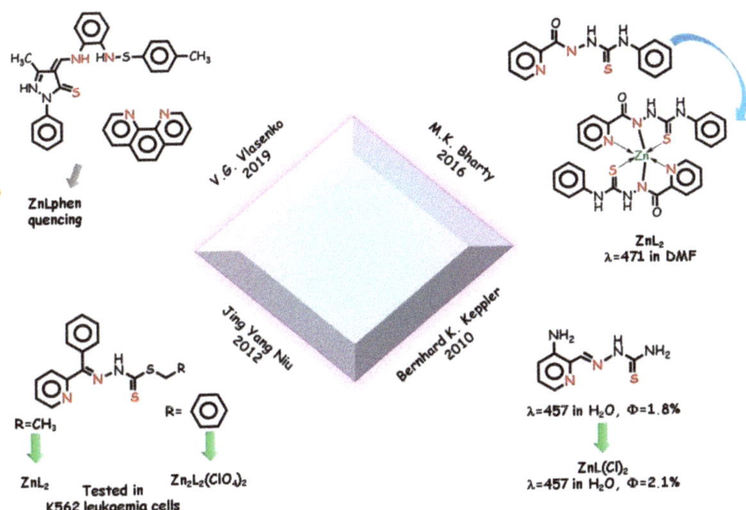

Figure 11. Selected example of *N,N,S* tridentate pincers and their complexes.

4.2. *N,S,O* Ligands

In 2015, N.K. Singh and coworkers synthetized two trinuclear Zn(II) complexes [129] from carboperthioate ligands (Figure 12). In both complexes, the middle zinc cation has a tetrahedral arrangement with two hydrazinic nitrogens and two sulfur atoms from two perthio ligands, structurally similar to the zinc finger protein. Both side zinc cations are five coordinated by one carbonyl oxygen, one hydrazinic nitrogen, and one sulfur from the carboperthioate ligand, which acts as a tridentate pincer toward the side cations. Two of the pyridinic nitrogen atom groups act as auxiliary ligands. Interestingly, from a structural point of view, the trimeric complexes generate self-assembly supramolecular structures in dependence on the different position of the pyridinic nitrogen atom of the ligand. The ligand is fluorescent silent while the complex with the 4-pyridyl substituent displays a blue emission at 470 nm in DMSO, predominantly ascribable to MLCT transitions. In this case, the mobility of the electron transfer in the backbone is enhanced and the electron transition energy of ILCT decreases due to back-coupling of π-bond between the metal and ligand. Moreover, the formation of a five-membered chelate between the coordination units and the central metal ion increases the π,π^* conjugation and the conformational coplanarity, consequently decreasing the energy gap between the π and π^* molecular orbitals of the ligand.

In 2016, a Schiff base ligand derived from 2-aminothiophenol was coordinated as an *N,S,O* tridentate ligand to different transition metal cations by Bita Shafaatian and coworkers [19]. The fluorescence properties of the ligand and of the dinuclear complexes (Figure 12) were examined. Interestingly, in all cases, the metal complexes in dichloromethane exhibit weak fluorescence in comparison to ligand. For Zn(II) complex, no emission observed was assigned to π,π^* IL transitions. PLQY decreases to about one third with a relevant blue shift in the emission maxima with respect to the free ligand. Finally, as an example of biological application of *N,S,O* complexes, in 2017, two novel triazole containing Schiff base ligands were employed with zinc cation and other transition metals by Sulekh Chandra and coworkers [130]. The ligands behave as binegative tridentate in the formation of 1:1 aqueous metal complex (Figure 12), which were employed in fluorescence quenching experiments of the strong emission band at 327 nm of BSA, revealing a zinc complex that was more promising due to its strong binding ability.

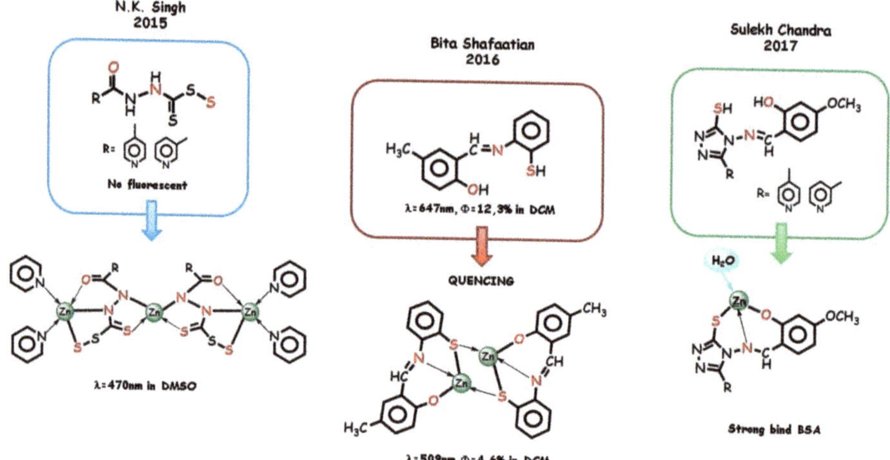

Figure 12. Selected example of *N,S,O* tridentate pincers and their complexes.

5. Conclusions

One of the biggest challenges of the modern era is sustainability. Over the last years, the high variety of *N*, *O*, *S* donor ligands moved a strong interest toward spectroscopic and applicative features of the related complexes. Tridentate ligands are simple, low cost, and easy synthesizable pincers able to guarantee good stability to the derived complexes. On the other hand, zinc cation has a unique mix of attractive properties, which potentially make it a smart "green" metal. Combinations of suitable tridentate ligands with zinc cation are the perfect union to achieve remarkable PL responses and targeted applications. As a d^{10} closed shell cation, zinc(II) plays a quite innocent role in the electronic and therefore spectroscopic pattern of the ligand, often guaranteeing a relevant CHEF effect. For this reason, zinc is a key issue in developing an alternative class of environmentally friendly and highly efficient fluorophores for display and lighting technologies. On the other hand, zinc plays a crucial role in many important biological processes and is a structural key component of proteins and enzymes. Versatile molecules able in the zinc(II) detection or in the zinc binding of fluorescence markers to specific biological substrates are recently proposed chemo- and biosensors.

As scientists, we presented here some examples of zinc tridentate complexes in an attempt to highlight some of their unique properties. We examined the PL emission tuning due to coordination and stating the most interesting applications. As researchers, we are committed to continuing the study of novel systems for the new technological frontier and we expect our research to be the subject of further interesting discoveries.

Author Contributions: Conceptualization, B.P.; data curation and formal analysis, R.D.; methodology, R.D. and B.P.; writing—original draft, B.P., writing—review & editing, B.P. All authors have read and agreed to the published version of the manuscript.

Funding: This research was funded by the Italian Ministry of Education, University and Research (MIUR) under grants PON PANDION 01_00375.

Conflicts of Interest: The authors declare no conflict of interest.

References

1. Mautner, F.A.; Berger, C.; Gspan, C.; Sudy, B.; Fischer, R.C.; Massoud, S.S. Pyridyl and triazole ligands directing the assembling of zinc(II) into coordination polymers with different dimensionality through azides. *Polyhedron* **2017**, *130*, 136–144. [CrossRef]

2. Caruso, U.; Panunzi, B.; Roviello, A.; Tuzi, A. Fluorescent metallopolymers with Zn(II) in a Schiff base/phenoxide coordination environment. *Inorg. Chem. Commun.* **2013**, *29*, 138–140. [CrossRef]
3. Günay Sezer, G.; Zafer Yeşilel, O.; Şahin, O.; Burrows, A.D. Zinc(II) and cadmium(II) coordination polymers containing phenylenediacetate and bis(imidazol-1-ylmethyl)benzene linkers: The effect of ligand isomers on the solid state structures. *J. Solid State Chem.* **2017**, *252*, 8–21. [CrossRef]
4. Gusev, A.; Shul'gin, V.; Braga, E.; Zamnius, E.; Kryukova, M.; Linert, W. Luminescent properties of Zn complexes based on tetradentate N2O2-donor pyrazolone schiff bases. *Dye. Pigm.* **2020**, *183*. [CrossRef]
5. Caruso, U.; Panunzi, B.; Roviello, A.; Tingoli, M.; Tuzi, A. Two aminobenzothiazole derivatives for Pd(II) and Zn(II) coordination: Synthesis, characterization and solid state fluorescence. *Inorg. Chem. Commun.* **2011**, *14*, 46–48. [CrossRef]
6. Matozzo, P.; Colombo, A.; Dragonetti, C.; Righetto, S.; Roberto, D.; Biagini, P.; Fantacci, S.; Marinotto, D. A Chiral Bis(salicylaldiminato)zinc(II) Complex with Second-Order Nonlinear Optical and Luminescent Properties in Solution. *Inorganics* **2020**, *8*, 25. [CrossRef]
7. Barbieri, A.; Accorsi, G.; Armaroli, N. Luminescent complexes beyond the platinum group: The d10 avenue. *Chem. Commun. (Camb.)* **2008**, *19*, 2185–2193. [CrossRef]
8. Alam, R.; Bhowmick, R.; Islam, A.S.M.; Chaudhuri, K.; Ali, M. A rhodamine based fluorescent trivalent sensor (Fe 3+, Al 3+, Cr 3+) with potential applications for live cell imaging and combinational logic circuits and memory devices. *New J. Chem.* **2017**, *41*, 8359–8369. [CrossRef]
9. Aragoni, M.C.; Arca, M.; Bencini, A.; Caltagirone, C.; Garau, A.; Isaia, F.; Light, M.E.; Lippolis, V.; Lodeiro, C.; Mameli, M. Zn 2+/Cd 2+ optical discrimination by fluorescent chemosensors based on 8-hydroxyquinoline derivatives and sulfur-containing macrocyclic units. *Dalton Trans.* **2013**, *42*, 14516–14530. [CrossRef]
10. Lee, J.J.; Lee, S.A.; Kim, H.; Nguyen, L.; Noh, I.; Kim, C. A highly selective CHEF-type chemosensor for monitoring Zn2+ in aqueous solution and living cells. *RSC Adv.* **2015**, *5*, 41905–41913. [CrossRef]
11. Lee, H.; Lee, H.S.; Reibenspies, J.H.; Hancock, R.D. Mechanism of "turn-on" fluorescent sensors for mercury(II) in solution and its implications for ligand design. *Inorg. Chem.* **2012**, *51*, 10904–10915. [CrossRef] [PubMed]
12. Tan, S.S.; Kim, S.J.; Kool, E.T. Differentiating between fluorescence-quenching metal ions with polyfluorophore sensors built on a DNA backbone. *J. Am. Chem. Soc.* **2011**, *133*, 2664–2671. [CrossRef] [PubMed]
13. Deems, J.C.; Reibenspies, J.H.; Lee, H.-S.; Hancock, R.D. Strategies for a fluorescent sensor with receptor and fluorophore designed for the recognition of heavy metal ions. *Inorg. Chim. Acta* **2020**, *499*. [CrossRef]
14. Lee, H.; Hancock, R.D.; Lee, H.S. Role of fluorophore-metal interaction in photoinduced electron transfer (PET) sensors: Time-dependent density functional theory (TDDFT) study. *J. Phys. Chem. A* **2013**, *117*, 13345–13355. [CrossRef] [PubMed]
15. Patra, L.; Das, S.; Gharami, S.; Aich, K.; Mondal, T.K. A new multi-analyte fluorogenic sensor for efficient detection of Al3+ and Zn2+ ions based on ESIPT and CHEF features. *New J. Chem.* **2018**, *42*, 19076–19082. [CrossRef]
16. Hancock, R.D. The pyridyl group in ligand design for selective metal ion complexation and sensing. *Chem. Soc. Rev.* **2013**, *42*, 1500–1524. [CrossRef]
17. Panunzi, B.; Diana, R.; Concilio, S.; Sessa, L.; Tuzi, A.; Piotto, S.; Caruso, U. Fluorescence pH-dependent sensing of Zn(II) by a tripodal ligand. A comparative X-ray and DFT study. *J. Lumin.* **2019**, *212*, 200–206. [CrossRef]
18. Wu, M.X.; Filley, S.J.; Hill, K.A. Cooperative binding of zinc to an aminoacyl-tRNA synthetase. *Biochem. Biophys. Res. Commun.* **1994**, *201*, 1079–1083. [CrossRef]
19. Shafaatian, B.; Mousavi, S.S.; Afshari, S. Synthesis, characterization, spectroscopic and theoretical studies of new zinc(II), copper(II) and nickel(II) complexes based on imine ligand containing 2-aminothiophenol moiety. *J. Mol. Struct.* **2016**, *1123*, 191–198. [CrossRef]
20. Bi, X.; Pang, Y. Optical Response of Terpyridine Ligands to Zinc Binding: A Close Look at the Substitution Effect by Spectroscopic Studies at Low Temperature. *J. Phys. Chem. B* **2016**, *120*, 3311–3317. [CrossRef]
21. Zhang, Y.; Yang, Y.; Zhou, S.; Ma, Z. Synthesis, structure, and thermal and photoluminescent properties of a zinc(II) sulfate 4′-phenyl-terpyridine compound. *Inorg. Nano-Met. Chem.* **2017**, *47*, 876–880. [CrossRef]
22. Mandal, H.; Chakrabartty, S.; Ray, D. Isothiocyanato and azido coordination induced structural diversity in zinc(ii) complexes with Schiff base containing tetrahydrofuran group: Synthesis, characterization, crystal structure and fluorescence study. *RSC Adv.* **2014**, *4*, 65044–65055. [CrossRef]
23. Borbone, F.; Caruso, U.; Causà, M.; Fusco, S.; Panunzi, B.; Roviello, A.; Shikler, R.; Tuzi, A. Series of O,N,O-tridentate ligands zinc(II) complexes with high solid-state photoluminescence quantum yield. *Eur. J. Inorg. Chem.* **2014**, *16*, 2695–2703. [CrossRef]

24. Borbone, F.; Caruso, U.; Concilio, S.; Nabha, S.; Panunzi, B.; Piotto, S.; Shikler, R.; Tuzi, A. Mono-, Di-, and Polymeric Pyridinoylhydrazone ZnII Complexes: Structure and Photoluminescent Properties. *Eur. J. Inorg. Chem.* **2016**, *2016*, 818–825. [CrossRef]
25. Panunzi, B.; Concilio, S.; Diana, R.; Shikler, R.; Nabha, S.; Piotto, S.; Sessa, L.; Tuzi, A.; Caruso, U. Photophysical Properties of Luminescent Zinc(II)-Pyridinyloxadiazole Complexes and their Glassy Self-Assembly Networks. *Eur. J. Inorg. Chem.* **2018**, *2018*, 2709–2716. [CrossRef]
26. Borbone, F.; Carella, A.; Caruso, U.; Roviello, G.; Tuzi, A.; Dardano, P.; Lettieri, S.; Maddalena, P.; Barsella, A. Large second-order NLO activity in poly(4-vinylpyridine) grafted with PdII and CuII chromophoric complexes with tridentate bent ligands containing heterocycles. *Eur. J. Inorg. Chem.* **2008**, *11*, 1846–1853. [CrossRef]
27. Caruso, U.; Centore, R.; Panunzi, B.; Roviello, A.; Tuzi, A. Grafting poly(4-vinylpyridine) with a second-order nonlinear optically active nickel(II) chromophore. *Eur. J. Inorg. Chem.* **2005**, *13*, 2747–2753. [CrossRef]
28. Li, G.; Wang, L.; Wu, L.; Guo, Z.; Zhao, J.; Liu, Y.; Bai, R.; Yan, X. Woven Polymer Networks via the Topological Transformation of a [2]Catenane. *J. Am. Chem. Soc.* **2020**, *142*, 14343–14349. [CrossRef]
29. Diana, R.; Panunzi, B.; Piotto, S.; Caruso, T.; Caruso, U. Solid-state fluorescence of two zinc coordination polymers from bulky dicyano-phenylenevinylene and bis-azobenzene cores. *Inorg. Chem. Commun.* **2019**, *110*. [CrossRef]
30. Diana, R.; Panunzi, B.; Shikler, R.; Nabha, S.; Caruso, U. Highly efficient dicyano-phenylenevinylene fluorophore as polymer dopant or zinc-driven self-assembling building block. *Inorg. Chem. Commun.* **2019**, *104*, 145–149. [CrossRef]
31. Cannizzo, A.; Blanco-Rodriguez, A.M.; El Nahhas, A.; Sebera, J.; Zalis, S.; Vlcek, A., Jr.; Chergui, M. Femtosecond fluorescence and intersystem crossing in rhenium(I) carbonyl-bipyridine complexes. *J. Am. Chem. Soc.* **2008**, *130*, 8967–8974. [CrossRef] [PubMed]
32. Stufkens, D. Ligand-dependent excited state behaviour of Re(I) and Ru(II) carbonyl–diimine complexes. *Coord. Chem. Rev.* **1998**, *177*, 127–179. [CrossRef]
33. Vlček, A., Jr. Ultrafast excited-state processes in Re(I) carbonyl-diimine complexes: From excitation to photochemistry. In *Topics in Organometallic Chemistry*; Lees, A.J., Ed.; Springer: Berlin/Heidelberg, Germany, 2010; Volume 29, pp. 73–114.
34. Zalis, S.; Milne, C.J.; El Nahhas, A.; Blanco-Rodriguez, A.M.; van der Veen, R.M.; Vlcek, A., Jr. Re and Br X-ray absorption near-edge structure study of the ground and excited states of [ReBr(CO)3(bpy)] interpreted by DFT and TD-DFT calculations. *Inorg. Chem.* **2013**, *52*, 5775–5785. [CrossRef]
35. Baková, R.; Chergui, M.; Daniel, C.; Vlček, A., Jr.; Záliš, S. Relativistic effects in spectroscopy and photophysics of heavy-metal complexes illustrated by spin–orbit calculations of [Re(imidazole)(CO)3(phen)]+. *Coord. Chem. Rev.* **2011**, *255*, 975–989. [CrossRef]
36. Świtlicka-Olszewska, A.; Klemens, T.; Nawrot, I.; Machura, B.; Kruszynski, R. Novel Re(I) tricarbonyl coordination compound of 5-amino-1,10-phenanthroline – Synthesis, structural, photophysical and computational studies. *J. Lumin.* **2016**, *171*, 166–175. [CrossRef]
37. Kurtz, D.A.; Brereton, K.R.; Ruoff, K.P.; Tang, H.M.; Felton, G.A.N.; Miller, A.J.M.; Dempsey, J.L. Bathochromic Shifts in Rhenium Carbonyl Dyes Induced through Destabilization of Occupied Orbitals. *Inorg. Chem.* **2018**, *57*, 5389–5399. [CrossRef]
38. Zarkadoulas, A.; Koutsouri, E.; Kefalidi, C.; Mitsopoulou, C.A. Rhenium complexes in homogeneous hydrogen evolution. *Coord. Chem. Rev.* **2015**, *304*, 55–72. [CrossRef]
39. Nganga, J.K.; Samanamu, C.R.; Tanski, J.M.; Pacheco, C.; Saucedo, C.; Batista, V.S.; Grice, K.A.; Ertem, M.Z.; Angeles-Boza, A.M. Electrochemical Reduction of CO2 Catalyzed by Re(pyridine-oxazoline)(CO)3Cl Complexes. *Inorg. Chem.* **2017**, *56*, 3214–3226. [CrossRef]
40. Martinez, J.F.; La Porte, N.T.; Wasielewski, M.R. Electron Transfer from Photoexcited Naphthalene Diimide Radical Anion to Electrocatalytically Active Re(bpy)(CO)3Cl in a Molecular Triad. *J. Phys. Chem. C* **2018**, *122*, 2608–2617. [CrossRef]
41. Hostachy, S.; Policar, C.; Delsuc, N. Re(I) carbonyl complexes: Multimodal platforms for inorganic chemical biology. *Coord. Chem. Rev.* **2017**, *351*, 172–188. [CrossRef]
42. Gabr, M.T.; Pigge, F.C. Rhenium tricarbonyl complexes of AIE active tetraarylethylene ligands: Tuning luminescence properties and HSA-specific binding. *Dalton Trans.* **2017**, *46*, 15040–15047. [CrossRef]
43. Ramdass, A.; Sathish, V.; Babu, E.; Velayudham, M.; Thanasekaran, P.; Rajagopal, S. Recent developments on optical and electrochemical sensing of copper(II) ion based on transition metal complexes. *Coord. Chem. Rev.* **2017**, *343*, 278–307. [CrossRef]

44. Ma, Z.; Lu, W.; Liang, B.; Pombeiro, A.J.L. Synthesis, characterization, photoluminescent and thermal properties of zinc(ii) 4′-phenyl-terpyridine compounds. *New J. Chem.* **2013**, *37*. [CrossRef]
45. Ma, Z.; Wei, L.; Alegria, E.C.; Martins, L.M.; Guedes da Silva, M.F.; Pombeiro, A.J. Synthesis and characterization of copper(II) 4′-phenyl-terpyridine compounds and catalytic application for aerobic oxidation of benzylic alcohols. *Dalton Trans.* **2014**, *43*, 4048–4058. [CrossRef]
46. Ma, Z.; Zhang, B.; Guedes da Silva, M.F.; Silva, J.; Mendo, A.S.; Baptista, P.V.; Fernandes, A.R.; Pombeiro, A.J. Synthesis, characterization, thermal properties and antiproliferative potential of copper(II) 4′-phenyl-terpyridine compounds. *Dalton Trans.* **2016**, *45*, 5339–5355. [CrossRef]
47. Winter, A.; Friebe, C.; Chiper, M.; Schubert, U.S.; Presselt, M.; Dietzek, B.; Schmitt, M.; Popp, J. Synthesis, characterization, and electro-optical properties of Zn(II) complexes with pi-conjugated terpyridine ligands. *ChemPhysChem* **2009**, *10*, 787–798. [CrossRef] [PubMed]
48. Ghosh, B.N.; Topic, F.; Sahoo, P.K.; Mal, P.; Linnera, J.; Kalenius, E.; Tuononen, H.M.; Rissanen, K. Synthesis, structure and photophysical properties of a highly luminescent terpyridine-diphenylacetylene hybrid fluorophore and its metal complexes. *Dalton Trans.* **2015**, *44*, 254–267. [CrossRef] [PubMed]
49. Hasegawa, Y.; Nakagawa, T.; Kawai, T. Recent progress of luminescent metal complexes with photochromic units. *Coord. Chem. Rev.* **2010**, *254*, 2643–2651. [CrossRef]
50. Yam, V.W.-W.; Lo, K.K.-W. Luminescent polynuclear d10 metal complexes. *Chem. Soc. Rev.* **1999**, *28*, 323–334. [CrossRef]
51. Ghosh, B.N.; Puttreddy, R.; Rissanen, K. Synthesis and structural characterization of new transition metal complexes of a highly luminescent amino-terpyridine ligand. *Polyhedron* **2020**, *177*. [CrossRef]
52. Medlycott, E.A.; Hanan, G.S. Designing tridentate ligands for ruthenium(II) complexes with prolonged room temperature luminescence lifetimes. *Chem. Soc. Rev.* **2005**, *34*, 133–142. [CrossRef] [PubMed]
53. Goodall, W.; Williams, J.A.G. A new, highly fluorescent terpyridine which responds to zinc ions with a large red-shift in emissionElectronic supplementary information (ESI) available: Details of synthetic details and characterisation and plots of: (i) the decrease in fluorescence as a function of pH for the two ligands L1 and L2, and (ii) the increase in the intensity of fluorescence of L1 at 620 nm as a function of added zinc, including the method used for determination of Kass. *Chem. Commun.* **2001**, *23*, 2514–2515. [CrossRef]
54. Diana, R.; Caruso, U.; Di Costanzo, L.; Bakayoko, G.; Panunzi, B. A novel DR/NIR T-shaped aiegen: Synthesis and x-ray crystal structure study. *Crystals* **2020**, *10*, 269. [CrossRef]
55. Kozhevnikov, V.N.; Shabunina, O.V.; Sharifullina, A.R.; Rusinov, V.L.; Chupakhin, O.N.; König, B. Aminomethyl bi- and terpyridines as luminescent probes for Zn2+ ions. *Mendeleev Commun.* **2005**, *15*, 8–9. [CrossRef]
56. Feng, Z.; Li, D.; Zhang, M.; Shao, T.; Shen, Y.; Tian, X.; Zhang, Q.; Li, S.; Wu, J.; Tian, Y. Enhanced three-photon activity triggered by the AIE behaviour of a novel terpyridine-based Zn(ii) complex bearing a thiophene bridge. *Chem. Sci.* **2019**, *10*, 7228–7232. [CrossRef]
57. Jiang, T.; Lu, N.; Hang, Y.; Yang, J.; Mei, J.; Wang, J.; Hua, J.; Tian, H. Dimethoxy triarylamine-derived terpyridine–zinc complex: A fluorescence light up sensor for citrate detection based on aggregation-induced emission. *J. Mater. Chem. C* **2016**, *4*, 10040–10046. [CrossRef]
58. Ghosh, K.; Kumar, P.; Tyagi, N. Synthesis, crystal structure and DNA interaction studies on mononuclear zinc complexes. *Inorg. Chim. Acta* **2011**, *375*, 77–83. [CrossRef]
59. Diana, R.; Caruso, U.; Concilio, S.; Piotto, S.; Tuzi, A.; Panunzi, B. A real-time tripodal colorimetric/fluorescence sensor for multiple target metal ions. *Dye. Pigm.* **2018**, *155*, 249–257. [CrossRef]
60. Otsuka, M.; Fujita, M.; Sugiura, Y.; Ishii, S.; Tsutomu, A.; Yamamoto, T.; Inoue, J.I. Novel zinc chelators which inhibit the binding of HIV-EP1 (HIV enhancer binding protein) to NF-kappa B recognition sequence. *J. Med. Chem.* **1994**, *37*, 4267–4269. [CrossRef]
61. Zamora, F.; Kunsman, M.; Sabat, M.; Lippert, B. Metal-Stabilized Rare Tautomers of Nucleobases. 6. Imino Tautomer of Adenine in a Mixed-Nucleobase Complex of Mercury(II). *Inorg Chem.* **1997**, *36*, 1583–1587. [CrossRef]
62. Ray, S.; Konar, S.; Jana, A.; Jana, S.; Patra, A.; Chatterjee, S.; Golen, J.A.; Rheingold, A.L.; Mandal, S.S.; Kar, S.K. Three new pseudohalide bridged dinuclear Zn(II), Cd(II) complexes of pyrimidine derived Schiff base ligands: Synthesis, crystal structures and fluorescence studies. *Polyhedron* **2012**, *33*, 82–89. [CrossRef]
63. Konar, S.; Jana, A.; Das, K.; Ray, S.; Chatterjee, S.; Kar, S.K. Complexes of a functionally modified pyrazole derived ligand—Mononuclear zinc(II), dinuclear nickel(II) and a rare pentanuclear cadmium(II) complex with a TBP core and their photoluminescence studies. *Polyhedron* **2012**, *47*, 143–150. [CrossRef]

64. Paira, M.K.; Dinda, J.; Lu, T.H.; Paital, A.R.; Sinha, C. Zn(II), Cd(II) and Hg(II) complexes of 8-aminoquinoline. *Polyhedron* **2007**, *26*, 4131–4140. [CrossRef]
65. Konar, S. Dicynamide bridged two new zig-zag 1-D Zn(II) coordination polymers of pyrimidine derived Schiff base ligands: Synthesis, crystal structures and fluorescence studies. *J. Mol. Struct.* **2015**, *1092*, 34–43. [CrossRef]
66. Blackman, A.G. The coordination chemistry of tripodal tetraamine ligands. *Polyhedron* **2005**, *24*, 1–39. [CrossRef]
67. Contreras, R.; Flores-Parra, A.; Mijangos, E.; Téllez, F.; López-Sandoval, H.; Barba-Behrens, N. From mono to polydentate azole and benzazole derivatives, versatile ligands for main group and transition metal atoms. *Coord. Chem. Rev.* **2009**, *253*, 1979–1999. [CrossRef]
68. Boča, M.; Jameson, R.F.; Linert, W. Fascinating variability in the chemistry and properties of 2,6-bis-(benzimidazol-2-yl)-pyridine and 2,6-bis-(benzthiazol-2-yl)-pyridine and their complexes. *Coord. Chem. Rev.* **2011**, *255*, 290–317. [CrossRef]
69. Li, L.-Q.; Li, M.; Zhang, H.; Li, S.; Nie, F.-M. Synthesis, structures, and fluorescent properties of two oxalato-bridged dinuclear zinc(II) complexes with tridentate and tetradentate polybenzimidazole ligands. *J. Coord. Chem.* **2014**, *67*, 847–856. [CrossRef]
70. Feng, R.; Huang, F.-F.; Yuan, J.-L.; Lu, Z.; Fang, T.; Nie, F.-M. Structures and fluorescent properties of picolinato zinc(II) and cadmium(II) complexes based on tridentate and tetradentate benzimidazole ligands. *J. Coord. Chem.* **2016**, *69*, 3776–3791. [CrossRef]
71. Concilio, S.; Bugatti, V.; Neitzert, H.C.; Landi, G.; De Sio, A.; Parisi, J.; Piotto, S.; Iannelli, P. Zn-complex based on oxadiazole/carbazole structure: Synthesis, optical and electric properties. *Thin Solid Film.* **2014**, *556*, 419–424. [CrossRef]
72. Borbone, F.; Tuzi, A.; Panunzi, B.; Piotto, S.; Concilio, S.; Shikler, R.; Nabha, S.; Centore, R. On-Off Mechano-Responsive Switching of ESIPT Luminescence in Polymorphic N-salicylidene-4-amino-2- methylbenzotriazole. *Cryst. Growth Des.* **2017**, *17*, 5517. [CrossRef]
73. Diana, R.; Caruso, U.; Di Costanzo, L.; Gentile, F.S.; Panunzi, B. Colorimetric recognition of multiple first-row transition metals: A single water-soluble chemosensor in acidic and basic conditions. *Dye. Pigm.* **2021**, *184*. [CrossRef]
74. Panunzi, B.; Diana, R.; Caruso, U. A highly efficient white luminescent zinc (II) based metallopolymer by RGB approach. *Polymers* **2019**, *11*, 1712. [CrossRef] [PubMed]
75. Diana, R.; Panunzi, B.; Tuzi, A.; Caruso, U. Two tridentate pyridinyl-hydrazone zinc(II) complexes as fluorophores for blue emitting layers. *J. Mol. Struct.* **2019**, *1197*, 672–680. [CrossRef]
76. Diana, R.; Panunzi, B.; Concilio, S.; Marrafino, F.; Shikler, R.; Caruso, T.; Caruso, U. The effect of bulky substituents on two π-conjugated mesogenic fluorophores. Their organic polymers and zinc-bridged luminescent networks. *Polymers* **2019**, *11*, 1379. [CrossRef] [PubMed]
77. Sun, Y.; Zuo, T.; Guo, F.; Sun, J.; Liu, Z.; Diao, G. Perylene dye-functionalized silver nanoparticles serving as pH-dependent metal sensor systems. *RSC Adv.* **2017**, *7*, 24215–24220. [CrossRef]
78. Li, C.; Krautler, B. Transition metal complexes of phyllobilins—A new realm of bioinorganic chemistry. *Dalton Trans.* **2015**, *44*, 10116–10127. [CrossRef]
79. Gil-Ramirez, G.; Leigh, D.A.; Stephens, A.J. Catenanes: Fifty years of molecular links. *Angew. Chem. Int. Ed. Engl.* **2015**, *54*, 6110–6150. [CrossRef]
80. Niu, Z.; Gibson, H.W. Polycatenanes. *Chem. Rev.* **2009**, *109*, 6024–6046. [CrossRef]
81. Wu, Q.; Rauscher, P.M.; Lang, X.; Wojtecki, R.J.; de Pablo, J.J.; Hore, M.J.A.; Rowan, S.J. Poly[n]catenanes: Synthesis of molecular interlocked chains. *Science* **2017**, *358*, 1434–1439. [CrossRef]
82. Li, H.; Zhu, Z.; Fahrenbach, A.C.; Savoie, B.M.; Ke, C.; Barnes, J.C.; Lei, J.; Zhao, Y.L.; Lilley, L.M.; Marks, T.J.; et al. Mechanical bond-induced radical stabilization. *J. Am. Chem. Soc.* **2013**, *135*, 456–467. [CrossRef]
83. Wang, X.Q.; Li, W.J.; Wang, W.; Wen, J.; Zhang, Y.; Tan, H.; Yang, H.B. Construction of Type III-C Rotaxane-Branched Dendrimers and Their Anion-Induced Dimension Modulation Feature. *J. Am. Chem. Soc.* **2019**, *141*, 13923–13930. [CrossRef] [PubMed]
84. Zhang, L.; Stephens, A.J.; Nussbaumer, A.L.; Lemonnier, J.F.; Jurcek, P.; Vitorica-Yrezabal, I.J.; Leigh, D.A. Stereoselective synthesis of a composite knot with nine crossings. *Nat. Chem.* **2018**, *10*, 1083–1088. [CrossRef]
85. Chui, S.S.; Lo, S.M.; Charmant, J.P.; Orpen, A.G.; Williams, I.D. A chemically functionalizable nanoporous material. *Science* **1999**, *283*, 1148–1150. [CrossRef] [PubMed]
86. Albrecht, M. "Let's twist again" double-stranded, triple-stranded, and circular helicates. *Chem. Rev.* **2001**, *101*, 3457–3497. [CrossRef] [PubMed]

87. Akbar Ali, M.; Livingstone, S.E. Metal complexes of sulphur-nitrogen chelating agents. *Coord. Chem. Rev.* **1974**, *13*, 101–132. [CrossRef]
88. Katsuki, T. Catalytic asymmetric oxidations using optically active (salen)manganese(III) complexes as catalysts. *Coord. Chem. Rev.* **1995**, *140*, 189–214. [CrossRef]
89. Singh, R.; Banerjee, A.; Colacio, E.; Rajak, K.K. Enantiopure tetranuclear iron(III) complexes using chiral reduced Schiff base ligands: Synthesis, structure, spectroscopy, magnetic properties, and DFT studies. *Inorg. Chem.* **2009**, *48*, 4753–4762. [CrossRef]
90. Bagai, R.; Datta, S.; Betancur-Rodriguez, A.; Abboud, K.A.; Hill, S.; Christou, G. Diversity of new structural types in polynuclear iron chemistry with a tridentate N,N,O ligand. *Inorg. Chem.* **2007**, *46*, 4535–4547. [CrossRef]
91. Naiya, S.; Wang, H.S.; Drew, M.G.; Song, Y.; Ghosh, A. Structural and magnetic studies of Schiff base complexes of nickel(II) nitrite: Change in crystalline state, ligand rearrangement and a very rare mu-nitrito-1kappaO:2kappaN:3kappaO′ bridging mode. *Dalton Trans.* **2011**, *40*, 2744–2756. [CrossRef]
92. Mukherjee, P.; Drew, M.G.; Gomez-Garcia, C.J.; Ghosh, A. The crucial role of polyatomic anions in molecular architecture: Structural and magnetic versatility of five nickel(II) complexes derived from A N,N,O-donor Schiff base ligand. *Inorg. Chem.* **2009**, *48*, 5848–5860. [CrossRef] [PubMed]
93. Basak, S.; Sen, S.; Banerjee, S.; Mitra, S.; Rosair, G.; Rodriguez, M.T.G. Three new pseudohalide bridged dinuclear Zn(II) Schiff base complexes: Synthesis, crystal structures and fluorescence studies. *Polyhedron* **2007**, *26*, 5104–5112. [CrossRef]
94. Ruiz, E.; Alemany, P.; Alvarez, S.; Cano, J. Toward the Prediction of Magnetic Coupling in Molecular Systems: Hydroxo- and Alkoxo-Bridged Cu(II) Binuclear Complexes. *J. Am. Chem. Soc.* **1997**, *119*, 1297–1303. [CrossRef]
95. Jelfs, K.E.; Wu, X.; Schmidtmann, M.; Jones, J.T.; Warren, J.E.; Adams, D.J.; Cooper, A.I. Large self-assembled chiral organic cages: Synthesis, structure, and shape persistence. *Angew. Chem. Int. Ed. Engl.* **2011**, *50*, 10653–10656. [CrossRef]
96. Diedrich, C.; Deeth, R.J. On the performance of ligand field molecular mechanics for model complexes containing the peroxido-bridged [Cu2O2]2+ center. *Inorg. Chem.* **2008**, *47*, 2494–2506. [CrossRef]
97. Holm, R.H.; Kennepohl, P.; Solomon, E.I. Structural and Functional Aspects of Metal Sites in Biology. *Chem. Rev.* **1996**, *96*, 2239–2314. [CrossRef] [PubMed]
98. Canali, L.; Sherrington, D.C. Utilisation of homogeneous and supported chiral metal(salen) complexes in asymmetric catalysis. *Chem. Soc. Rev.* **1999**, *28*, 85–93. [CrossRef]
99. Chen, H.-Y.; Tang, H.-Y.; Lin, C.-C. Ring-Opening Polymerization of Lactides Initiated by Zinc Alkoxides Derived from NNO-Tridentate Ligands. *Macromolecules* **2006**, *39*, 3745–3752. [CrossRef]
100. Morris, G.A.; Zhou, H.; Stern, C.L.; Nguyen, S.T. A general high-yield route to bis(salicylaldimine) zinc(II) complexes: Application to the synthesis of pyridine-modified salen-type zinc(II) complexes. *Inorg. Chem.* **2001**, *40*, 3222–3227. [CrossRef] [PubMed]
101. Hamada, Y.; Sano, T.; Fujita, M.; Fujii, T.; Nishio, Y.; Shibata, K. Blue Electroluminescence in Thin Films of Azomethin-Zinc Complexes. *Jpn. J. Appl. Phys.* **1993**, *32*, L511–L513. [CrossRef]
102. De La Durantaye, L.; McCormick, T.; Liu, X.Y.; Wang, S. Interaction of 2-(2′-pyridyl)benzimidazolyl derivative ligands with group 12 metal ions: Coordination, structures and luminescence. *Dalton Trans.* **2006**, *48*, 5675–5682. [CrossRef] [PubMed]
103. Maiti, M.; Thakurta, S.; Sadhukhan, D.; Pilet, G.; Rosair, G.M.; Nonat, A.; Charbonnière, L.J.; Mitra, S. Thermally stable luminescent zinc–Schiff base complexes: A thiocyanato bridged 1D coordination polymer and a supramolecular 1D polymer. *Polyhedron* **2013**, *65*, 6–15. [CrossRef]
104. Shit, S.; Nandy, M.; Saha, D.; Zhang, L.; Schmitt, W.; Rizzoli, C.; Row, T.N.G. Synthesis, crystal structure and fluorescence properties of two dinuclear zinc(II) complexes incorporating tridentate (NNO) Schiff bases. *J. Coord. Chem.* **2016**, *69*, 2403–2414. [CrossRef]
105. Konar, S.; Jana, A.; Das, K.; Ray, S.; Golen, J.A.; Rheingold, A.L.; Kar, S.K. A rare pentanuclear cadmium(II) complex and two new mononuclear zinc(II) complexes of pyrazole derived ditopic ligands—Synthesis, crystal structures and spectral studies. *Inorg. Chim. Acta* **2013**, *397*, 144–151. [CrossRef]
106. Caruso, U.; Panunzi, B.; Diana, R.; Concilio, S.; Sessa, L.; Shikler, R.; Nabha, S.; Tuzi, A.; Piotto, S. AIE/ACQ effects in two DR/NIR emitters: A structural and DFT comparative analysis. *Molecules* **2018**, *23*, 1947. [CrossRef] [PubMed]

107. Mondal, S.; Chakraborty, M.; Mondal, A.; Pakhira, B.; Mukhopadhyay, S.K.; Banik, A.; Sengupta, S.; Chattopadhyay, S.K. Crystal structure, spectroscopic, DNA binding studies and DFT calculations of a Zn(ii) complex. *New J. Chem.* **2019**, *43*, 5466–5474. [CrossRef]
108. Jayendran, M.; Begum, P.M.S.; Kurup, M.R.P. Structural, spectral and biological investigations on Cu(II) and Zn(II) complexes derived from NNO donor tridentate Schiff base: Crystal structure of a 1D Cu(II) coordination polymer. *J. Mol. Struct.* **2020**, *1206*. [CrossRef]
109. Li, M.; Xing, Y.; Zou, Y.; Chen, G.; You, J.; Yu, F. Imaging of the mutual regulation between zinc cation and nitrosyl via two-photon fluorescent probes in cells and in vivo. *Sens. Actuators B Chem.* **2020**, *309*. [CrossRef]
110. Phatangare, K.R.; Lanke, S.K.; Sekar, N. Fluorescent coumarin derivatives with viscosity sensitive emission–synthesis, photophysical properties and computational studies. *J. Fluoresc.* **2014**, *24*, 1263–1274. [CrossRef]
111. Li, D.; Zhao, W.; Sun, X.; Zhang, J.; Anpo, M.; Zhao, J. Photophysical properties of coumarin derivatives incorporated in MCM-41. *Dye. Pigm.* **2006**, *68*, 33–37. [CrossRef]
112. Ray, D.; Bharadwaj, P.K. A coumarin-derived fluorescence probe selective for magnesium. *Inorg. Chem.* **2008**, *47*, 2252–2254. [CrossRef] [PubMed]
113. Verma, S.K.; Savani, C.; Singh, V.K. Synthesis, Photophysical, Thermal and Crystallographic Studies of 3-Aminocoumarin Based Monobasic κ3-O,N,O-tridentate/κ2-N,O-bidentate Schiff Base Divalent Complexes. *ChemistrySelect* **2019**, *4*, 14244–14252. [CrossRef]
114. Aslkhademi, S.; Noshiranzadeh, N.; Sadjadi, M.S.; Mehrani, K.; Farhadyar, N. Synthesis, crystal structure and investigation of the catalytic and spectroscopic properties of a Zn(II) complex with coumarin-hydrazone ligand. *Polyhedron* **2019**, *160*, 115–122. [CrossRef]
115. Gusev, A.N.; Braga, E.V.; Kryukova, M.A.; Lyubomirskii, N.V.; Zamnius, E.A.; Shul'gin, V.F. Photoluminescence of the Coordination Zinc Compounds with 3-Methyl-4-Formyl-1-Phenylpyrazol-5-one Acylhydrazones. *Russ. J. Coord. Chem.* **2020**, *46*, 251–259. [CrossRef]
116. Argeri, M.; Borbone, F.; Caruso, U.; Causà, M.; Fusco, S.; Panunzi, B.; Roviello, A.; Shikler, R.; Tuzi, A. Color tuning and noteworthy photoluminescence quantum yields in crystalline mono-/dinuclear ZnII complexes. *Eur. J. Inorg. Chem.* **2014**, *2014*, 5916–5924. [CrossRef]
117. Costa, R.D.; Orti, E.; Bolink, H.J.; Monti, F.; Accorsi, G.; Armaroli, N. Luminescent ionic transition-metal complexes for light-emitting electrochemical cells. *Angew. Chem. Int. Ed. Engl.* **2012**, *51*, 8178–8211. [CrossRef]
118. Dianu, M.L.; Kriza, A.; Musuc, A.M. Synthesis, spectral characterization, and thermal behavior of mononuclear Cu(II), Co(II), Ni(II), Mn(II), and Zn(II) complexes with 5-bromosalycilaldehyde isonicotinoylhydrazone. *J. Therm. Anal. Calorim.* **2012**, *112*, 585–593. [CrossRef]
119. Borbone, F.; Caruso, U.; Palma, S.D.; Fusco, S.; Nabha, S.; Panunzi, B.; Shikler, R. High solid state photoluminescence quantum yields and effective color tuning in polyvinylpyridine based zinc(II) metallopolymers. *Macromol. Chem. Phys.* **2015**, *216*, 1516–1522. [CrossRef]
120. Diana, R.; Panunzi, B.; De Simone, B.; Borbone, F.; Tuzi, A.; Caruso, U. RGB emission of three charged O,N,O-chelate zinc (II) complexes in pyridine solution. *Inorg. Chem. Commun.* **2020**, *113*. [CrossRef]
121. Sun, X.; Wang, Y.; Lei, Y. Fluorescence based explosive detection: From mechanisms to sensory materials. *Chem. Soc. Rev.* **2015**, *44*, 8019–8061. [CrossRef]
122. Taraba, L.; Krizek, T.; Kozlik, P.; Hodek, O.; Coufal, P. Protonation of polyaniline-coated silica stationary phase affects the retention behavior of neutral hydrophobic solutes in reversed-phase capillary liquid chromatography. *J. Sep. Sci.* **2018**, *41*, 2886–2894. [CrossRef] [PubMed]
123. Hou, S.-S.; Fan, N.-S.; Tseng, Y.-C.; Jan, J.-S. Self-Assembly and Hydrogelation of Coil–Sheet Poly(l-lysine) -block-poly(l-threonine) Block Copolypeptides. *Macromolecules* **2018**, *51*, 8054–8063. [CrossRef]
124. Miecznikowski, J.R.; Lo, W.; Lynn, M.A.; Jain, S.; Keilich, L.C.; Kloczko, N.F.; O'Loughlin, B.E.; DiMarzio, A.P.; Foley, K.M.; Lisi, G.P.; et al. Syntheses, characterization, density functional theory calculations, and activity of tridentate SNS zinc pincer complexes based on bis-imidazole or bis-triazole precursors. *Inorg. Chim. Acta* **2012**, *387*, 25–36. [CrossRef]
125. Li, M.X.; Zhang, L.Z.; Chen, C.L.; Niu, J.Y.; Ji, B.S. Synthesis, crystal structures, and biological evaluation of Cu(II) and Zn(II) complexes of 2-benzoylpyridine Schiff bases derived from S-methyl- and S-phenyldithiocarbazates. *J. Inorg. Biochem.* **2012**, *106*, 117–125. [CrossRef]
126. Bharti, A.; Bharati, P.; Singh, N.K.; Bharty, M.K. NNS tridentate thiosemicarbazide and 1,3,4-thiadiazole-2-amine complexes of some transition metal ions: Syntheses, structure and fluorescence properties. *J. Coord. Chem.* **2016**, *69*, 1258–1271. [CrossRef]

127. Kowol, C.R.; Trondl, R.; Arion, V.B.; Jakupec, M.A.; Lichtscheidl, I.; Keppler, B.K. Fluorescence properties and cellular distribution of the investigational anticancer drug triapine (3-aminopyridine-2-carboxaldehyde thiosemicarbazone) and its zinc(II) complex. *Dalton Trans.* **2010**, *39*, 704–706. [CrossRef]
128. Vlasenko, V.G.; Garnovskii, D.A.; Aleksandrov, G.G.; Makarova, N.I.; Levchenkov, S.I.; Trigub, A.L.; Zubavichus, Y.V.; Uraev, A.I.; Koshchienko, Y.V.; Burlov, A.S. Electrochemical synthesis, structural, spectral studies and DFT calculations of heteroleptic metal-chelates bearing N, N, S tridentate tosylamino functionalized pyrazole containing Schiff base and 1,10-phenathroline. *Polyhedron* **2019**, *157*, 6–17. [CrossRef]
129. Bharati, P.; Bharti, A.; Chaudhari, U.K.; Bharty, M.K.; Kashyap, S.; Singh, U.P.; Singh, N.K. Trinuclear supramolecular Zn(II) complexes derived from N′-(pyridine carbonyl) hydrazine carboperthioates: Synthesis, structural characterization, luminescent properties and metalloaromaticity. *Inorg. Chim. Acta* **2015**, *425*, 100–107. [CrossRef]
130. Tyagi, P.; Tyagi, M.; Agrawal, S.; Chandra, S.; Ojha, H.; Pathak, M. Synthesis, characterization of 1,2,4-triazole Schiff base derived 3d-metal complexes: Induces cytotoxicity in HepG2, MCF-7 cell line, BSA binding fluorescence and DFT study. *Spectrochim. Acta A Mol. Biomol. Spectrosc.* **2017**, *171*, 246–257. [CrossRef]

Publisher's Note: MDPI stays neutral with regard to jurisdictional claims in published maps and institutional affiliations.

Review

Zinc (II) and AIEgens: The "Clip Approach" for a Novel Fluorophore Family. A Review

Rosita Diana and Barbara Panunzi *

Department of Agricultural Sciences, University of Naples Federico II, 80055 Portici, Italy; rosita.diana@unina.it
* Correspondence: barbara.panunzi@unina.it; Tel.: +39-081-674-170

Abstract: Aggregation-induced emission (AIE) compounds display a photophysical phenomenon in which the aggregate state exhibits stronger emission than the isolated units. The common term of "AIEgens" was coined to describe compounds undergoing the AIE effect. Due to the recent interest in AIEgens, the search for novel hybrid organic–inorganic compounds with unique luminescence properties in the aggregate phase is a relevant goal. In this perspective, the abundant, inexpensive, and nontoxic d^{10} zinc cation offers unique opportunities for building AIE active fluorophores, sensing probes, and bioimaging tools. Considering the novelty of the topic, relevant examples collected in the last 5 years (2016–2021) through scientific production can be considered fully representative of the state-of-the-art. Starting from the simple phenomenological approach and considering different typological and chemical units and structures, we focused on zinc-based AIEgens offering synthetic novelty, research completeness, and relevant applications. A special section was devoted to Zn(II)-based AIEgens for living cell imaging as the novel technological frontier in biology and medicine.

Keywords: AIE; zinc complex; fluorescence

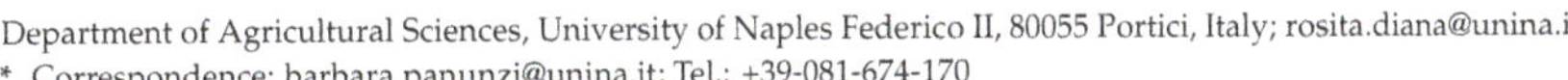

Citation: Diana, R.; Panunzi, B. Zinc (II) and AIEgens: The "Clip Approach" for a Novel Fluorophore Family. A Review. *Molecules* **2021**, *26*, 4176. https://doi.org/10.3390/molecules26144176

Academic Editors: Jorge Bañuelos Prieto and Ugo Caruso

Received: 16 June 2021
Accepted: 6 July 2021
Published: 9 July 2021

1. Introduction

1.1. Activation of the Fluorescence Channel in AIEgens

A luminescent material is a material able to emit light in the process of returning from the electronic or vibrational excited state to the ground state after being excited by external energy. The term "luminescence" was introduced in 1888 by the physicist and historian Eilhard Wiedemann to describe light emission not simply related to an increase in temperature. Photo-induced luminescence or photoluminescence (PL), often referred simply as "luminescence", is the light emission in the optical range of visible, ultraviolet, or infrared light. According to the mode of excitation and relaxation, luminescence can be classified into various types, including fluorescence and phosphorescence.

Specifically, "fluorescence" is a word coined from the mineral form of CaF_2 known as fluorite. In 1819, E. D. Clarke reported that some crystals of green fluorite deeply emitted a blue colour under the illumination of a UV lamp. Contrary to phosphorescence, fluorescence is a form of photoluminescence in which the excitation–emission process occurs very quickly [1,2]. After a molecule absorbs energy from a light source and becomes excited, fluorescence occurs within nanoseconds. Upon excitation with electromagnetic energy at the correct wavelength, an electron in the fluorescent molecule is promoted to an upper level, and finally, the energy is released in the form of a photon (fluorescence emission) while the electron moves back down to the lower energy level. In most cases, fluorescence requires a longer excitation wavelength—and so lower energy—than the absorbed radiation. Therefore, fluorescence can occur when the absorbed radiation is in the ultraviolet (UV)-visible region of the spectrum, while the emitted light is in the visible region. Fluorescent dyes are compounds that strongly emit in the visible region so that they show a naked-eye-perceivable colour when exposed to a UV-visible light source [1,3–6].

Over the last century, the rapid development of molecular science moved to study the working mechanisms of fluorescence at the molecular level. Since 1800, fluorescent organic dyes highly emissive in diluted solutions were examined. In the middle of the nineteenth century, the chemist Adolf von Baeyer developed fluorescein as a synthetic organic material highly fluorescent in aqueous solutions in daylight. Early fluorescent materials found only rare applications, such as tracers for the detection of underground waterways or optical whitening agents in paper, textiles, and marking inks [7,8]. Even less attention was paid to the photoluminescence of aggregate molecules until the last century. The first report about the concentration effect on the fluorescence of aromatic compounds in solutions is due to J. B. Perrin, who in 1923 found that the photoluminescence of uranine (disodium salt form of fluorescein) decreased as the concentration increased. This fluorescence quenching effect at high concentrations was clarified in 1955 by Th. Förster, who proposed the concentration-quenching effect upon aggregation due to strong π–π stacking interactions in the organic dyes. Such effect is now commonly known as the ACQ (aggregation-caused quenching) effect [9]. In 1970, the ACQ effect was told as "common to most aromatic hydrocarbons and their derivatives" by J. B. Birks, in his book entitled *Photophysics of Aromatic Molecules* [10].

With the rise and development of new technologies, the twentieth century opened the doors to a new perspective about luminescent materials. In most modern applications, such as optoelectronic devices (as laser, optical storage, and OLEDs) and biomedical tools, fluorescence materials are used as solids and/or aggregates [11–19]. The need for materials suitable for such applications led to a growing interest in the design and synthesis of aggregate emitters.

The first pivotal report containing the concept of aggregation-induced emission materials is due to Tang and coworkers [20]. Aggregation-induced emission (AIE) compounds display a photophysical phenomenon in which molecular aggregates exhibit stronger emission than single molecules. In the following articles, the research group presented a silole derivative (hexaphenylsilole) non-emissive in solutions up to 80% water fraction. Above this threshold, strong fluorescence was observed due to the strong aggregation of the molecules. Contrary to the traditional ACQ effect, this effect was coined as "AIE", and the molecules undergoing the AIE effect were termed "AIEgens" [21,22].

The unique AIE behaviour changed researchers' way of thinking. The AIE operating mechanism is now under examination from a new applicative perspective due to the growing demand for advanced luminescent materials [23–25]. Milestones for AIE theoretical development are several scientific publications from 2001 up to now. First, the origins of the phenomenon were investigated, and several mechanisms proposed for an explanation of the AIE effect. In 2003, again Tang and coworkers proposed the restriction of intramolecular rotations (RIR) mechanism based on the study of hexaphenylsilole [26] and proved it as applicable to most AIEgens. When the single AIE molecule is dissolved in solution, the dynamic intramolecular rotations cause a fluorescence quenching by dissipating the exciton energy. Contrarily, upon aggregation, the restriction of the intramolecular motions suppresses the radiationless decay pathway. Despite this, the RIR process cannot explain some experimental evidence, and other effects were added in the mechanistic study of the phenomenon. The intramolecular vibrations of flexible moieties have been considered [27], generating an additional model named the "intramolecular vibrations (RIV) effect". A generalised mechanism for AIEgens named "RIM" (restriction of intramolecular motions) was introduced in 2014 by combining RIR and RIV effects, so including rotation, vibration, bending, flapping, twisting, and other intramolecular motions concurring to the radiationless decay pathways [28–32]. A theoretical confirmation of the RIM mechanism was achieved by time-dependent density-functional theory [25].

Additional mechanistic models and theories were proposed to investigate the photo-induced process occurring in different AIE systems. Specifically, to predict AIEgens behaviour, other energetic parameters are involved and must be considered. The Duschinsky rotation energy [33–38] (due to the difference between the ground state and excited-state potential energy surfaces, calculated by the harmonic oscillator model); the reorganisation energy [39] (required to relax the structure and environment upon electron transfer); the formation of J-aggregates (causing a bathochromic shift in the absorption due to π–π stacking of the aromatic moieties) [22]. Finally, the restricted access to a conical intersection (RACI) model to analyse the global potential energy surface topology was introduced in 2013 by Blancafort and coworkers [38]. Conical intersections were described as regions of the potential energy surface where the ground and excited states are degenerate, and the probability of non-radiative internal conversion is maximal. RACI model generally leads to a forecast on the fluorescence yield (PLQY, Φ). This fundamental parameter is defined as the ratio of the number of photons emitted to the number of photons absorbed and derives from evaluating both the radiative and the non-radiative channels activated in the AIE molecule [40–44].

To conclude, a correct theoretical setting of the chemical system is crucial for the actual prediction of the photoluminescence yield of the AIE fluorescent dye. The observation of the effects involved in the activation of emission in the aggregate state gave scientists new ideas about AIEgens design.

1.2. *AIEgens as the Novel Scientific Frontier for Cutting-Edge Technologies*

The great potential of AIEgens is documented by the number of publications and citations about the topic, exponentially increasing since the first AIE report in 2001. To date, more than 2000 papers (articles and many reviews) have been published under the keyword "AIEgen", and AIE topic was ranked no. 2 and no. 3 in the list of "Top 100 Research Fronts in Chemistry and Materials Science" by Thomson Reuters, respectively, in 2015 and 2013. In 2016, Nature Journal placed AIE-based dots between the four key materials for the novel lightening nanotechnology [45]. In the last 20 years, AIEgens have been deeply explored and their unique advantages over conventional ACQ molecules fully recognised. The implementation in AIEgens study is moved by the cutting-edge technologies demand, ranging from optoelectronic to sensing and bioimaging [46].

Many optoelectronic techniques greedily require emissive solid layers. The earliest application of AIEgen in optoelectronic devices was reported in 2001 as blue-emissive AIE siloles with excellent external quantum efficiency [47]. Several AIEgens have been designed with the precise aim of obtaining solid-state emissions. The nature of AIEgens matches the structural tunability and manipulability required to produce an optical device. In addition, the whole emission colour display, up to the NIR region, can be easily obtained by structural modifications in the AIEgen skeleton.

Due to their versatility, AIEgens can be easily obtained in a polymeric- or macro-structured form by reaction of functionalised fluorophores or by doping the fluorophores in a polymeric matrix. Macromolecular materials are ideal candidates for optoelectronic applications based on solid-state layers, such as organic light-emitting diodes (OLEDs, optical waveguides, and luminescent solar concentrators, or even soft-matter-based devices such as light-emitting electrochemical cells (LECs) and liquid-crystal displays [48–52].

Other significant applications of AIEgens are fluorescence chemosensors and multiple-stimuli (such as pH, electromagnetic radiation, morphology) responsive materials [53–56]. AIEgens can be utilised as fluorescent indicators/markers to characterise supramolecular interactions and macromolecular motions in the solid state, in an aggregate gel phase or even in a water-concentrated solution. In the presence of the target analyte, the AIE channel can be activated or deactivated, resulting in an optical recordable signal. In 2005, Tang and coworkers produced chemosensors based on silole as sensors for the regioisomers of nitroanilines [57] and in 2010, Park and coworkers promoted AIEgens as stimuli–responsive materials. To date, several AIE-based chemosensors were reported for the detection of ions, metals, small organic molecules, and biological targets [58–66].

Porous crystalline solids such as inorganic nanoparticles (NPs), metal–organic frameworks (MOFs), covalent organic frameworks (COFs), and carbon nanotubes (CNTs) are materials where the choice of suitable AIE active fragments can produce highly engineered composite materials with unique properties [59–61]. In the composite material, the AIE moieties interact with a regular structure resulting in tunable electronic and optical properties. Low self-quenching and a highly modulable inner chemical environment can be achieved. Thanks to the porous architecture, such nanostructured AIEgens are promising candidates for fluorescence sensing with high sensitivity and selectivity toward specific analytes [61–66].

Finally, photoluminescent devices for biomedical uses take advantage of the fluorescence characteristics of AIEgens [63]. The incorporation of AIEgens in bio-compatible materials through various methods (covalent or coordinate bonds, noncovalent interactions) provides AIEgen-based composite biomaterials for several uses. Bio-engineering of nano-sized probes based on AIEgens produced highly specific and responsive systems. In 2012, Tang and Liu and coworkers developed AIEgen bioprobes for cell-imaging and monitoring of biological processes [64]. Biocompatible AIEgen-based NPs were designed still by Tang for biological and biomedical applications [65]. Due to their permeability and retention, biocompatible AIE-based dots and NPs were also applied for in vivo cancer imaging and diagnostics [65,66]. Again, Tang and coworkers, in 2018, reported a simple approach to link AIEgens to bio-relevant species for biological labelling and monitoring [67].

Lessons learned from these advanced tools provide new ways of thinking about the most relevant in vivo applications.

1.3. *Metal-Containing AIEgens and The Role of Zinc (II) Ion*

In the following years, many purely organic, aromatic and heteroaromatic RIM undergoing AIEgens were described. Specifically, AIEgens bearing cores such as tetraphenylethene, thiophene-triphenylamine, tetraphenylpyrazine, quinoline, 9,10-distrylanthracene, dithiole, and derivatives have been extensively reviewed in recent articles [21,25,63,68–74].

Despite the obvious advantages of organic fluorophores, which are singlet emitters, heavy atoms such as transition metals display triplet emission. The related spin–orbit coupling leads to efficient singlet–triplet state mixing so that the presence of a heavy atom can improve the photophysical properties of π-conjugated ligands [75–78]. The formation of metal complexes represents a strategy for obtaining new luminescent materials with long luminescence lifetimes, large Stoke's shifts, and high PLQYs in the visible region. Metal complexes are potentially unique luminophores with highly tunable structures and photophysics. Specifically, the d-block metal complexes are rich in different charge-transfer electronic states (MC, Metal Centered; ICT, Intramolecular Charge Transfer; ILCT, Intraligand Charge Transfer; LC, Ligand Centred; MLCT, Metal to Ligand Charge Transfer; LL'CT Ligand to Ligand Charge Transfer) and local π–π^* transitions on the ligands [78–84].

The origin of the AIE effect in transition metal complexes can be explained based on the competing radiative and non-radiative de-excitation paths. The presence of the coordinated metal leads to a selective activation or blocking of these channels. Besides, blocking the non-radiative deactivation pathway, the presence of the coordinated metal can also produce a more efficient emitting state of the organic part in its aggregate state. In fact, the electronic charge in the complex can be transferred between different molecular moieties, involving electron transfer between metal cation and ligands. This transfer causes an alteration in the energy levels and in the emitting state. The AIE effect is potentially ascribable to simultaneous changes of the emitting state and activation of RIM and/or RACI mechanisms. Obviously, the activated AIE channel can be predicted only with an in-depth theoretical analysis [85,86].

Several transition metals such as Ir(III), Pt(II), Re(I), Ru(II), Os(IV), Au(I), and Pd(II) are reported to cause the activation of the AIE channel [78]. The development of AIE-active complexes seems to be a promising strategy for many AIE systems where the metal is strictly involved in the emission process. In all cases, ML and/or LM transfer must be expected in the emissive complexes.

Among the wide variety of metal-based luminophores, experience as researchers moves us to recognise a unique role of zinc (II) cation [87]. In the growing demand for high-performance devices, sustainability is a relevant parameter. Zinc, as a small eco-friendly cation, is a good alternative to other hazardous or expansive metals. Zinc complexes claim unique peculiarities which endorse the use in the production of AIE undergoing materials.

Specifically, zinc (II) complexes exhibit fluorescence intensity and/or colour tuning in dependence on the electronic pattern of the ligands and the coordination pattern imposed by the ligands. In fact, in the d^{10} closed-shell zinc (II) ion, d–d electronic transitions are not expected. The lowest energy excited states are mainly LCT transitions (such as ICT and ILCT bands) and/or LL'CT transitions, rarely LMCT due to s or p empty orbitals of the metal [7].

That is what we mean as a "clip approach" (Scheme 1). In most cases, the cation acts as a constraint for the ligand by locking it into a favourable emissive conformation, with few electronic effects mostly due to π–π^* LCT transitions. Chelation enhanced fluorescence (CHEF mechanism) is often involved in zinc (II) binding as the result of the stabilisation of the excited state in poorly emissive ligands [14]. Therefore, the role of zinc (II) cation can be to turn a non-emissive ligand in an AIEgen molecule upon coordination. Alternatively, zinc (II) can tune and/or increase the emissive properties of the AIE ligand through suitable assembly of AIE portions. In addition to the "optically innocent clip role", zinc cation offers the ability to give rise to the most varied architectures through coordination of organic ligands, self-assembly of two and three-dimensional macrostructures, aggregation of nanomaterials, and production of regular or amorphous structures [88–94].

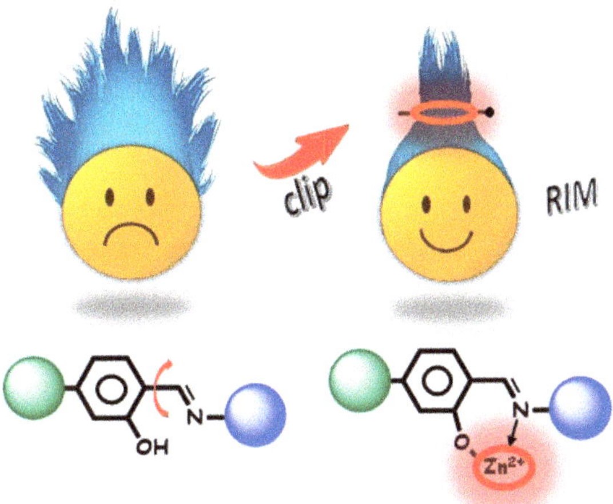

Scheme 1. The "clip approach" in Zn AIEgens.

Despite several studies encompassing the topic of AIEgens, in this review, we will focus our attention on the role of Zn(II) in the design of AIE active complexes and materials. Our main aim is to give a unified overview of AIEgens involving zinc (II) cation (abbreviated as Zn AIEgens). With the term "Zn AIEgens", we mean: the zinc (II) complexes exhibiting AIE properties; the AIEgens giving a fluorescence response in the presence of zinc (II) cation; the zinc (II) complexes exploiting an AIE mechanism to detect specific analytes. In short, with this review, we wanted to answer the question: what zinc (II) is dealing with AIEgens.

Considering the novelty of the topic, we cut the last 5 years (years 2016–2021) as representative of the recent scientific production. Selected examples were reported to elucidate the state-of-the-art, starting from the simple phenomenological approach and considering different typological and chemical units and structures. We will focus on the recent Zn AIEgens based on the synthetic novelty, research completeness, fluorescence response, theoretical deepening, and relevant applications. Specifically, we reviewed the literature and organised the discussion in the following sections:

1. Newly developed Zn AIEgens;
2. Zn AIEgens for OLEDs and other optical technologies;

3. The role of zinc (II) in AIEgen chemosensing;
4. Zn AIEgens for cell imaging.

Scientific literature, including reviews, were consulted to understand the structure, the optical response, and the operating mode of Zn AIEgens. Finally, we underlined the representative applications for the targeted biological and medical scope. By use of a quick glance representation which should be considered an integral part of the discussion, we propose an intuitive overview of groups of structures examined according to the application area.

2. Newly Developed Zn AIEgens

The examples reported in this section can be considered the first level in the study and development of novel targeted Zn AIEgens. The starting point is represented by the design and synthesis of simple small organic molecules, even non-emissive, acting toward Zn(II) as ligands or chelators. In the formation of the complex, the organic units are subjected to constraints (RIM), and therefore, the final complex undergoes the AIE effect. The zinc complexes are expected to meet specific requirements, such as high PLQYs in the aggregate phase, photostability, large lifetime, and Stokes shifts. The study of the solvent effects is relevant. An easy test to assess the AIE nature of the compound consists of recording emission intensity as the ratio of non-solvent/solvent increases [95]. Emission is expected to grow as the aggregation increases. Many solid-state emitters are *de facto* AIEgens. A complete analysis of the solid-state emission properties achieved by the AIE mechanism and the relationship with the structural data is a central point to these articles. Relevant information is inferred by a supporting theoretical analysis of the energetic transitions involved in the AIE response [28,69,96–99].

The synthetic Zn AIEgens reported in this section are referred to as "potentially useful" materials for optoelectronic applications and/or optical techniques. On the other hand, articles including targeted applications will be reported in Section 3.

In Section 2.1, we will report the most relevant examples of novel Zn AIEgens obtained by encumbered nitrogen and/or oxygen donor-based low-molecular-weight ligands (RIM effect undergoing complexes). Our discussion will be articulated based on structural similarities and differences and on the AIE activated channel. The related structures are summarised in Figure 1, accordingly with discussion. In Section 2.2, we referred to a few relevant examples of metal–organic frameworks acting as highly fluorescent Zn AIEgens. The related structures are reported in Figure 2.

2.1. RIM Designed Zn AIEgens

The most common and easy approach to obtain Zn AIEgens is to enhance steric hindrance by adding bulky substituents to the chelating ligands. The ligand undergoes a conformational block upon coordination, producing the RIM fragment. A selection of relevant examples of synthetic Zn AIEgens undergoing the RIM effect is grouped based on chemical/structural features. As it will be discussed, some peculiar skeletons able to cause the RIM effect are recurrent.

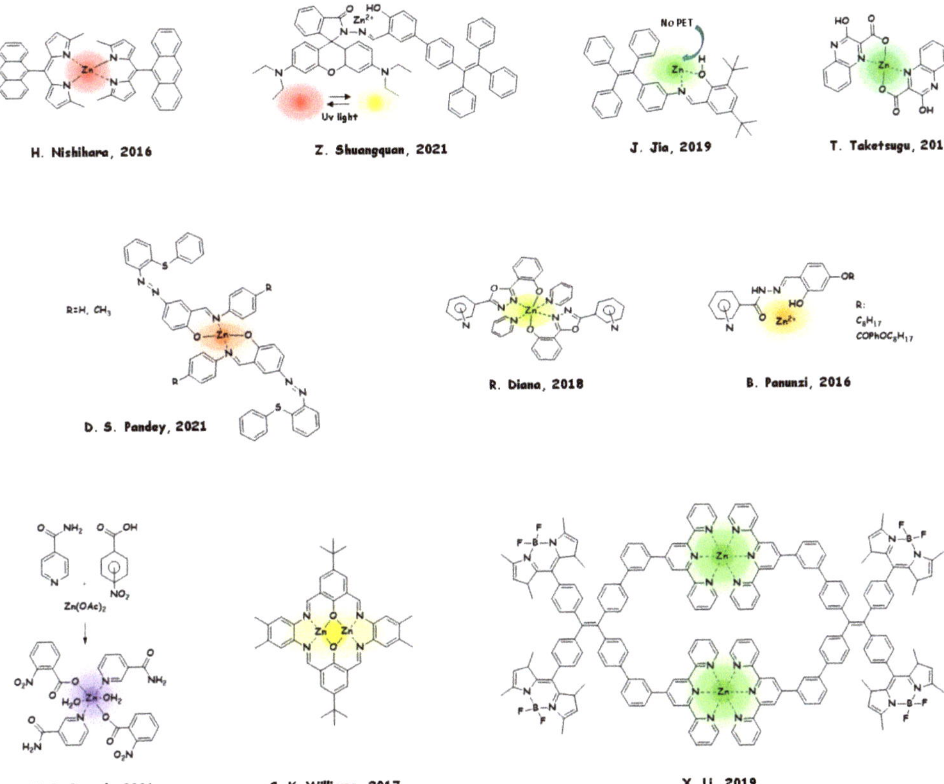

Figure 1. Zn AIEgens obtained by encumbered nitrogen and/or oxygen-donor-based ligands.

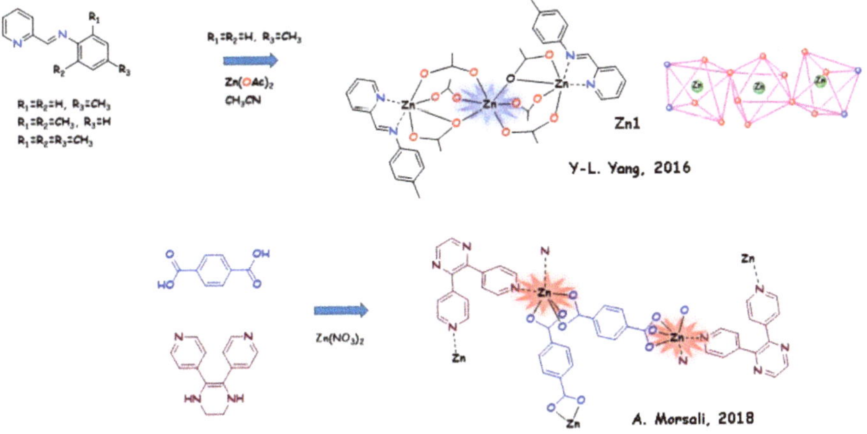

Figure 2. MOFs acting as highly fluorescent Zn AIEgens.

In 2016, Hiroshi Nishihara [100] and coworkers studied the solid-state PL emission of seven encumbered tetradentate bis(dipyrrinato)zinc (II) complexes. The bulky effect [101–103] due to the meso-aryl group in the dipyrrin ligands proved to affect the emissive properties of the dipyrrin Zn(II) complexes. The conformation of the meso-aryl group in the excited state greatly affects the nonradiative decay pathways. In solution, PLQYs increased as the solution turned into aggregate state PLQYs by adding a non-solvent. In the solid-state, each complex showed different emission wavelengths and spectra depending on the crystal packing structure. With respect to the solution behaviour, the different PLQYs and lifetimes recorded on the crystalline complexes proved the activation of different photophysics in the solid state. A *O,N,O* hydrazone tridentate ligand was synthesised by Zang Shuangquan and coworkers [104] in 2020, starting from Rhodamine B 2-Hydroxy-5-tetraphenylethene-benzaldehyde hydrazone. A typical AIE moiety, tetraphenylethene (TPE) group, was introduced as a bulky substituent, able to undergo fluorescence resonance energy transfer (FRET) [105] process from TPE moiety to rhodamine B moiety. As a result, the complex obtained by reacting the composed ligand with zinc (II) cation is a Zn AIEgen and exhibits reversible fluorescence colour and intensity changes (red/yellow) upon UV light photo-induced irradiation.

Common *salen* type ligands with encumbered substituents are excellent candidates in building Zn AIEgens, zinc cation acting as a constraint of the structure. In some cases, intriguing fluorescence behaviour can be achieved by specific substituents onto the simple Schiff-base *O,N* moiety. In 2019 Junhui Jia and coworkers [106] synthesised a tetraphenylethylene-functioned Schiff-base ligand (TPEMO) with multifunctional behaviour. The molecule itself is an AIEgen and shows reversible mechano-fluorochromism after grinding. It was found responsive towards Zn^{2+} (LOD = 8.05×10^{-8} M) and CO_3^{2-} ions. The sensing mechanism of the molecule with Zn^{2+} was ascribed to the inhibition of PET transfer and consequently of the ESIPT mechanism. Very recently, Daya Shankar Pandey and coworkers [107] studied two novel azo-based Schiff-base functionalised ligands. Again, the ligands themselves are AIEgens, potentially useful as solid-state fluorescent sensors for acidic vapours. Upon irradiation with UV light, the N=N azo bridge exhibits *cis–trans* isomerisation both in the ligands and in their zinc (II) complexes. Comparative photoswitching studies and DFT analysis was performed on ligands and complexes based on X-ray analysis. In 2019, Tetsuya Taketsugu and coworkers [108] performed a complete theoretical study on a Zn AIEgen derived from 3-hydroxy-2-quinoxalinecarboxylate showing ESIPT emission in the solid state, even though the ligand does not undergo ESIPT emission. TD-DFT analysis was employed to explore the role of the zinc atom in the emission mechanism, resulting in a structural change of the ligand from the lactam form (3,4-dihydro-3-oxo-2-quinoxalinecarboxylic acid) to the enol form (3-hydroxy-2-quinoxalinecarboxylic acid) in the presence of zinc (II). Emission was imputed to LC transitions of the ligand.

A systematic study of isomeric Schiff-base zinc complexes showing relevant differences in their crystalline pattern and emission behaviour was published in 2018 by Rosita Diana and coworkers [109]. Three bidentate *N,O* Schiff-base ligands contained a 2,5-disubstituted-oxadiazole unit and an *ortho/meta/para*-substituted pyridine unit. The reactions of the ligands with zinc (II) acetate in pyridine gave crystalline complexes with the general formula $Zn(L)_2py_2$. In the solid phase, PLQYs (ranging from 16 to 75%) and emission colour (in the red–green–blue or RGB shape) were recorded in dependence on the structural pattern. The reaction with zinc cation self-assemblies the *meta* derivative ligand in a stable glassy network, with $Zn(L)_2$ general stoichiometry and PLQY = 44%. The same research group (2016, Barbara Panunzi and coworkers [110]) produced aroyl- and acylhydrazones ligands acting as differently substituted *O,N,O* tridentate pincers to zinc. The photophysical properties of the tridentate complexes in the crystalline phase were explored. Additionally, in this case, the one-dimensional crystalline coordination polymer obtained from the *meta* derivatives showed high PLQY (74%) in the solid state.

The results achieved in both articles suggested that the polymeric or reticulated macrostructures obtained by zinc-driven self-assembly provide emission enhancement. Both in the case of an amorphous and crystalline material, the tight structure with optically "innocent" zinc nodes could guarantee a more efficient electron hopping in the whole macrostructure.

The role of the metal in assembly together non-fluorescent components was examined in 2020 by Jubaraj B. Baruah and coworkers [111]. By zinc-assembly reaction of nitrobenzoate and pyridine-3-(or-4)carboxamide, five different complexes were obtained. The role of MLCT in the resulting Zn AIEgens was explored by DFT calculations. The frontier molecular energy levels of different combinations of the positional isomeric complexes were ascertained. In such cases, the d^{10} cation resulted in contributing to the emission through CT transitions. The highest PLQY (13.56%) was recorded for the non-ionic complex (di-aqua)bis(pyridine-3-carboxamide)di(2-nitrobenzoato)zinc (II) monohydrate at 439 nm. The AIE effect caused by adding water to a DMSO solution was studied. As expected, the participation of water molecules to form aggregates with the complexes caused increased emission intensity up to a characteristic limit. Above that concentration, emission quenching was recorded due to the equilibrium between the complex and the hydrate cation.

Other structures causing the AIE effect by steric constrain are metallo–supramolecular architectures, achieved by zinc-driven self-assembling reaction. Specifically, emissive or non-emissive polyfunctional ligands can be assembly in macrocycles or metallocycles by reaction with zinc cation. The final metal complex undergoes restrictions of intramolecular motion increased with respect to the starting organic moieties and are expected to be a Zn AIEgen. Charlotte K. Williams and coworkers in 2017 [112] investigate the solid-state light emission of seven zinc *salphen* (*salphen* = *N*,*N*′-bis(salicylidene)-1,2-phenylenediamine) complexes in a macrocycle pattern. By gradual addition of a non-solvent to the complexes dissolved in chloroform, a visible aggregation was induced, and emission enhancement was recorded. The complexes showed yellow to orange–red emission in the solid state (PLQE = 1–5%), displaying up to a 75-fold increase in peak emission intensity upon aggregation, in the best case.

Terpyridine (TPY) is a tridentate ligand with three coordination sites belonging to *N*-heteroaromatic rings. Owing to the strong chelating ability, TPY derivatives can form stable complexes with several different transition metal ions. Xiaopeng Li and coworkers in 2019 [113] studied self-assembly of emissive metallocycles with tetraphenylethylene (TPE), boron-dipyrromethene (BODIPY), and terpyridine (TPY). Combining the three blocks into one system by zinc coordination driven self-assembly reaction, a metallocycle dimer with typical AIE characteristics was produced. Interestingly, by combining the three fluorophores into a metallo–supramolecular architecture, emissive properties were recorded both in solution and in the aggregated state.

2.2. *AIE Zn-MOFs*

Metal-organic frameworks (MOFs) are hybrid organic–inorganic crystalline materials derived by the regular array of metal cations (or clusters) surrounded by organic linkers. The metal ions act as nodes binding the organic fragments into a regular cage-like structure. They are a subclass of self-assembly coordination polymers (CPs) with a tridimensional feature. Due to this hollow structure, MOFs have a porous internal surface area. The small-sized internal cavity of the materials is nanoscaled (in at least one dimension, between 1 and 100 nm). Therefore, MOF research adopts a scientific approach to nanotechnology. Professor Omar Yaghi at UC Berkeley in the late 1990s ("Design and synthesis of an exceptionally stable and highly porous metal-organic framework") presented MOFs for the first time, and they rapidly become a cutting-edge research field. The synergistic effects of structures and compositions make MOFs fascinating examples of unique structural design and tunable properties. Different metal atoms and organic linkers lead to MOFs selectively absorbing targeted molecules / ions. Tailored MOFs offer great potential due to their optical, electronic, opto-electronic, or sensing properties [114–116].

In recent decades, MOFs with luminescence characteristics (luminescent MOFs, LMOFs) has been considered as the new pathway for the fabrication of solid-state luminescent nanomaterials. AIEgen ligands can be employed as building blocks in the MOFs construction. To retain the fluorescence of AIE linkers or activate the AIE channel, the closed-shell zinc

(II) ion with low-lying d-orbital energy can be preferred to classic heavy metals. Zn(II) nodes afford MOFs with strong luminescence due to constraining effect on the AIE ligands.

In this section, we reported a few relevant examples of LMOFs containing zinc (II) and acting as highly active AIEgens (we will use the term "AIE Zn-MOFs"). In Figure 2, we reported a schematic representation focusing on the structural feature.

The Schiff-base pattern also demonstrated to be a winner strategy in building LMOFs. In 2016, Yu-Lin Yang and coworkers [117] used five Schiff-base ligands containing a *N*-(pyridine-2-yl) fragment with different alkyl substitutions on the phenyl ring for the synthesis of Zn(II)/Cd(II) complexes. The structures of the ligands can direct the formation of 3D supramolecular LMOFs due to hydrogen bonds and π–π interactions. Nine Zn(II)/Cd(II) complexes were obtained, displaying deep blue emissions (401−436 nm) in acetonitrile solution and light blue/bluish green emissions (485−575 nm) in the solid state. Structural analysis gave information on the crystalline packing of the complexes. PL measurements recorded in CH_3CN/H_2O mixtures gave evidence of an effective AIE behaviour for the Zn1 (in Figure 2) aggregate complex.

Ali Morsali and coworkers [118] in 2018 synthesised a turn-on AIE-based MOF by in situ ligand fabrication and employing different valence-shell cations, with the aim of investigating the effect of metal nodes. The coordination of 5,6-di(pyridin-4-yl)-1,2,3,4-tetrahydropyrazine (AIE fluorophore) and terephthalic acid ligand with Zn(II), Co(II), or Cd(II) leads to the preparation of strongly luminescent MOFs (named "TMU-40(Zn)", "TMU-40(Cd)", and "TMU-40(Co)"). The formation of rigid frameworks improved the fluorescence of the organic parts more with zinc than with the other two metals (PLQY = 38.2% with respect to 31.17% and 11.69%, respectively).

3. Zn AIEgens for OLEDs and Other Optical Technologies

The basic research on RIM undergoing Zn AIEgens was a platform to produce effective technological devices. Solid-state emitters have a great potential for the development of novel optoelectronic technologies, such as OLEDs and other optical tools. The growing demand for low-cost light-emitting devices focused attention on the abundant, nontoxic, versatile zinc cation. Zinc (II) complexes are a potential alternative to expensive transition metals complexes of iridium, osmium, and platinum, and can represent excellent luminescent and electron-accepting/transporting materials. They can be employed as emitter layers in a variety of forms: solid powders, dyes dispersed in host solid and/or gel matrixes, and metallo–polymers. Since the first OLED based on a Schiff-base zinc (II) complex developed by Hamada in 1993 [119], the research on highly efficient emissive zinc (II) complexes continued. Zinc complexes also have a potential for the construction of white OLEDs (WOLEDs) due to the tunable emission colour. To date, reports on effective zinc (II) complex-based OLED are rare, and there is no commercialisation. Nonetheless, the study of Zn AIE systems makes room for new ideas, and Zn AIEgens claim a role in the novel technological frontier.

In this section, we will discuss the most relevant examples of opto-applications of Zn AIEgens, from OLEDs and LECs to other cutting-edge technologies. In Figure 3, we collected the related structures, referring to the most PL active architectures.

Wai-Yeung Wong and coworkers in 2017 [120] presented four novel *salphen* complexes bearing pyridyl functionalised ligands with dendritic structures. OLED devices based on the novel zinc complexes were fabricated by a solution process. The electroluminescence (EL) properties achieved by zinc (II) complexes were checked: peak luminance (L_{max}) of 3589 cd m^{-2}, maximal external quantum efficiency (η_{ext}) of 1.46%, maximum current efficiency (η_L) of 4.1 cd A^{-1}, and maximal power efficiency (η_P) of 3.8 lm W^{-1}. Garry S. Hanan and coworkers in 2018 [121] synthesised a dinuclear complex of zinc (II) with 4-bromo-*N,N'*-diphenylbenzamidinate N-oxide. This compound is a Zn AIEgen and was used as a dopant in a co-host matrix of the emissive layer used in the fabrication of a solution-processed white–green WOLED. A luminance efficiency and power efficiency of 1.12 cd/A and 0.30 lm/W, respectively, were obtained.

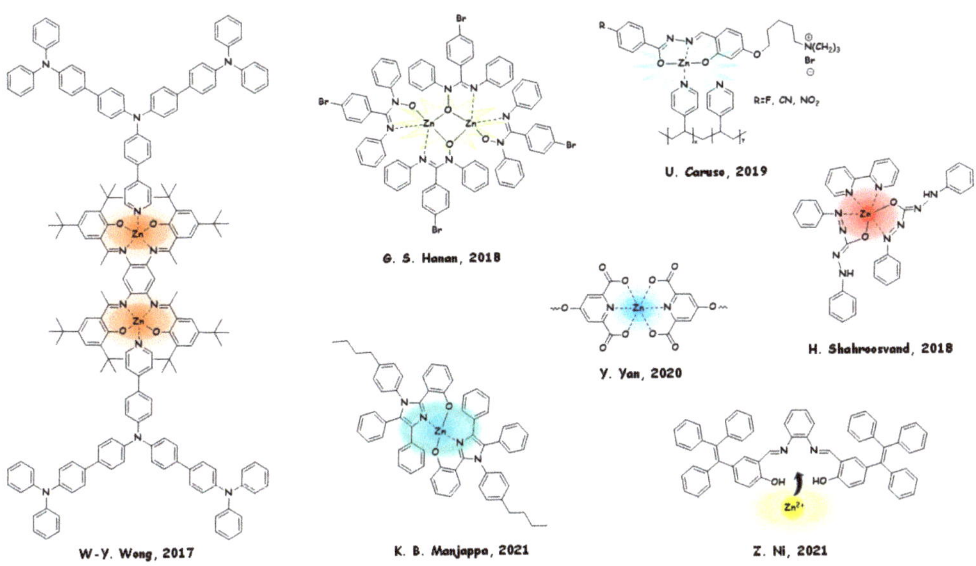

Figure 3. Zn AIEgens for optical applications.

RGB (red, green, blue) metallopolymer emitters with potential in WOLEDs preparation were examined by Ugo Caruso and coworkers in 2019 [70]. Three aryl-hydrazone *O,N,O* tridentate ligands with a different electron-withdrawing substituent and a charged moiety were prepared by grafting ligand-Zn(II) coordination fragments onto commercial poly-(4-vinylpyridine). Three RGB emissive polymers were obtained. By grafting a suitable mix of the three different coordination moieties, an efficient single-component white-light blue emissive metallopolymer (CIE: 0.30, 0.31) was prepared. The same material resulted emissive both in the solid phase and mixed with an ionic liquid, so resulting employable in WOLEDs and in LECs preparation. A rare example of LEC was prepared by Hashem Shahroosvand and coworkers in 2018 [122], based on a blend of the cationic complex $[Ru(bpy)_3]^{2+}$ and a neutral zinc (II) complex derived from diphenylcarbazone ligands. The crystal structure of the Zn fluorescent complex was examined, and the assignment of ground- and excited-state transitions was achieved by TD-DFT analysis. The deep red LEC shows a brightness (740 cd m^{-2}) and luminous efficiency of 0.39 cd/A at a low turn-on voltage of 2.5 V.

Recently, Kiran B. Manjappa and coworkers [123] performed a comprehensive structural, spectroscopic, and theoretical analysis of a bisimidazolyl zinc AIEgen complex. This complex emits cyan fluorescence with a high quantum efficiency (83%), leading to two different applications. First, it was used for the fabrication of an OLED, with a maximum brightness of 15,000 cd/m^2 (16 V), and external quantum efficiency close to 3.8% (comparable with state-of-art of the purely fluorescent non-thermally activated OLEDs). Moreover, the complex was employed in the fabrication of a latent fingerprint (LFPs) investigation tool. Fingerprint detection is a crucial area in forensic science and is based on fluorescent dyes. Under UV light, the finger marks on the surfaces can be photographed by a digital camera resulting in more clearly visible and examinable fingerprints. Dispersed in poly(methyl methacrylate), the dye can also be used as a straightforward security ink for anticounterfeiting applications. Moved by a similar approach, Yun Yan and coworkers in 2020 [124] reported a water-based polyion anticounterfeiting micellar ink displaying full-spectral RGB emission colours by a combination of AIEgens and luminescence of rare earth metals. Blue emission was achieved from the AIE fluorophore tetraphenylethylene (TPE) linked by zinc cation into coordination supramolecular polymers. Recently, Zhonghai Ni [125] developed an ESIPT-based AIEgen Schiff-base compound containing a tetraphenylethene skeleton preventing the ACQ effect. Both the absorption and fluorescence spectra are water- and Zn(II)-responsive. The multifunctional ligand sensor was successfully applied as an inkless rewritable paper and as a colourimetric/fluorescent dual-channel sensor for zinc (II) ion.

4. The Role of Zinc (II) in AIEgen Chemosensing

Some of the established methods for trace elements or molecules detection require sophisticated spectrometry and spectroscopy as accurate and sensitive-but-expansive analytical techniques. Today, many sought-after substitutes are low-cost luminescent probes for on-site determinations.

Fluorescence sensors for zinc detection are a relevant research area. Zinc (II) ion plays a crucial role in many chemical, biological, and environmental processes. From a physiological point of view, zinc (II) cation is an essential metal ion for many biological processes, such as the formation of zinc finger proteins, the neural signal transmission/modulation [9,10], the regulation of gene expression [11], and various catalytic activity [8]. The mechanism of sensing fluorescence turn-on of most zinc (II) chemosensors is based on MLCT, CHEF effect, PET, and C=N isomerisation. Up to now, many classic organic fluorophores (such as coumarin, fluorescein, rhodamine, and cyanine) used for sensing applications undergo ACQ effect at high concentrations or in the aggregated state. Contrarily, organic AIE receptors can be advantageously employed in sensing zinc (II). According to a common strategy, when an organic AIEgen is dissolved in solution, a turn-off of fluorescence is expected. The addition of zinc can induce a turn-on of the fluorescence by different mechanisms: the restriction of intramolecular motions (chelation-aided rigidification, CAR); the decrease of solubility of the sensor, which causes the formation of emissive aggregates (cleavage-triggered aggregation, CTA); the metal-caused blocking of nonradiative pathways (metal-bridged crosslinking, MBC); the metal-caused blocking of intramolecular motion (coordination-induced complexation; CIC) [126]. Therefore, the sensing mechanism involves the active role of the metal in the PL turn-on of the solution. A linear relationship between PL increase and zinc concentration can be exploited to achieve quantitative information.

On the other hand, integrates Zn-containing chemosensors claim a role in the wide field of fluorescence sensing probes. Probes based upon the AIE strategy proved to be attractive and versatile tools for sensing biologically relevant small molecules or ions. AIE chemosensors acting in water solution shows potential in imaging techniques due to their high sensitivity, fast response, low cost, and technical smartness. Fluorescent probes exhibiting AIE behaviour thanks to the presence of zinc (II) cation in a structural key position are an interesting analytical application of Zn AIEgens. The role of the zinc (II) ion is crucial in the turn-on (or turn-off) of the sensor emission. Undergoing coordination-decoordination equilibria, zinc (II) dissolve and reform AIE-active compounds depending on the presence of the analyte.

In this chapter, we will refer to the most relevant examples of AIEgens sensors involving zinc (II) cation as an analyte or as a part of the sensor. Specifically, our discussion will be articulated based on the target analyte as follows:

- AIE sensors designed for selective detection of zinc (II) ion (Section 4.1)
- Zinc (II) containing AIE systems able to detect several target analytes (Sections 4.2 and 4.3).

We will explore and discuss single-analyte or multiple-targeted sensors on the basis of the structural pattern and the sensing mechanism.

4.1. AIEgens for Zinc (II) Sensing

In this section, we will refer to a selection of relevant AIEgens acting as mono-channel zinc sensors (Figure 4), as multi-channel zinc sensors (Figure 5), and macro-structured AIEgens acting as zinc (II) sensors (Figure 6).

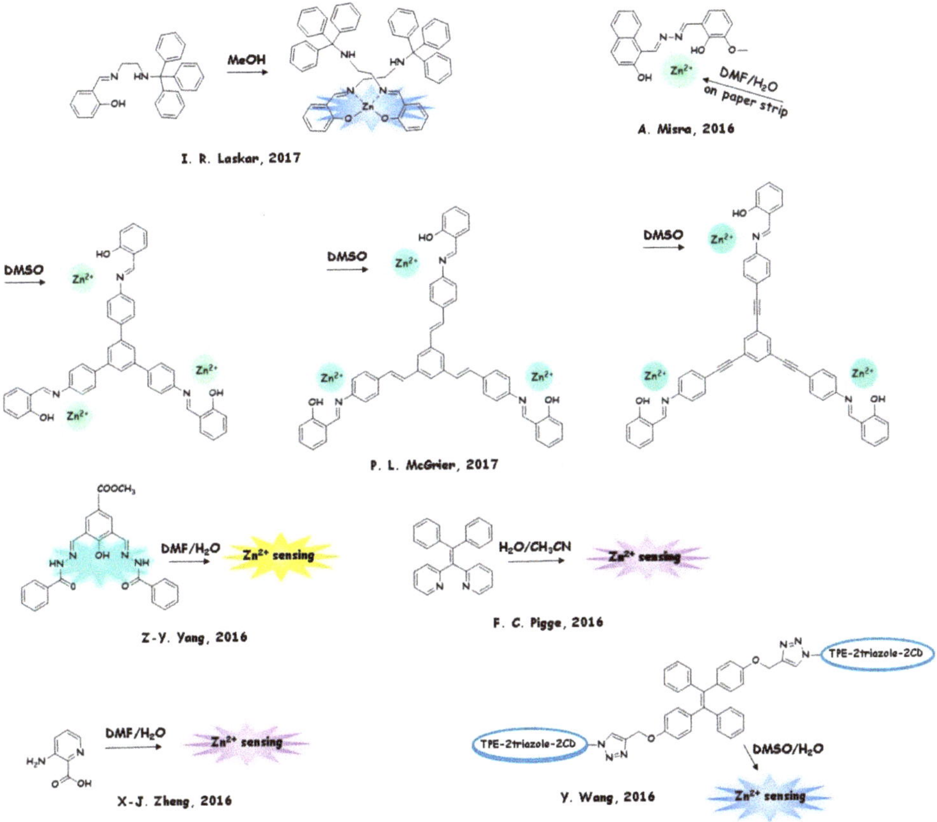

Figure 4. AIEgens acting as zinc (II) mono-channel sensor.

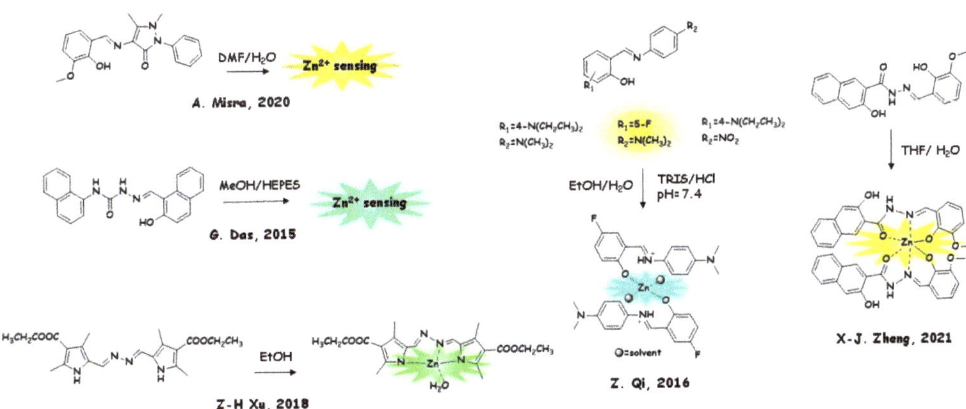

Figure 5. AIEgens acting as zinc (II) multi-channel sensor.

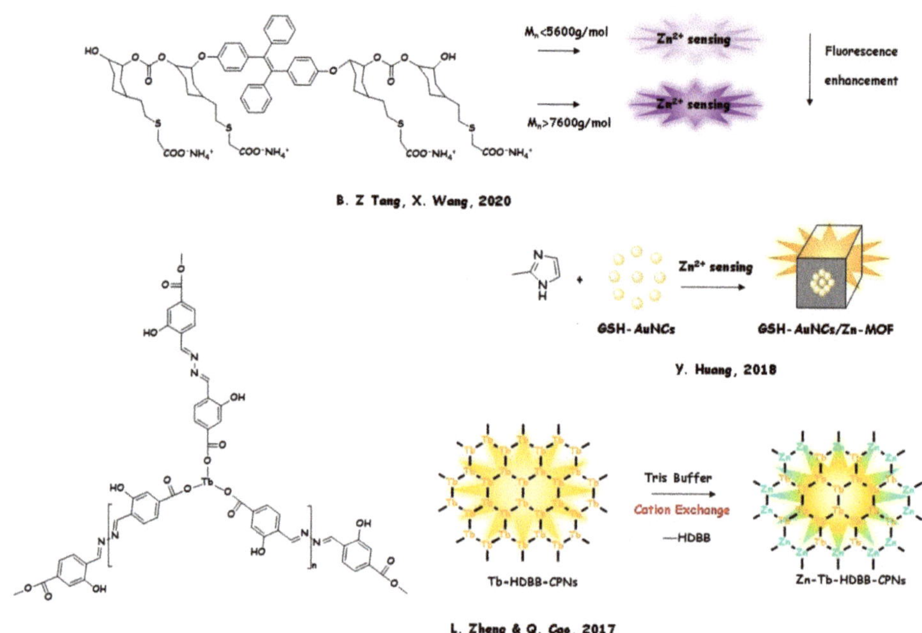

Figure 6. Macro-structured AIEgens acting as zinc (II) sensor.

Schiff-base ligands are good candidates in building zinc sensitive AIEgens. Inamur Rahaman Laskar and coworkers in 2017 [127] synthesised an RIR undergoing chelate, with J-aggregate in the crystalline phase. The sensor [2-(2-(tritylamino)ethylideneamino) methyl phenol] exhibits irreversible mechanoluminescence with an emission colour change from blue to green upon grinding of the powder sample. The ligand proved to be a selective sensor showing PL turn-on (23 folds the unbonded sensor) with LOD = 0.064 ppm upon zinc (II) addition. Three analogue AIE Schiff-base ligands containing 1,3,5-triarylbenzene, -tristyrylbenzene, and -tris(arylethynyl)-benzene cores were published in 2016 by Psaras L. McGrier and coworkers [128]. Their AIE behaviour was attributed to an ESIPT process, and the ligands are turn-on responsive to zinc (II) and copper (II) cations in DMSO solution. Ajay Misra and coworkers in 2016 [129] proposed an AIE fluorescence probe, 1-(2-hydroxynaphthylmethylene)-2-(3-methoxy-2-hydroxybenzylidene) hydrazine, which resulted in being highly selective toward zinc (II) based on a CHEF/AIE feature. Fluorescence turn-on was detected in DMF/H_2O. Zheng-yin Yang and coworkers in 2016 [130] designed a bis Schiff-base AIEgen which detected Zn^{2+} ions in a ratiometric way (a ratiometric fluorescence sensor measures emission intensities at two or more wavelengths to accurately detect changes to the local environment/analyte), due to an ESIPT process.

In addition to classical Schiff-base type ligands, differently structured AIE chemosensors were successfully employed, with a fluorescence turn-on performance toward zinc (II). F. Christopher Pigge and coworker in 2016 [131] proposed three AIEgens, 2,2′-bis(pyridyl) tetraarylethylenes type ligands, for their ability to selectively detect zinc (II) ions in aqueous solution. In the same year, Xiang-Jun Zheng and coworkers [132] produced a 3-aminopyridine-2-carboxylic acid ligand that is AIE-active and zinc (II) responsive. RIR and RIV effects were activated by hydrogen bonds, resulting in rigidity enhancement of the molecule. Still, in 2016, Yong Wang and coworkers [133] produced a tetra-phenyl ethylene triazole-bridged sensor (TPE-2triazole-2CD). The sensor contains a cyclodextrin moiety which has a unique hydrophobic cavity able to encapsulate different guest molecules and a hydrophilic surface due to the hydroxyl moieties on the rings. The elaborate sensor shows a highly selective turn-on florescence response towards Zn^{2+} in neutral environments and potential biocompatibility.

AIE-based multi-responsive zinc (II) probes (Figure 5) are a growing research field. Multianalyte AIE sensors, or generally multifunctional AIE sensors, are highly desirable tools for many applications [134–136]. Ajay Misra and coworkers in 2020 [137] produced an antipyrine-based fluorescent probe, 4-[(2-hydroxy-3-methoxy-benzylidene)-amino]-1,5-dimethyl-2-phenyl-1,2-dihydro-pyrazol-3-one exhibiting turn-on selective fluorescence sensing toward Al^{3+} and Zn^{2+}. In 2016 Zhengjian Qi and coworkers [138] synthesised three Schiff-base derivatives with electron acceptor and donor substituents exhibiting AIE properties due to an ESIPT process. Their response toward both zinc (II) cation and pH parameters was assessed, and the proposed binding mode at pH = 7.4 is illustrated in Figure 5. Another ESIPT-based turn-on AIE sensor was recently synthesised by Xiang-Jun Zheng and coworkers [139]. N-(3-methoxy-2-hydroxybenzylidene)-3-hydroxy-2-naphthahydrazone ligand demonstrated its ability to differently coordinate Zn^{2+} and Cd^{2+} ions. The sensor selectively recognises the metals due to different emission peaks (at 560 and 645 nm, respectively in solution). Interestingly, the ions were recognised quantitatively in water, and qualitatively in the solid state via test paper (light yellow emission).

A more elaborate multianalyte performance was realised in 2015 by Gopal Das and coworkers [140]. They synthesised an AIE probe with both an anion and a cation binding site. The ligand resulted in being turn-on fluorescence responsive and selective toward Al^{3+} and Zn^{2+}. It can also detect Cu^{2+} in mixed buffer medium, and F^- in acetonitrile through colorimetric response. Zhi-Hong Xu and coworkers in 2018 [136] proposed an AIE active bis-hydrazone selectively detecting Cu^{2+}, Zn^{2+}, and Hg^{2+} by monitoring absorption (for copper ion) and fluorescence (for zinc and mercury ions) spectral patterns.

Examples of polymeric AIEgen sensors for zinc (II) are rare and intriguing due to the relationship between structural pattern and AIE response (Figure 6). Ben Zhong Tang and Xianhong Wang and coworkers [141], in 2020, synthesised conjugated polyelectrolytes with adjustable molecular weight (M_n) through a combination of controllable polymerisation and aggregation-induced emission (AIE) techniques. The AIE-active polyelectrolytes are responsive to Zn^{2+} ions with a fluorescence turn-on response associated with M_n. Specifically, in the polymers with $M_n < 5600$ gmol^{-1}, the aggregation of the nanoparticles is due to the diffusion of the cation into the interstitial space between the COO− groups. On the other hand, the long-wrapped chains of polymers with $M_n > 7600$ gmol^{-1} hinder the diffusion of Zn^{2+}, which in turn can coordinate between different nanoparticles, enlarging the size of the aggregates. As expected, the fluorescence increases much more in the second case. The observed phenomena provide a relationship between AIE behaviour and structural parameters, assuming a zinc ion detection of nanoparticle sizes.

Coordination polymers obtained by self-assembly of binding organic ligands with metal cations can be obtained in the nanoscale. CPs derived by nanoparticles (CPNs, sized between 1 and 100 nm) can be formed by self-assembly of a metal ion and organic ligands and provide a unique platform for designing nanosized AIEgen. A relevant example of zinc (II)-responsive CPN was reported in 2017 by Liyan Zheng and Qiue Cao and coworkers [142]. The fluorescent probe for Zn^{2+} based on CPNs was prepared from AIE fluorophore HDBB molecules with Tb^{3+} ion (Tb-HDBB-CPNs). Interestingly, a fluorescence turn-on was observed in the presence of Zn-HDBB-CPNs after the cation exchange process of Tb-HDBB-CPNs with zinc (II). Tb-HDBB-CPNs and Zn-HDBB-CPNs emit different wavelength fluorescence in aqueous solution upon excitation; therefore, the system acts as a ratiometric fluorescent sensor linearly dependent on zinc (II) concentration, from 100 nM to 60 μM.

Yuming Huang and coworkers in 2018 [143] reported an Au nanocluster (AuNC) detecting zinc (II) by AIE effect. Glutathione capped AuNCs (GSH-AuNCs) was synthesised by reduction of Au^{3+} by glutathione. Fluorescence enhance GSH-AuNCs upon addition of 2-methylimidazole in the presence of zinc (II) ion was attributed to the formation of the Zn-MOF (PLQY = 36.6%, about nine times that of AuNCs). GSH-AuNCs resulted in an AIE active sensor that was reported as zinc (II) with a linear range from 12.3 nM to 24.6 mM and LOD = 6 nM. Interestingly, the nanoscaled sensor was successfully applied to assay the content of zinc in human serum, water, milk, and zinc sulfate syrup oral solution samples.

4.2. *Sensing by Molecular Zn AIEgens*

Integrates Zn-containing molecular chemosensors will be discussed in this section, with a focus on the sensing mechanism (Figure 7). As will be detailed described, sensing can take place through an alternation of complexation–decomplexation equilibria involving zinc (II) ions.

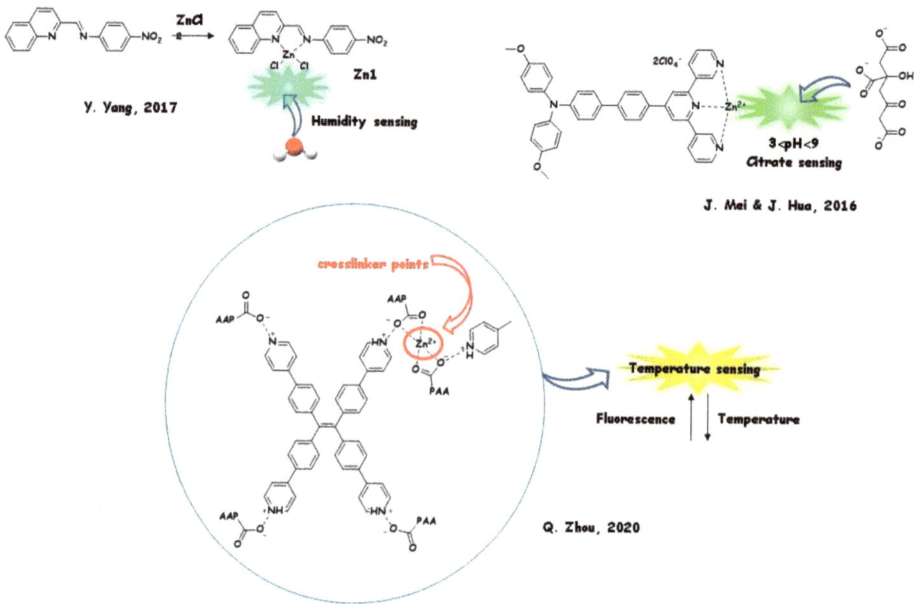

Figure 7. Relevant examples of zinc (II)-based AIE active sensors.

Ju Mei and Jianli Hua and coworkers in 2016 [144] developed a new citrate-sensitive, highly selective fluorescence chemosensor (DTPA-TPY-Zn) by integrating an AIE moiety (DTPA-TPY, where TPY guarantees the AIE behaviour) with zinc (II) ion. The probe detects citrate by an immediate significant fluorescence enhancement and LOD = 3.5×10^{-7} M. The fluorescence turn-on mechanism is ascribed to the complexation of DTPA-TPY-Zn. Specifically, after coordination to zinc (II) ion, the sensor DTPA-TPY undergoes a decrease in fluorescence emission, but after citrate coordinates to the metal cation of DTPA-TPY-Zn, the derived aggregate state in aqueous solution promotes a fluorescence recovery. The effect of pH on the fluorescence of DTPA-TPY-Zn was investigated, finding a stable response in the pH range from 3 to 9. As citrate is a vital metabolite in the Krebs cycle of aerobic cells, its determination is of biological interest. A quantification of citrate in the artificial urine was successfully performed, proving the potential of DTPA-TPY-Zn in real biological samples. The next year, Yulin Yang and coworkers [145] developed several sensors for the detection of the molecule of life: water. Six novel Schiff-base Zn(II) complexes were synthesised, which exhibit sensing ability in water detection. Remarkable characteristics are the wide linear range in water traces sensing (most can reach 0–94%, v/v), LODs (0.2%, v/v), and fast response time (8 s) in various organic solvents. In addition, the sensors demonstrated their application for humidity (42–80%) detection. A mechanism was proposed and validated by experimental and theoretical analysis. Thanks to the water-activated, hydrogen-bonding crosslinking AIE mechanism (WHCAIE), the six zinc (II) complexes bearing chlorine groups/nitro groups can interact with water through intermolecular hydrogen bonding, forming cross-linking networks in the presence of water. The resulting water-activated aggregates of Zn(II) complexes are AIE moieties, whit a superior performance gained by Zn1 (see Figure 7).

An engineered supramolecular system with a key role for zinc (II) ion was designed by Qiong Zhou and coworkers in 2020 [146]. An AIE undergoing hydrogel consisting of a supramolecular network with zinc (II) crosslinking nodes was found to be temperature responsive. Due to the RIM effect in the polymeric segments and the increase of ionic interactions, the hydrogel undergoes a blue shift and an enhancement in fluorescence emission as the temperature decreases.

4.3. Sensing by AIE Active Zn-Nanoprobes and Zn-MOFs

We previously described some zinc-based LMOFs as fascinating materials with structural and spectral tunability. The AIE effect has been revealed as a new route for LMOFs fabrication for a wide range of application areas. In contrast to the ACQ effect, AIE-based MOFs display poor emission in dilute solution and higher fluorescence in the aggregated states due to RIM effects undergoing the organic frameworks. LMOFs, where the AIE channel can be selectively activated by a specific analyte, are powerful tools in chemo and biosensing. AIE Zn-MOFs and other nanosized zinc probes are good candidates for sensing of targeted analytes due to the unique vehiculation and the specificity of the porous architectures achievable by zinc (II) cation. Usually, such probes are formed by the reaction of AIE organic ligands and zinc (II). In addition, the replacement of classical heavy metals—often toxic and hardly disposable—with zinc (II) makes Zn AIE nanoprobes sustainable biocompatible tools.

Selected examples of such nanosized probes will be reported in this section, based on their mono- or multifunctional abilities, starting from chemical targets to biological analytes. Their structures and sensing mode are summarised in Figure 8.

Xiliang Luo and coworkers 2019 [147] produced a sensitive method for the detection of inorganic pyrophosphate (PPi) and pyrophosphatase activity (PPase), which play vital roles in biological systems. Dual-ligand functionalised Au NCs nanoclusters (NCs, dimension between 1nm and 100 nm) exhibiting AIE characteristics were used for fluorescence detection of PPi and PPase, based on a Zn(II)-regulated strategy. Very recently, Zhihong Liu and coworkers [148] developed a similar method for the detection of PPi and PPase. Glutathione stabilised water-soluble alloy Au/Ag nanoclusters (NCs, dimension between 1 nm and 100 nm) exhibiting AIE characteristics were used, still based on a Zn(II)-regulated strategy. The on–off–on mechanism employed in both cases involves the aggregation of Au NCs (or Au/Ag NCs) in the presence of zinc (II) due to electrostatic interaction between Zn^{2+} and the NCs. The fluorescence enhancement is due to the activation of RIR and RIV channels. PPi, containing two phosphate groups, competitively coordinates to zinc (II) ion, subtracting the metal to the previous NCs aggregates. Therefore, a decrease in emission (turn-off) is expected, and the reduction of PL emission can be used to determine the concentration of PPi. On the other hand, PPase catalyzes the hydrolysis of PPi to a single phosphate group and hence break down the PPi-Zn^{2+} non-emissive aggregates. By re-forming of NCs aggregates with zinc (II), the emission turns on. A relationship between the recovery of fluorescence intensity and the amount of PPase leads to a quantitative analysis of PPase activity in water.

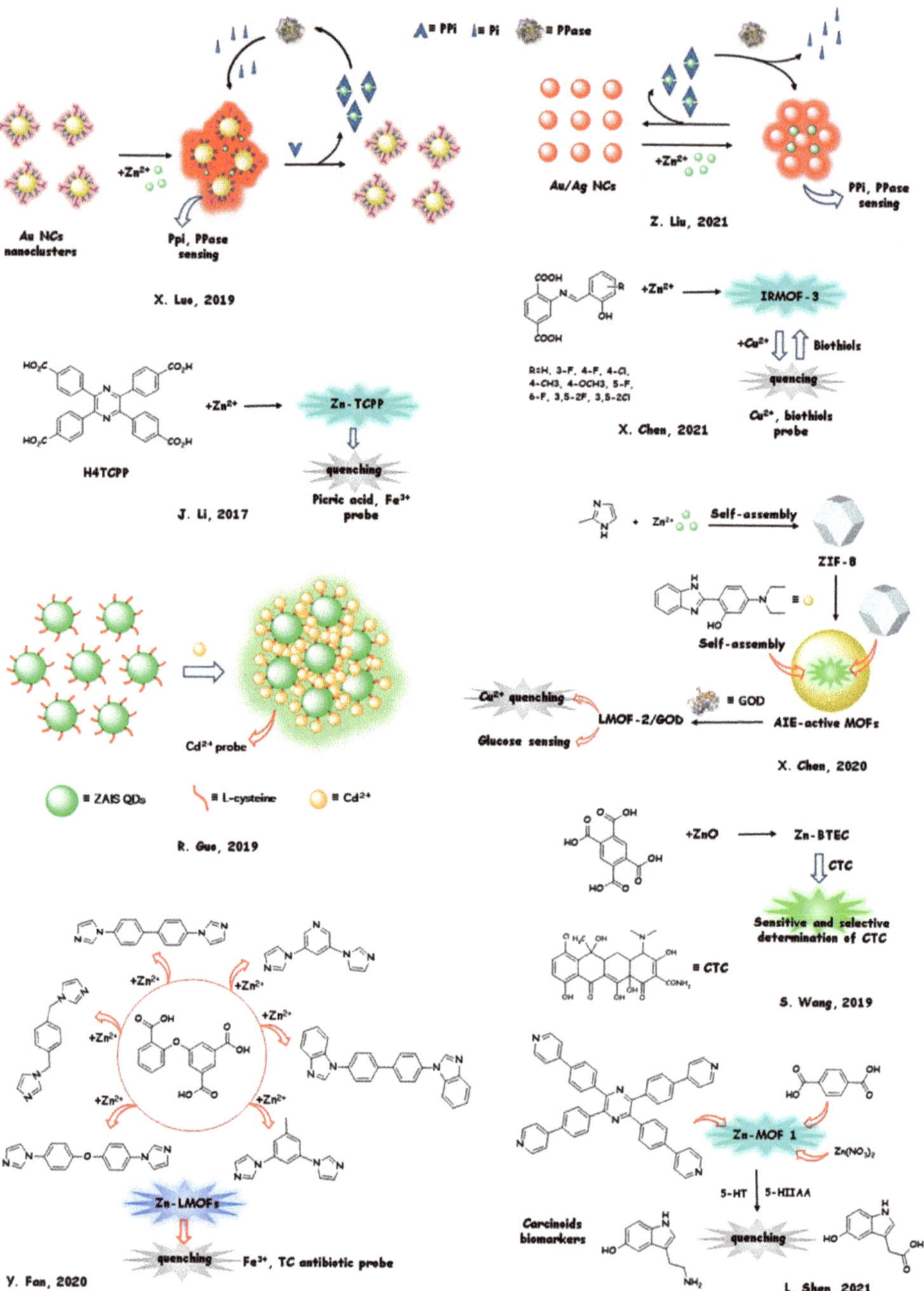

Figure 8. Structures and sensing mode of relevant nanosized Zn-AIEgens.

Jiyang Li and coworkers in 2017 [149] produced a new rigid symmetric tetracarboxylic ligand 2,3,5,6-tetrakis(4-carboxyphenyl)pyrazine (H4TCPP) with AIE properties as a ligand for building the multifunctional Zn-TCPP MOF. The three-dimensional channel structure of Zn-TCPP displays bright blue luminescence in the presence of CO_2 and a quenching effect in the presence of picric acid and Fe^{3+} ions. A micrometre-sized phase of Zn-TCPP (named "Zn-TCPP") exhibiting the best sensing performance was obtained by the solvothermal reaction of H4TCPP and $Zn(NO_3)_2{\cdot}6H_2O$. Recently, Xiaoqing Chen and coworkers [150] produced a series of functional IRMOF-3 frameworks with solid-state luminescence and tuneable light emission (from 490 to 608 nm). The MOFs were obtained by reaction of AIE-active Schiff-base ligands with zinc. The AIE-active ligands formed a "flower-like" morphology due to the formation of J-aggregates. A coordination site of the precursors was designed to bind copper (II) ions. IRMOFs were found PL responsive toward Cu(II) and thiols.

Suhua Wang and coworkers in 2019 [151] developed a Zn-LMOF of pyromellitic acid (Zn-BTEC) used for selective and sensitive AIE detection of chlortetracycline (CTC), a broad-spectrum antibiotic, with LOD = 28 nM. CTC molecules defusing into the rigid MOF structure produce additional aggregation leading to a fluorescence enhancement. Zn-BTEC sensors were successfully applied for the sensitive and selective determination of CTC in fish and urine samples. In 2020, Yuhua Fan and coworkers [152] prepared six Zn-LMOFs, starting from flexible or semiflexible moieties (i.e., 5-(2-carboxylphenoxy)isophthalic acid, 11,4-bis(imidazol-l-ylmethyl)benzene, 4,4-bis(imidazolyl)diphenyl ether) combined with rigid moieties (i.e., 4,4-bis(benzoimidazo-1-ly)biphenyl, 4,4-bis(imidazolyl)biphenyl, bis(1-imidazoly)pyridine, and 11,3-bis(l-imidazoly)toluene). The Zn-LMOFs gave a fluorescence quenching sensing response towards Fe(III) ions and TC antibiotic with LOD = 0.35 mM and 0.15 mM, respectively.

Xiaoqing Chen and coworkers in 2020 [153] reported a series of AIE-active LMOFs with good photostability and excellent dispersibility. The emission of the LMOFs could be tuned by varying the substituents on AIEgens. A MOF of the series was used as a fluorescent probe for detecting copper (II) ions in a wide concentration range (1–100 nM) with LOD = 550 pM and for on-site determination of glucose in real serum samples. Recently, Liang Shen and coworkers [154] produced a non-interpenetrated three-dimensional tetraphenylpyrazine (TPP)-based LMOF (denoted as Zn-MOF1) starting from TPyTPP = tetrakis(4-(pyridin-4-yl)phenyl)pyrazine and H2BDC = 1,4-benzenedicarboxylic acid. Zn-MOF1 displays excellent luminescence-quenching sensitivity toward 5-hydroxytryptamine (5-HT) and 5-hydroxyindole-3-acetic acid (5-HIAA) with LODs of 0.50 μM for 5-HT and 0.57 μM for 5-HIAA. Zn-MOF1 performs as efficient sensing of carcinoid biomarkers.

Quantum dots (QDs) are semiconductor nanoparticles with specific optical and electronic properties with the ability to produce PL emission. Ruiqian Guo and coworkers in 2019 [155] designed a nanosized fluorescence probe for Cd^{2+} based on the AIE active QDs containing several inorganic elements, including zinc (ZAIS QDs, i.e., Zn–Ag–In–S QDs). ZAIS QDs were synthesised starting from L-cysteine ligands with an average size of 3.2 nm and an emission peak at 520 nm. The probe was able to detect cadmium (II) ion with LOD = 1.56 mM by activation of the AIE channel due to the interaction between Cd^{2+} and thiol groups on the surface of QDs, which in turn causes electrostatic aggregation. A year later, Jiye Jin and coworkers [156] produced CdSe/ZnS-based QDs applied as ECL (electrochemiluminescence) hydro-phobic dyes acting as AIE active fluorophores for the determination of H_2O_2 with LOD = 9.2×10^{-8} M.

5. Zn AIEgens for Cell Imaging

The scientific research about sensitive and selective fluorescent probes opens new horizons for biological parameters detection. Nowadays, non-invasive and in situ operating biosensors for analysis in living cells are highly required. By modern microscopy and imaging techniques in combination with fluorescence sensor, sharper cytologic details can be visualised and recorded. Cell organisation can be analysed with clarity and precision by employin the sensing tool combined with confocal microscopy and computer integration in a three-dimensional space. From the first fluorescence microscope (due to Heimstadt in 1911 and Lehmann in 1913) to modern bioimaging techniques, more than a century has passed, and research made great strides from the early fluorophores (as rhodamine, fluorescein, flavin derivatives) to the modern highly engineered fluorogenic dyes. Several fluorescent dyes for staining live cells, labelling subcellular organelles, live cell tracking, and carrying in live cells are now commercially available. Nonetheless, the search for new marking systems is constantly evolving. Fluorescence bioimaging continuously requires new real-time tools for bioanalysis and molecular dynamic biological processes with high sensitivity, flexibility, and biocompatibility.

Time-resolved imaging is a modern microscopy technique whereby fast kinetic and PL decay parameters (decay times and the corresponding resolved amplitudes) are directly and simultaneously measured throughout an image in an optical microscope. Thus far, most fluorescent dyes utilised in sensing commonly suffer from severe concentration or the ACQ effect, and their fluorescence lifetimes can be shortened to nanoseconds in the aggregate phase. This behaviour is unwelcomed for the time-resolved fluorescence imaging technique. The employ of AIE active sensors or markers can solve the ACQ problems due to the high fluorescence of the aggregates in aqueous media and short fluorescence lifetimes.

In this context, Zn AIEgens have drawn considerable attention, thanks to the synthetic feasibility, tuneable PL pattern, and biocompatibility.

Again, the involvement of zinc (II) cation with nanosized AIE sensors can be as a biologically relevant analyte or as a part of the sensor itself. In the next section (Section 5.1), we will discuss the monitoring of the zinc (II) level in living biological substrates. The structures and the sensing/marking mechanism are reported in Figure 9.

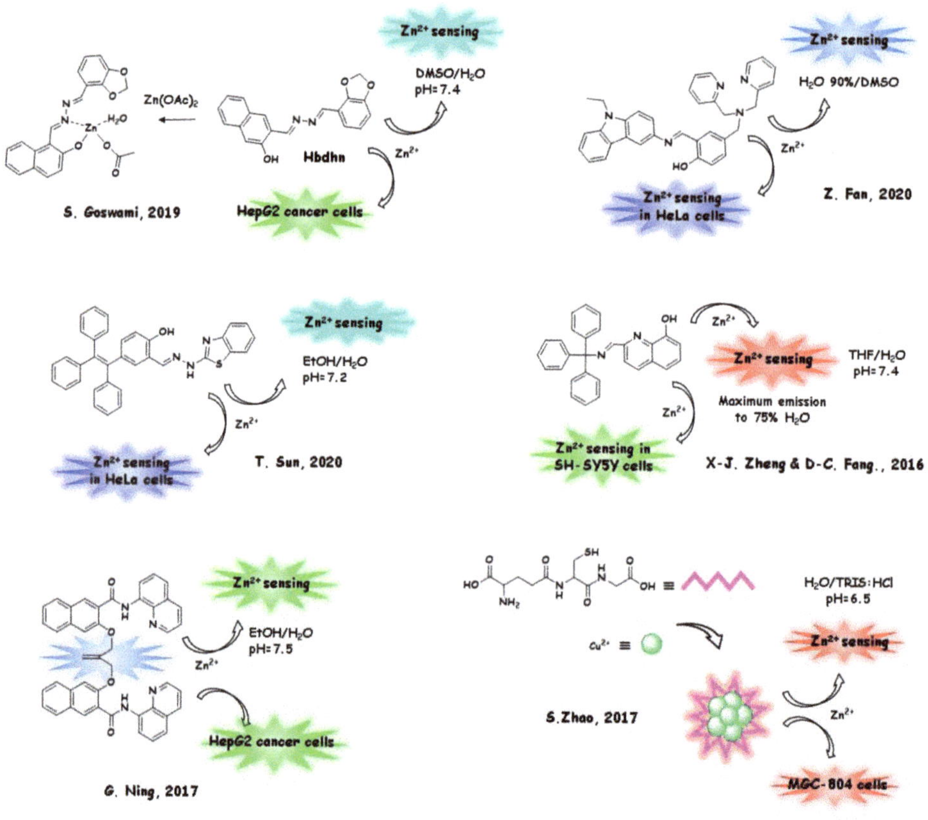

Figure 9. Structures and sensing/marking pattern of relevant AIEgens for zinc (II) detection in living cells.

In Section 5.2, we will report a selection of the most relevant biosensors and markers containing zinc in their architecture and employed for their sensing/marking ability in living cells (Figure 10). The latter bio-probes can be considered as a class of innovative and desirable biocompatible tools, with a strong technological impact on diagnostic imaging methods.

5.1. AIEgens for Zinc (II) Detection in Living Cells

Even for in vivo tests, probes based on Schiff-base moieties give efficient and selective responses. Sanchita Goswami and coworkers in 2019 [157] reported a Schiff-base chemosensor 1-(benzo(1,3)dioxol-4-ylmethylene-hydrazonomethyl)-naphthalen-2-ol (Hbdhn). The ligand is a zinc (II)-responsive, long-lived fluorescence AIEgen. The sensor was also checked by test paper. A photographic image of Hbdhn on a paper strip in the absence and presence of Zn^{2+} in natural light and under UV light (λ_{ex} = 366 nm) evidenced a potential employment of the sensor in the solid state. Moreover, high efficiency of Hbdhn to monitor zinc (II) cation under physiological conditions was evaluated by fluorescence images of HepG2 cancer cells, detecting a colorimetric turn-on of the fluorescence response in the presence of zinc (II) ion. Zhefeng Fan and coworkers in 2020 [158] synthesised a Schiff-base-chelating probe containing di(2-picolyl)amine (DPA) acting as an AIE chelating agent for Zn^{2+}. The probe shows good biocompatibility and a large Stokes shift upon coordination with naked-eye perceivable emission colour. Again, a probe-based test filter paper evidenced a good naked-eye response onto a solid support. A good linear dependence in fluorescence increase was detected between 0.2 and 18 μM, and LOD was 1.3×10^{-7} M. The low toxicity and the excellent membrane permeability enable the probe for the application in confocal fluorescence images of HeLa cells. Tongming Sun and coworkers in 2020 [159] produced an AIE probe based on a hydrazonoic skeleton, able to detect Zn^{2+} with LOD = 2.1×10^{-8} M and highly sensitive in a pH range of 4.8–12.0. The probe L obtained by condensation of 4-hydroxy-5-aldehyde tetrastyrene and 2-hydrazine benzothiazole, activates by blocking of PET process. It shows good membrane permeability and low cytotoxicity and can be applied in HeLa living cells for endogenous and exogenous zinc (II) detection.

Among the fluorescent AIE probes developed in recent years, quinoline-based probes attracted the interest of researchers due to their strong binding ability towards metal cations and excellent fluorescent properties [160]. A quinoline derivative was employed by Xiang-Jun Zheng and De-Cai Fang and coworkers 2016 [161]. They explored the PL properties of the zinc (II) complex obtained from 2-(trityliminomethyl)-quinolin-8-ol ligand (L) acting as a *N,O* pincer. ZnL_2 units formed stacking by the coordination bonds and intermolecular π–π interactions (J-aggregates). The ligands show no fluorescence, and the complex is emissive in the aggregate phase due to the restriction of the ESIPT process. Therefore, ligand L is an ESIPT-based AIE chemosensor for Zn(II) in an aqueous medium. To prove the ability of the ligand to detect Zn^{2+} in living cells, bioimaging experiments were carried out in SH-SY5Y cells incubated with L 20μM for 16h and treated with perchloride zinc (II) salt 20 μM for 30 min, recording a bright green fluorescence when complex formed. In 2017 Guiling Ning and coworkers [162] synthesised two organic AIE active fluorophores containing bis-naphthylamide and quinoline moieties. An isobutylene group was introduced in one case to increase rigidity and to reduce intramolecular molecular rotations (RIR). As expected, the more encumbered ligand shows a higher fluorescence response towards Zn^{2+}. The probe exhibits a highly ratiometric fluorescence response towards zinc (II) ion and excellent cell permeability in human hepatoma cancer cells HepG2, with a binding constant of 5×10^{-7} M^{-1} and LOD = 2.6 nM.

Glutathione (GSH)-capped nanoclusters were reported in Section 4 for their sensing ability. In 2017, Shulin Zhao and coworkers [163] prepared glutathione (GSH)-capped copper nanoclusters (CuNCs) with red emission for Zn^{2+} detecting and explored their potential by confocal fluorescence images in MGC-803 cells. The AIE mechanism was based on zinc (II) triggered aggregation of CuNCs, inducing the increase of PLQY from 1.3% to

6.2%. The probe showed a fast response in the detection range from 4.68 to 2240 μM and LOD = 1.17 μM.

5.2. Zn AIEgens for Biological Targets in Living Cells

TPY, as a *N,N,N*-tridentate ligand able to form stable complexes with zinc (II), is the building block of several zinc-based AIE biological probes. In 2016, Duobin Chao and Shitan Ni [164] developed two TPY-based Zn(II) complexes for selective nanomolar PPi detection in a water medium, with LOD = 5.37 nM. The sensors are based on AIE and ICT effects due to the formation of nanoaggregates. The probes were successfully employed for nucleus staining in living cells. Specifically, confocal fluorescence images of HeLa cells incubated with the probes were recorded, showing potential applications in PPi detection by cell imaging and diagnosis. In 2019 Yupeng Tian and coworkers [165] presented a multiphoton bioimaging TPY-based zinc (II) complex bearing a thiophene bridge (DZ1). Due to the AIE effect, the probe emits bright yellow-green fluorescence under physiological conditions. DZ1 showed an excellent response to RNA, compared with commercially multiphoton probes, due to the self-aggregation ability under physiological conditions and excellent photostability. The spectral changes of DZ1 in binding RNA result in a twofold enhancement of the three-photon action cross-sections located at the second near-infrared window (1700 nm). Fluorescence microscopy images of HeLa cells stained with different concentrations of DZ1 (1.0–10 mM) and in aqueous phosphate buffered were presented.

The apoptosis level plays a crucial role in cell growth, and its dysregulation may result in cancers and neurodegenerative diseases. In 2016, Ben Zhong Tang and coworkers [166] synthesised a novel AIE-active fluorescent probe (TPE–Zn_2BDPA) comprised of two components: tetraphenylethene (TPE) as the AIEgen, and 2,6-bis(zinc (II)-dipicolylamine)phenoxide (Zn_2BDPA) as the targeting group for phosphatidylserine (PS). PS has emerged as an attractive indicator for apoptosis detection. It shows fluorescence response to apoptotic cells and differentiates the early and late stages of cell apoptosis. Light-blue emissive TPE–Zn_2BDPA with low cytotoxicity and excellent biocompatibility resulted in being comparable with two commonly used probes for apoptosis detection (Annexin V-FITC and propidium iodide, PI) commonly employed to image HeLa and HepG2 cells. Fafu Yang and coworkers in 2020 [167] synthesised an AIE probe based on a hydrazono-bis-tetraphenylethylene (Bis-TPE) multiple-detecting adenosine triphosphate (ATP) and two biologically relevant metal cations. Bis-TPE exhibited good AIE fluorescence at 550–700 nm based on the highly conjugated electron effect. The probe display sensing abilities for copper (II) ion in a fluorescence turn-off mode. The strong red fluorescence in the Bis-TPE system, quenched by Cu^{2+}, can be recovered by adding ATP. On the other hand, the addition of zinc (II) to Bis-TPE produces a relevant blue shift, and the orange fluorescence of the Bis-TPE-Zn^{2+} system can be quenched by copper (II) and then recovered by adding ATP. The "allochroic off–on" orange fluorescence in Bis-TPE-Zn^{2+}-Cu^{2+} combined system was applied to sense copper (II) and zinc (II) cations and ATP both on test paper and by confocal fluorescence images of MCF-7 cancer cells.

Jintao Zhu and coworkers in 2018 [168] developed a metallo–supramolecular assembly (Z/E-TPE2CyZn-PV) consisting of a tetraphenylethene (TPE)-based dinuclear Zn^{2+}–cyclen complex and pyrocatechol violet (PV). The PV fragment can bind with TPE2CyZn and quenched the background fluorescence. Z/E-TPE2CyZn-PV acted as a fluorescence turn-on chemosensor for selective detection of the T-rich nucleic acid in aqueous buffers, discriminating the T-rich single-stranded DNA (ssDNA) from double-stranded DNA (dsDNA). Z-TPE2CyZn-PV proved to be an indicator displacement assay (IDAs) inside living cells by fluorescence microscopy images of B16–F10 cells.

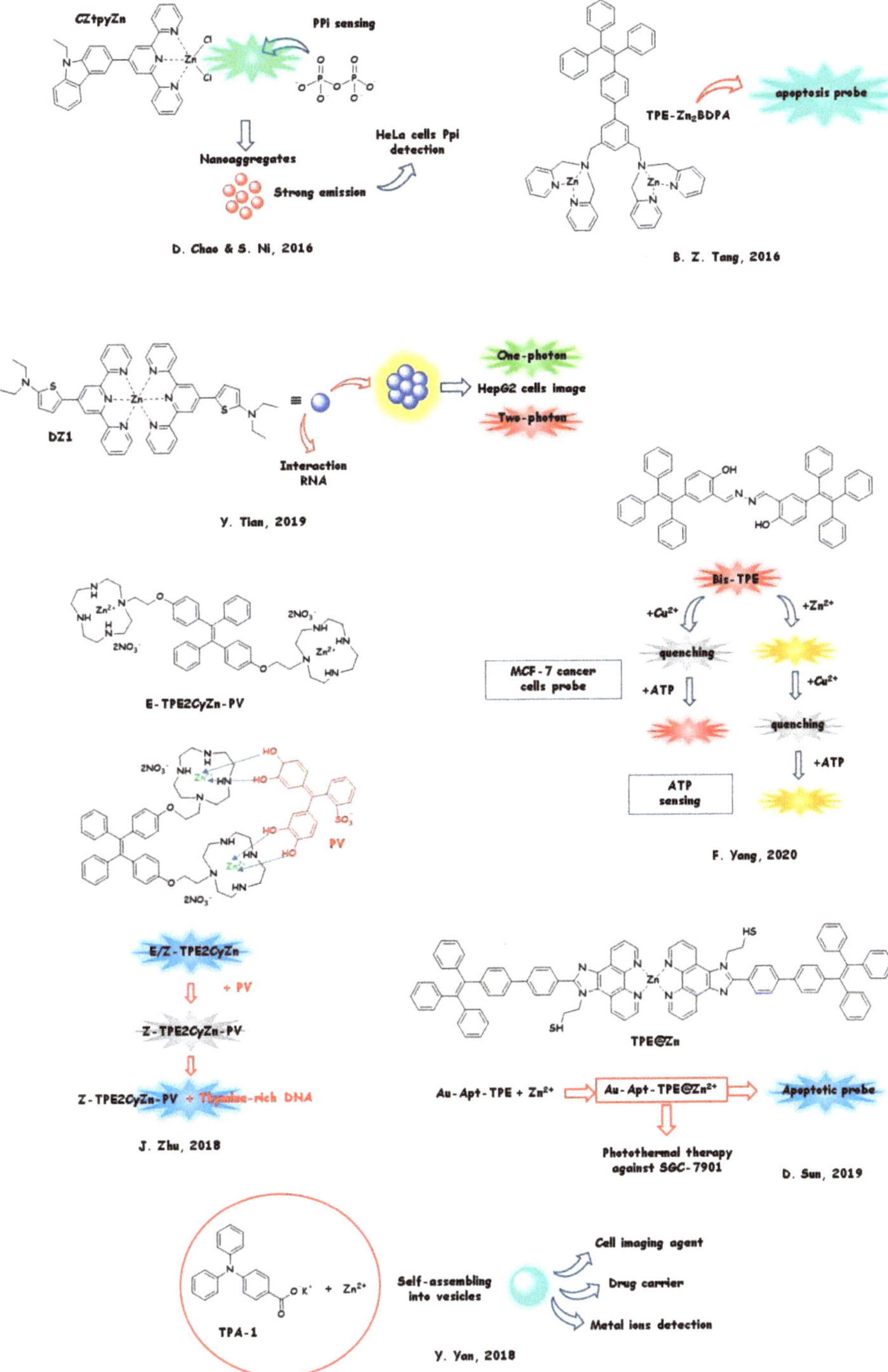

Figure 10. Structures and sensing/marking pattern of zinc-based AIE active biosensors/markers.

As reported in previous sections, nanoprobes found their ultimate application in luminescence sensing. Detection of biological species in physiological conditions by employing nanoparticles are drawing more and more attention by virtue of their selectivity, low cytotoxicity, and cell permeability. Multifunctional nanosized probes for cell imaging, drug-carrying, and metal ion detection are effective cutting-edge in scientific research. Recently, hybrid organic–inorganic nanoprobes based on the bio-friendly zinc (II) cation [169–173] have been explored as highly engineered sensors and markers for bioimaging [63,126]. Rare articles have been produced in which the link between the AIE properties and the sensing ability has been highlighted and explored.

Dongdong Sun and coworkers in 2019 [174] fabricated dual-targeted gold nanoprisms, whereby bare gold nanoprisms (AuNPR) were conjugated to phenanthroline derivatives tetraphenylethene (TPE) functionalised and exhibiting AIE characteristics. The probe was further stabilised with target peptide aptamers via Au-S bonds (Au-Apt-TPE). The free nitrogen atoms in Au-Apt-TPE can chelate zinc (II) ions (obtaining Au-Apt-TPE@Zn) which were used for selectively monitoring early stage apoptotic cells. Remarkably, Au-Apt-TPE@Zn under NIR irradiation showed effective penetration and photothermal therapy against SGC-7901 human gastric carcinoma cells growth. In vivo studies revealed deep penetration of Au-Apt-TPE@Zn under NIR irradiation and dual-model imaging application for cancer-targeted fluorescence imaging. In 2018, Yun Yan and coworkers [175] reported a metallo–organic fluorescent vesicle based on the triarylamine carboxylate AIE molecules (TPA-1). When zinc (II) ions electrostatically interact with the TPA-1 molecule, the complex $Zn(TPA\text{-}1)_2$ forms, which further self-assembles into vesicles. TPA-1 molecules confined in the vesicle membrane display AIE behaviour. The vesicles are excellent cell imaging agents (as tested in Hela cells) by two-photon emission. The probes can be a powerful tool for cell imaging, drug-carrying, and metal ion detection.

6. Conclusions

The review briefly summarised the development history, the AIE-channel activation, and the recent achievements of Zn AIEgens research. During the past 10 years, the development in AIEgens research has undergone rapid growth. Novel AIE active systems have been designed and synthesised; their properties and mechanism have been deeply investigated by experimental verification and theoretical models. The basic study gave the boost to novel AIE systems for more and more applications, from optics to sensing and bioimaging. The search for hybrid inorganic–organic compounds with unique photophysical properties is crucial in developing novel AIEgens. In this perspective, the abundant, inexpensive, and nontoxic zinc (II) metal cation offers many opportunities. The benefits of using zinc complexes instead of conventional luminescent AIE active organic materials or in place of other metal cations have been examined. Acting as a "chemical clip", zinc (II) ion imposes a constraint to the ligand or assembles the ligands in AIE active macro-architectures, with a scarce contribution to the electronic pattern and a large contributor to the RIM effect.

The most relevant examples of Zn AIEgens from the last 5 years (2016–2021) were collected, and the discussion articulated with the level increasing. The structure–properties relationship and the AIE triggered applications were the thread running through the text. In brief, the review displays an overview, from the novel fully characterised Zn AIEgens to the Zn AIEgens involved in chemo and biosensing. As zinc (II) cation is of great importance to humans and other biological systems, the sensing ability of AIEgens toward the metal was a relevant section of our discussion. Finally, we reported applications of Zn AIEgens targeted for living cell imaging as materials designed for the novel technological frontier.

We trust that the comprehensive and informative approach of this review could trigger new ideas because ideas open new synthetic strategies, promote new ways of projecting, and address new avenues for the next generation of lighting and imaging devices.

Author Contributions: Conceptualization, writing—original draft preparation, review, and editing, B.P. and R.D. Both authors read and agreed to the published Version of the manuscript.

Funding: This research wos funded by the Italian Ministry of Education, University and Research (MIUR) under grants PON PANDION 01_00375.

Institutional Review Board Statement: Not applicable.

Informed Consent Statement: Not applicable.

Conflicts of Interest: The authors declare no conflict of interest.

References

1. Langhals, H. Fluorescence and fluorescent dyes. *Phys. Sci. Rev.* **2020**, *5*, 1–26. [CrossRef]
2. Valeur, B.; Berberan-santos, N. Emergence of Quantum Theory. *J. Chem. Educ.* **2011**, *88*, 731–738. [CrossRef]
3. Caruso, U.; Panunzi, B.; Diana, R.; Concilio, S.; Sessa, L.; Shikler, R.; Nabha, S.; Tuzi, A.; Piotto, S. AIE/ACQ Effects in Two DR/NIR Emitters: A Structural and DFT Comparative Analysis. *Molecules* **2018**, *23*, 1947. [CrossRef] [PubMed]
4. Concilio, S.; Ferrentino, I.; Sessa, L.; Massa, A.; Iannelli, P.; Diana, R.; Panunzi, B.; Rella, A.; Piotto, S.P. A novel fluorescent solvatochromic probe for lipid bilayers. *Supramol. Chem.* **2017**, *29*, 887–895. [CrossRef]
5. Tang, J.; Robichaux, M.A.; Wu, K.-L.; Pei, J.; Nguyen, N.T.; Zhou, Y.; Wensel, T.G.; Xiao, H. Single-Atom Fluorescence Switch: A General Approach toward Visible-Light-Activated Dyes for Biological Imaging. *J. Am. Chem. Soc.* **2019**, *141*, 14699–14706. [CrossRef]
6. Umezawa, K.; Nakamura, Y.; Makino, H.; Citterio, D.; Suzuki, K. Bright, Color-Tunable Fluorescent Dyes in the Visible—Near-Infrared Region. *J. Am. Chem. Soc.* **2008**, *130*, 1550–1551. [CrossRef]
7. Christie, R.M. Pigments, Dyes and Fluorescent Brightening Agents for Plastics: An Qverview. *Polym. Int.* **1994**, *34*, 351–361. [CrossRef]
8. Hussain, M.; Shamey, R.; Hinks, D.; El-Shafei, A.; Ali, S.I. Synthesis of novel stilbene-alkoxysilane fluorescent brighteners, and their performance on cotton fiber as fluorescent brightening and ultraviolet absorbing agents. *Dyes Pigments* **2012**, *92*, 1231–1240. [CrossRef]
9. Förster, T.; Kasper, K. Ein Konzentrationsumschlag der Fluoreszenz des Pyrens. *Ber. Bunsenges. Phys. Chem.* **1955**, *59*, 976–980.
10. Birks, J.B. *Photophysics of Aromatic Molecules*; Wiley-Interscience: London, UK, 1970; ISBN 0471074209 9780471074205.
11. Dumur, F. Zinc complexes in OLEDs: An overview. *Synth. Met.* **2014**, *195*, 241–251. [CrossRef]
12. Diana, R.; Panunzi, B.; Tuzi, A.; Piotto, S.; Concilio, S.; Caruso, U. An Amphiphilic Pyridinoyl-hydrazone Probe for Colorimetric and Fluorescence pH Sensing. *Molecules* **2019**, *24*, 3833. [CrossRef]
13. Huang, J.; Nie, H.; Zeng, J.; Zhuang, Z.; Gan, S.; Cai, Y.; Guo, J.; Su, S.-J.; Zhao, Z.; Tang, B.Z. Highly Efficient Nondoped OLEDs with Negligible Efficiency Roll-Off Fabricated from Aggregation-Induced Delayed Fluorescence Luminogens. *Angew. Chem. Int. Ed.* **2017**, *56*, 12971–12976. [CrossRef]
14. Kwok, R.T.K.; Leung, C.W.T.; Lam, J.W.Y.; Tang, B.Z. Biosensing by luminogens with aggregation-induced emission characteristics. *Chem. Soc. Rev.* **2015**, *44*, 4228–4238. [CrossRef] [PubMed]
15. Wang, Z.; Wang, R.; Mi, Y.; Lu, K.; Liu, Y.; Yang, C.; Zhang, J.; Liu, X.; Wang, Y.; Shuai, Z.; et al. Creating Side Transport Pathways in Organic Solar Cells by Introducing Delayed Fluorescence Molecules. *Chem. Mater.* **2021**, *33*, 4578–4585. [CrossRef]
16. De Girolamo Del Mauro, A.; Diana, R.; Grimaldi, I.A.; Loffredo, F.; Morvillo, P.; Villani, F.; Minarini, C. Polymer solar cells with inkjet-printed doped-PEDOT: PSS anode. *Polym. Compos.* **2013**, *34*, 1493–1499. [CrossRef]
17. Lee, H.L.; Lee, K.H.; Lee, J.Y.; Lee, H.J. Molecular design opening two emission pathways for high efficiency and long lifetime of thermally activated delayed fluorescent organic light-emitting diodes. *J. Mater. Chem. C* **2021**, *9*, 7328–7335. [CrossRef]
18. Yu, R.; Song, Y.; Chen, M.; He, L. Green to blue-green-emitting cationic iridium complexes with a CF3-substituted phenyl-triazole type cyclometalating ligand: Synthesis, characterization and their use for efficient light-emitting electrochemical cells. *Dalton Trans.* **2021**, *50*, 8084–8095. [CrossRef] [PubMed]
19. Petdee, S.; Chaiwai, C.; Benchaphanthawee, W.; Nalaoh, P.; Kungwan, N.; Namuangruk, S.; Sudyoadsuk, T.; Promarak, V. Imidazole-based solid-state fluorophores with combined ESIPT and AIE features as self-absorption-free non-doped emitters for electroluminescent devices. *Dyes Pigments* **2021**, *193*, 109488. [CrossRef]
20. Luo, J.D.; Xie, Z.L.; Lam, J.W.Y.; Cheng, L.; Tang, B.Z.; Chen, H.Y.; Qiu, C.F.; Kwok, H.S.; Zhan, X.W.; Liu, Y.; et al. Aggregation-induced emission of 1-methyl-1,2,3,4,5-pentaphenylsilole. *Chem. Commun.* **2001**, 1740–1741. [CrossRef]
21. Mei, J.; Leung, N.; Kwok, R.T.K.; Lam, J.W.Y.; Tang, B.Z. Aggregation-Induced Emission: Together We Shine, United We Soar! *Chem. Rev.* **2015**, *115*, 11718–11940. [CrossRef]
22. Zhao, Z.; Zhang, H.; Lam, J.W.Y.; Tang, B.Z. Aggregation-Induced Emission: New Vistas at the Aggregate Level. *Angew. Chem. Int. Ed.* **2020**, *59*, 9888–9907. [CrossRef]
23. Dou, Y.; Zhu, Q.; Du, K. Recent Advances in Two-Photon AIEgens and Their Application in Biological Systems. *ChemBioChem* **2021**, *22*, 1871–1883. [CrossRef] [PubMed]
24. Suzuki, S.; Sasaki, S.; Sairi, A.S.; Iwai, R.; Tang, B.Z.; Konishi, G. Principles of Aggregation-Induced Emission: Design of Deactivation Pathways for Advanced AIEgens and Applications. *Angew. Chem. Int. Ed.* **2020**, *59*, 9856–9867. [CrossRef]

25. Khan, I.M.; Niazi, S.; Khan, M.K.I.; Pasha, I.; Mohsin, A.; Haider, J.; Iqbal, M.W.; Rehman, A.; Yue, L.; Wang, Z. Recent advances and perspectives of aggregation-induced emission as an emerging platform for detection and bioimaging. *TrAC Trends Anal. Chem.* **2019**, *119*, 115637. [CrossRef]
26. Chen, J.; Law, C.C.W.; Lam, J.W.Y.; Dong, Y.; Lo, S.M.F.; Williams, I.D.; Zhu, D.; Tang, B.Z. Synthesis, Light Emission, Nanoaggregation, and Restricted Intramolecular Rotation of 1,1-Substituted 2,3,4,5-Tetraphenylsiloles. *Chem. Mater.* **2003**, *15*, 1535–1546. [CrossRef]
27. Leung, N.L.C.; Xie, N.; Yuan, W.Z.; Liu, Y.; Wu, Q.; Peng, Q.; Miao, Q.; Lam, J.W.Y.; Tang, B.Z. Restriction of Intramolecular Motions: The General Mechanism behind Aggregation-Induced Emission. *Chem. A Eur. J.* **2014**, *20*, 15349–15353. [CrossRef] [PubMed]
28. Hong, Y.; Lam, J.W.Y.; Tang, B.Z. Aggregation-induced emission: Phenomenon, mechanism and applications. *Chem. Commun.* **2009**, 4332–4353. [CrossRef]
29. Niu, Y.; Peng, Q.; Deng, C.; Gao, X.; Shuai, Z. Theory of Excited State Decays and Optical Spectra: Application to Polyatomic Molecules. *J. Phys. Chem. A* **2010**, *114*, 7817–7831. [CrossRef]
30. Peng, Q.; Niu, Y.; Deng, C.; Shuai, Z. Vibration correlation function formalism of radiative and non-radiative rates for complex molecules. *Chem. Phys.* **2010**, *370*, 215–222. [CrossRef]
31. Shuai, Z.; Peng, Q. Excited states structure and processes: Understanding organic light-emitting diodes at the molecular level. *Phys. Rep.* **2014**, *537*, 123–156. [CrossRef]
32. Li, K.; Liu, Y.; Feng, Q.; Hou, H.; Tang, B.Z. 2,5-bis(4-alkoxycarbonylphenyl)-1,4-diaryl-1,4-dihydropyrrolo[3,2-b]pyrrole (AAPP) AIEgens: Tunable RIR and TICT characteristics and their multifunctional applications. *Chem. Sci.* **2017**, *8*, 7258–7267. [CrossRef] [PubMed]
33. Han, T.; Yan, D.; Wu, Q.; Song, N.; Zhang, H.; Wang, D. Aggregation-Induced Emission: A Rising Star in Chemistry and Materials Science. *Chin. J. Chem.* **2021**, *39*, 677–689. [CrossRef]
34. Zhang, T.; Zhu, G.; Lin, L.; Mu, J.; Ai, B.; Li, Y.; Zhuo, S. Cyano substitution effect on the emission quantum efficiency in stilbene derivatives: A computational study. *Org. Electron.* **2019**, *68*, 264–270. [CrossRef]
35. Shuai, Z.; Wang, D.; Peng, Q.; Geng, H. Computational Evaluation of Optoelectronic Properties for Organic/Carbon Materials. *Acc. Chem. Res.* **2014**, *47*, 3301–3309. [CrossRef]
36. Zhang, T.; Jiang, Y.; Niu, Y.; Wang, D.; Peng, Q.; Shuai, Z. Aggregation Effects on the Optical Emission of 1,1,2,3,4,5-Hexaphenylsilole (HPS): A QM/MM Study. *J. Phys. Chem. A* **2014**, *118*, 9094–9104. [CrossRef]
37. Li, Q.; Blancafort, L. A conical intersection model to explain aggregation induced emission in diphenyl dibenzofulvene. *Chem. Commun.* **2013**, *49*, 5966–5968. [CrossRef] [PubMed]
38. Crespo-Otero, R.; Li, Q.; Blancafort, L. Exploring Potential Energy Surfaces for Aggregation-Induced Emission—From Solution to Crystal. *Chem. Asian J.* **2019**, *14*, 700–714. [CrossRef] [PubMed]
39. Hu, L.; Farrokhnia, M.; Heimdal, J.; Shleev, S.; Rulíšek, L.; Ryde, U. Reorganization Energy for Internal Electron Transfer in Multicopper Oxidases. *J. Phys. Chem. B* **2011**, *115*, 13111–13126. [CrossRef] [PubMed]
40. Li, W.; Zhang, Y.-M.; Zhang, T.; Zhang, W.; Li, M.; Zhang, S.X.-A. Tunable RGB luminescence of a single molecule with high quantum yields through a rational design. *J. Mater. Chem. C* **2016**, *4*, 1527–1532. [CrossRef]
41. Nayak, S.; Sinha, R.K.; Lewis, P.M.; Kulkarni, S.D.; Gaonkar, S.L. Synthesis, characterization, DFT and photophysical studies of new class of 1,3,4-oxadiazole-isobenzofuran hybrids. *J. Lumin.* **2021**, *238*, 118212. [CrossRef]
42. Diana, R.; Caruso, U.; Di Costanzo, L.; Bakayoko, G.; Panunzi, B. A Novel DR/NIR T-Shaped AIEgen: Synthesis and X-Ray Crystal Structure Study. *Crystals* **2020**, *10*, 269. [CrossRef]
43. Keshri, S.K.; Mandal, K.; Kumar, Y.; Yadav, D.; Mukhopadhyay, P. Naphthalenediimides with High Fluorescence Quantum Yield: Bright-Red, Stable, and Responsive Fluorescent Dyes. *Chem.-Eur. J.* **2021**, *27*, 6954–6962. [CrossRef] [PubMed]
44. Wang, J.; Liang, B.; Wei, J.; Li, Z.; Xu, Y.; Yang, T.; Li, C.; Wang, Y. Highly Efficient Electrofluorescence Material Based on Pure Organic Phosphor Sensitization. *Angew. Chem. Int. Ed.* **2021**, *60*, 15335–15339. [CrossRef] [PubMed]
45. Lim, B.Y.X. The nanoscale rainbow. *Nat. NV* **2016**, *531*, 26–28. [CrossRef] [PubMed]
46. Roy, E.; Nagar, A.; Chaudhary, S.; Pal, S. Advanced Properties and Applications of AIEgens-Inspired Smart Materials. *Ind. Eng. Chem. Res.* **2020**, *59*, 10721–10736. [CrossRef]
47. Tang, B.Z.; Zhan, X.; Yu, G.; Sze Lee, P.P.; Liu, Y.; Zhu, D. Efficient blue emission from siloles. *J. Mater. Chem.* **2001**, *11*, 2974–2978. [CrossRef]
48. Li, J.; Bisoyi, H.K.; Lin, S.; Guo, J.; Li, Q. 1,2-Dithienyldicyanoethene-Based, Visible-Light-Driven, Chiral Fluorescent Molecular Switch: Rewritable Multimodal Photonic Devices. *Angew. Chem. Int. Ed.* **2019**, *58*, 16052–16056. [CrossRef]
49. Puthanveedu, A.; Shanmugasundaram, K.; Yoon, S.; Choe, Y. Thenil and furil-imidazole-based efficient ionic green emitters with high color purity for non-doped light-emitting electrochemical cells. *J. Mater. Chem. C* **2021**. [CrossRef]
50. Morvillo, P.; Grimaldi, I.A.; Diana, R.; Loffredo, F.; Villani, F. Study of the microstructure of inkjet-printed P3HT:PCBM blend for photovoltaic applications. *J. Mater. Sci.* **2012**, *48*, 2920–2927. [CrossRef]
51. Emami, M.; Shahroosvand, H.; Bikas, R.; Lis, T.; Daneluik, C.; Pilkington, M. Synthesis, Study, and Application of Pd(II) Hydrazone Complexes as the Emissive Components of Single-Layer Light-Emitting Electrochemical Cells. *Inorg. Chem.* **2021**, *60*, 982–994. [CrossRef]

52. Liu, Y.; Xiao, X.; Ran, Y.; Bin, Z.; You, J. Molecular design of thermally activated delayed fluorescent emitters for narrowband orange–red OLEDs boosted by a cyano-functionalization strategy. *Chem. Sci.* **2021**. [CrossRef]
53. Park, J.M.; Nam, S.H.; Hong, K.-I.; Jeun, Y.E.; Ahn, H.S.; Jang, W.-D. Stimuli-responsive fluorescent dyes for electrochemically tunable multi-color-emitting devices. *Sens. Actuators B Chem.* **2021**, *332*, 129534. [CrossRef]
54. Sayed, M.; Tom, D.M.; Pal, H. Multimode binding and stimuli responsive displacement of acridine orange dye complexed with p-sulfonatocalix[4/6]arene macrocycles. *Phys. Chem. Chem. Phys.* **2020**, *22*, 13306–13319. [CrossRef] [PubMed]
55. Diana, R.; Caruso, U.; Tuzi, A.; Panunzi, B. A Highly Water-Soluble Fluorescent and Colorimetric pH Probe. *Crystals* **2020**, *10*, 83. [CrossRef]
56. Liu, Z.; Jiang, Z.; Xu, C.; Chen, B.; Zhu, G. Fluorenyl-difluoroboron-β-diketonates with multi-stimuli fluorescent response behavior and their applications in a thermochromic logic gate device. *Dyes Pigments* **2021**, *186*, 108990. [CrossRef]
57. Li, Z.; Dong, Y.; Mi, B.; Tang, Y.; Häussler, M.; Tong, H.; Dong, Y.; Lam, J.W.Y.; Ren, Y.; Sung, H.H.Y.; et al. Structural Control of the Photoluminescence of Silole Regioisomers and Their Utility as Sensitive Regiodiscriminating Chemosensors and Efficient Electroluminescent Materials. *J. Phys. Chem. B* **2005**, *109*, 10061–10066. [CrossRef]
58. Yoon, S.-J.; Chung, J.W.; Gierschner, J.; Kim, K.S.; Choi, M.-G.; Kim, D.; Park, S.Y. Multistimuli Two-Color Luminescence Switching via Different Slip-Stacking of Highly Fluorescent Molecular Sheets. *J. Am. Chem. Soc.* **2010**, *132*, 13675–13683. [CrossRef]
59. Zhu, L.; Zhu, B.; Luo, J.; Liu, B. Design and Property Modulation of Metal–Organic Frameworks with Aggregation-Induced Emission. *ACS Mater. Lett.* **2021**, *3*, 77–89. [CrossRef]
60. Jin, J.; Xue, J.; Liu, Y.; Yang, G.; Wang, Y.-Y. Recent progresses in luminescent metal–organic frameworks (LMOFs) as sensors for the detection of anions and cations in aqueous solution. *Dalton Trans.* **2021**, *50*, 1950–1972. [CrossRef]
61. Wang, S.; Zhang, C.-H.; Zhang, P.; Chen, S.; Song, Z.-L.; Chen, J.; Zeng, R. Rational design of a HA-AuNPs@AIED nanoassembly for activatable fluorescence detection of HAase and imaging in tumor cells. *Anal. Methods* **2021**, *13*, 2030–2036. [CrossRef]
62. Deshpande, N.U.; Virmani, M.; Jayakannan, M. An AIE-driven fluorescent polysaccharide polymersome as an enzyme-responsive FRET nanoprobe to study the real-time delivery aspects in live cells. *Polym. Chem.* **2021**, *12*, 1549–1561. [CrossRef]
63. Li, Y.; Zhong, H.; Huang, Y.; Zhao, R. Recent Advances in AIEgens for Metal Ion Biosensing and Bioimaging. *Molecules* **2019**, *24*, 4593. [CrossRef] [PubMed]
64. Shi, H.; Liu, J.; Geng, J.; Tang, B.Z.; Liu, B. Specific Detection of Integrin αvβ3 by Light-Up Bioprobe with Aggregation-Induced Emission Characteristics. *J. Am. Chem. Soc.* **2012**, *134*, 9569–9572. [CrossRef] [PubMed]
65. Qin, W.; Ding, D.; Liu, J.; Yuan, W.Z.; Hu, Y.; Liu, B.; Tang, B.Z. Biocompatible Nanoparticles with Aggregation-Induced Emission Characteristics as Far-Red/Near-Infrared Fluorescent Bioprobes for In Vitro and In Vivo Imaging Applications. *Adv. Funct. Mater.* **2011**, *22*, 771–779. [CrossRef]
66. Guo, B.; Huang, Z.; Shi, Q.; Middha, E.; Xu, S.; Li, L.; Wu, M.; Jiang, J.; Hu, Q.; Fu, Z.; et al. Organic Small Molecule Based Photothermal Agents with Molecular Rotors for Malignant Breast Cancer Therapy. *Adv. Funct. Mater.* **2019**, *30*, 1–11. [CrossRef]
67. Hu, X.; Zhao, X.; He, B.; Zhao, Z.; Zheng, Z.; Zhang, P.; Shi, X.; Kwok, R.T.K.; Lam, J.W.Y.; Qin, A.; et al. A Simple Approach to Bioconjugation at Diverse Levels: Metal-Free Click Reactions of Activated Alkynes with Native Groups of Biotargets without Prefunctionalization. *Research* **2018**, *2018*, 1–12. [CrossRef]
68. Chen, M.; Xie, W.; Li, D.; Zebibula, A.; Wang, Y.; Qian, J.; Qin, A.; Tang, B.Z. Utilizing a Pyrazine-Containing Aggregation-Induced Emission Luminogen as an Efficient Photosensitizer for Imaging-Guided Two-Photon Photodynamic Therapy. *Chem.-Eur. J.* **2018**, *24*, 16603–16608. [CrossRef]
69. Gondia, N.; Sharma, S. Comparative optical studies of naphthalene based Schiff base complexes for colour tunable application. *Mater. Chem. Phys.* **2019**, *224*, 314–319. [CrossRef]
70. Panunzi, B.; Diana, R.; Caruso, U. A Highly Efficient White Luminescent Zinc (II) Based Metallopolymer by RGB Approach. *Polymers* **2019**, *11*, 1712. [CrossRef]
71. Diana, R.; Panunzi, B.; Shikler, R.; Nabha, S.; Caruso, U. A symmetrical azo-based fluorophore and the derived salen multipurpose framework for emissive layers. *Inorg. Chem. Commun.* **2019**, *104*, 186–189. [CrossRef]
72. Han, X.; Bai, Q.; Yao, L.; Liu, H.; Gao, Y.; Li, J.; Liu, L.; Liu, Y.; Li, X.; Lu, P.; et al. Highly Efficient Solid-State Near-Infrared Emitting Material Based on Triphenylamine and Diphenylfumaronitrile with an EQE of 2.58% in Nondoped Organic Light-Emitting Diode. *Adv. Funct. Mater.* **2015**, *25*, 7521–7529. [CrossRef]
73. Panunzi, B.; Diana, R.; Concilio, S.; Sessa, L.; Shikler, R.; Nabha, S.; Tuzi, A.; Caruso, U.; Piotto, S. Solid-State Highly Efficient DR Mono and Poly-dicyano-phenylenevinylene Fluorophores. *Molecules* **2018**, *23*, 1505. [CrossRef]
74. Zhou, Z.; Li, W.; Hao, X.; Redshaw, C.; Chen, L.; Sun, W.-H. 6-Benzhydryl-4-methyl-2-(1H-benzoimidazol-2-yl)phenol ligands and their zinc complexes: Syntheses, characterization and photoluminescence behavior. *Inorg. Chim. Acta* **2012**, *392*, 345–353. [CrossRef]
75. Lozano, G. The Role of Metal Halide Perovskites in Next-Generation Lighting Devices. *J. Phys. Chem. Lett.* **2018**, *9*, 3987–3997. [CrossRef] [PubMed]
76. Bizzarri, C.; Spuling, E.; Knoll, D.M.; Volz, D.; Bräse, S. Sustainable metal complexes for organic light-emitting diodes (OLEDs). *Coord. Chem. Rev.* **2018**, *373*, 49–82. [CrossRef]
77. Caruso, U.; Diana, R.; Panunzi, B.; Roviello, A.; Tingoli, M.; Tuzi, A. Facile synthesis of new Pd(II) and Cu(II) based metallomesogens from ligands containing thiophene rings. *Inorg. Chem. Commun.* **2009**, *12*, 1135–1138. [CrossRef]

78. Alam, P.; Climent, C.; Alemany, P.; Laskar, I.R. "Aggregation-induced emission" of transition metal compounds: Design, mechanistic insights, and applications. *J. Photochem. Photobiol. C Photochem. Rev.* **2019**, *41*, 100317. [CrossRef]
79. Castellano, F.; Pomestchenko, I.E.; Shikhova, E.; Hua, F.; Muro, M.L.; Rajapakse, N. Photophysics in bipyridyl and terpyridyl platinum(II) acetylides. *Coord. Chem. Rev.* **2006**, *250*, 1819–1828. [CrossRef]
80. Lamansky, S.; Djurovich, P.; Murphy, D.; Abdel-Razzaq, F.; Lee, H.-E.; Adachi, C.; Burrows, P.E.; Forrest, S.R.; Thompson, M. Highly Phosphorescent Bis-Cyclometalated Iridium Complexes: Synthesis, Photophysical Characterization, and Use in Organic Light Emitting Diodes. *J. Am. Chem. Soc.* **2001**, *123*, 4304–4312. [CrossRef]
81. Caruso, U.; Panunzi, B.; Roviello, A.; Tingoli, M.; Tuzi, A. Two aminobenzothiazole derivatives for Pd(II) and Zn(II) coordination: Synthesis, characterization and solid state fluorescence. *Inorg. Chem. Commun.* **2011**, *14*, 46–48. [CrossRef]
82. Chen, C.; Xu, Y.; Wan, Y.; Fan, W.; Si, Z. Aggregation-Induced Phosphorescent Emission from ReIComplexes: Synthesis and Property Studies. *Eur. J. Inorg. Chem.* **2016**, *2016*, 1340–1347. [CrossRef]
83. You, Y.; Nam, W. Photofunctional triplet excited states of cyclometalated Ir(iii) complexes: Beyond electroluminescence. *Chem. Soc. Rev.* **2012**, *41*, 7061–7084. [CrossRef] [PubMed]
84. Complex, I.I.I.; Seder, T.A.; Church, S.P.; Ouderkirk, A.J. Excited-state properties of a triply ortho-metalated iridium (III) complex. *J. Am. Chem. Soc.* **1985**, *107*, 1431–1432.
85. Wu, Q.; Deng, C.; Peng, Q.; Niu, Y.; Shuai, Z. Quantum chemical insights into the aggregation induced emission phenomena: A QM/MM study for pyrazine derivatives. *J. Comput. Chem.* **2012**, *33*, 1862–1869. [CrossRef] [PubMed]
86. Causà, M.; D'Amore, M.; Garzillo, C.; Gentile, F.; Savin, A. The Bond Analysis Techniques (ELF and Maximum Probability Domains) Application to a Family of Models Relevant to Bio-Inorganic Chemistry. *Fam. Med.* **2012**, *150*, 119–141. [CrossRef]
87. Diana, R.; Panunzi, B. The Role of Zinc(II) Ion in Fluorescence Tuning of Tridentate Pincers: A Review. *Molecules* **2020**, *25*, 4984. [CrossRef] [PubMed]
88. Xie, Y.-Z.; Shan, G.-G.; Li, P.; Zhou, Z.-Y.; Su, Z.-M. A novel class of Zn(II) Schiff base complexes with aggregation-induced emission enhancement (AIEE) properties: Synthesis, characterization and photophysical/electrochemical properties. *Dyes Pigments* **2013**, *96*, 467–474. [CrossRef]
89. Pan, Y.; Wang, J.; Guo, X.; Liu, X.; Tang, X.; Zhang, H. A new three-dimensional zinc-based metal-organic framework as a fluorescent sensor for detection of cadmium ion and nitrobenzene. *J. Colloid Interface Sci.* **2018**, *513*, 418–426. [CrossRef]
90. Diana, R.; Panunzi, B.; Shikler, R.; Nabha, S.; Caruso, U. Highly efficient dicyano-phenylenevinylene fluorophore as polymer dopant or zinc-driven self-assembling building block. *Inorg. Chem. Commun.* **2019**, *104*, 145–149. [CrossRef]
91. Diana, R.; Panunzi, B.; Tuzi, A.; Caruso, U. Two tridentate pyridinyl-hydrazone zinc(II) complexes as fluorophores for blue emitting layers. *J. Mol. Struct.* **2019**, *1197*, 672–680. [CrossRef]
92. Panunzi, B.; Diana, R.; Concilio, S.; Sessa, L.; Tuzi, A.; Piotto, S.; Caruso, U. Fluorescence pH-dependent sensing of Zn(II) by a tripodal ligand. A comparative X-ray and DFT study. *J. Lumin.* **2019**, *212*, 200–206. [CrossRef]
93. Zhang, G.; Chen, Q.; Zhang, Y.; Kong, L.; Tao, X.; Lu, H.; Tian, Y.; Yang, J. Bulky group functionalized porphyrin and its Zn (II) complex with high emission in aggregation. *Inorg. Chem. Commun.* **2014**, *46*, 85–88. [CrossRef]
94. Landi, G.; Fahrner, W.R.; Concilio, S.; Sessa, L.; Neitzert, H.C. Electrical Hole Transport Properties of an Ambipolar Organic Compound With Zn-Atoms on a Crystalline Silicon Heterostructure. *IEEE J. Electron. Devices Soc.* **2014**, *2*, 179–181. [CrossRef]
95. Suman, G.R.; Pandey, M.; Chakravarthy, A.J. Review on new horizons of aggregation induced emission: From design to development. *Mater. Chem. Front.* **2021**, *5*, 1541–1584. [CrossRef]
96. Gentile, F.S.; Salustro, S.; Di Palma, G.; Causà, M.; D'Arco, P.; Dovesi, R. Hydrogen, boron and nitrogen atoms in diamond: A quantum mechanical vibrational analysis. *Theor. Chem. Acc.* **2018**, *137*, 154. [CrossRef]
97. Copéret, C. Single-Sites and Nanoparticles at Tailored Interfaces Prepared via Surface Organometallic Chemistry from Thermolytic Molecular Precursors. *Acc. Chem. Res.* **2019**, *52*, 1697–1708. [CrossRef] [PubMed]
98. Gentile, F.S.; Salustro, S.; Desmarais, J.K.; Ferrari, A.M.; D'Arco, P.; Dovesi, R. Vibrational spectroscopy of hydrogens in diamond: A quantum mechanical treatment. *Phys. Chem. Chem. Phys.* **2018**, *20*, 11930–11940. [CrossRef]
99. Han, H.; Chen, Y.; Jin, Q.; Wang, Y.; Ji, J. The rational design of a gemcitabine prodrug with AIE-based intracellular light-up characteristics for selective suppression of pancreatic cancer cells. *Chem. Commun.* **2015**, *51*, 17435–17438. [CrossRef]
100. Tsuchiya, M.; Sakamoto, R.; Shimada, M.; Yamanoi, Y.; Hattori, Y.; Sugimoto, K.; Nishibori, E.; Nishihara, H. Bis(dipyrrinato)zinc(II) Complexes: Emission in the Solid State. *Inorg. Chem.* **2016**, *55*, 5732–5734. [CrossRef]
101. Diana, R.; Panunzi, B.; Concilio, S.; Marrafino, F.; Shikler, R.; Caruso, T.; Caruso, U. The Effect of Bulky Substituents on Two π-Conjugated Mesogenic Fluorophores. Their Organic Polymers and Zinc-Bridged Luminescent Networks. *Polymers* **2019**, *11*, 1379. [CrossRef]
102. Chen, Q.; Xiang, Y.; Yin, X.; Hu, K.; Li, Y.; Cheng, X.; Liu, Y.; Xie, G.; Yang, C. Highly efficient blue TADF emitters incorporating bulky acridine moieties and their application in solution-processed OLEDs. *Dyes Pigments* **2021**, *188*, 109157. [CrossRef]
103. Stolte, M.; Schembri, T.; Süß, J.; Schmidt, D.; Krause, A.-M.; Vysotsky, M.O.; Würthner, F. 1-Mono- and 1,7-Disubstituted Perylene Bisimide Dyes with Voluminous Groups at Bay Positions: In Search for Highly Effective Solid-State Fluorescence Materials. *Chem. Mater.* **2020**, *32*, 6222–6236. [CrossRef]
104. Qin, Y.; Peng, Q.; Chen, F.; Liu, Y.; Li, K.; Zang, S. AIE Ligand Constructed Zn(II) Complex with Reversible Photo-induced Color and Emission Changes. *Chem. Res. Chin. Univ.* **2021**, *37*, 123–128. [CrossRef]

105. Wu, L.; Huang, C.; Emery, B.P.; Sedgwick, A.C.; Steven, D.B.; He, X.-P.; Tian, H.; Yoon, J.; Sessler, J.L.; James, T.D. Förster resonance energy transfer (FRET)-based small-molecule sensors and imaging agents. *Chem. Soc. Rev.* **2020**, *49*, 5110. [CrossRef] [PubMed]
106. Jia, J.; Zhao, H. A multi-responsive AIE-active tetraphenylethylene-functioned salicylaldehyde-based schiff base for reversible mechanofluorochromism and Zn^{2+} and ${CO_3}^{2-}$ detection. *Org. Electron.* **2019**, *73*, 55–61. [CrossRef]
107. Kumar, Y.; Singh, V.D.; Dwivedi, B.K.; Singh, N.K.; Pandey, D.S. Solid state emissive azo-Schiff base ligands and their Zn(ii) complexes: Acidochromism and photoswitching behaviour. *New J. Chem.* **2021**, *45*, 199–207. [CrossRef]
108. Ebina, M.; Kondo, Y.; Iwasa, T.; Taketsugu, T. Low-Lying Excited States of hqxcH and Zn–hqxc Complex: Toward Understanding Intramolecular Proton Transfer Emission. *Inorg. Chem.* **2019**, *58*, 4686–4698. [CrossRef] [PubMed]
109. Panunzi, B.; Concilio, S.; Diana, R.; Shikler, R.; Nabha, S.; Piotto, S.; Sessa, L.; Tuzi, A.; Caruso, U. Photophysical Properties of Luminescent Zinc(II)-Pyridinyloxadiazole Complexes and their Glassy Self-Assembly Networks. *Eur. J. Inorg. Chem.* **2018**, *2018*, 2709–2716. [CrossRef]
110. Borbone, F.; Caruso, U.; Concilio, S.; Nabha, S.; Panunzi, B.; Piotto, S.; Shikler, R.; Tuzi, A. Mono-, Di-, and Polymeric Pyridinoylhydrazone ZnIIComplexes: Structure and Photoluminescent Properties. *Eur. J. Inorg. Chem.* **2016**, *2016*, 818–825. [CrossRef]
111. Brahma, R.; Baruah, J.B. Self-Assemblies of Zinc Complexes for Aggregation-Induced Emission Luminogen Precursors. *ACS Omega* **2020**, *5*, 3774–3785. [CrossRef]
112. Marafie, J.A.; Bradley, D.D.C.; Williams, C.K. Thermally Stable Zinc Disalphen Macrocycles Showing Solid-State and Aggregation-Induced Enhanced Emission. *Inorg. Chem.* **2017**, *56*, 5688–5695. [CrossRef] [PubMed]
113. Yan, Y.; Yin, G.-Q.; Khalife, S.; He, Z.-H.; Xu, C.; Li, X. Self-assembly of emissive metallocycles with tetraphenylethylene, BODIPY and terpyridine in one system. *Supramol. Chem.* **2019**, *31*, 597–605. [CrossRef] [PubMed]
114. Yaghi, O.M.; O'Keeffe, M.; Ockwig, N.W.; Chae, H.K.; Eddaoudi, M.; Kim, J. Reticular synthesis and the design of new materials. *Nat. Vol.* **2003**, *423*, 705–714. [CrossRef] [PubMed]
115. Yaghi, O.M.; Li, H.; Davis, C.; Richardson, D.; Groy, T. Synthetic Strategies, Structure Patterns, and Emerging Properties in the Chemistry of Modular Porous Solids. *Acc. Chem. Res.* **1998**, *31*, 474–484. [CrossRef]
116. Eddaoudi, M.; Moler, D.B.; Li, H.; Chen, B.; Reineke, T.M.; O'Keeffe, M.; Yaghi, O.M. Modular Chemistry: Secondary Building Units as a Basis for the Design of Highly Porous and Robust Metal−Organic Carboxylate Frameworks. *Acc. Chem. Res.* **2001**, *34*, 319–330. [CrossRef] [PubMed]
117. Dong, Y.-W.; Fan, R.-Q.; Wang, X.-M.; Wang, P.; Zhang, H.-J.; Wei, L.-G.; Chen, W.; Yang, Y.-L. (E)-N-(Pyridine-2-ylmethylene)arylamine as an Assembling Ligand for Zn(II)/Cd(II) Complexes: Aryl Substitution and Anion Effects on the Dimensionality and Luminescence Properties of the Supramolecular Metal–Organic Frameworks. *Cryst. Growth Des.* **2016**, *16*, 3366–3378. [CrossRef]
118. Rouhani, F.; Morsali, A.; Retailleau, P. Simple One-Pot Preparation of a Rapid Response AIE Fluorescent Metal–Organic Framework. *ACS Appl. Mater. Interfaces* **2018**, *10*, 36259–36266. [CrossRef]
119. Hamada, Y.; Sano, T.; Fujita, M.; Fujii, T.; Nishio, Y.; Shibata, K. Blue Electroluminescence in Thin Films of Azomethin-Zinc Complexes. *Jpn. J. Appl. Phys.* **1993**, *32*, L511–L513. [CrossRef]
120. Zhao, J.; Dang, F.; Liu, B.; Wu, Y.; Yang, X.; Zhou, G.; Wu, Z.; Wong, W.-Y. Bis-ZnII salphen complexes bearing pyridyl functionalized ligands for efficient organic light-emitting diodes (OLEDs). *Dalton Trans.* **2017**, *46*, 6098–6110. [CrossRef]
121. Cibian, M.; Shahalizad, A.; Souissi, F.; Castro, J.; Ferreira, J.G.; Chartrand, D.; Nunzi, J.-M.; Hanan, G.S. A Zinc(II) Benzamidinate N -Oxide Complex as an Aggregation-Induced Emission Material: Toward Solution-Processable White Organic Light-Emitting Devices. *Eur. J. Inorg. Chem.* **2018**, *2018*, 4322–4330. [CrossRef]
122. Shahroosvand, H.; Heydari, L.; Bideh, B.N.; Pashaei, B.; Tarighi, S.; Notash, B. Low-Turn-On-Voltage, High-Brightness, and Deep-Red Light-Emitting Electrochemical Cell Based on a New Blend of $[Ru(bpy)_3]^{2+}$ and Zn–Diphenylcarbazone. *ACS Omega* **2018**, *3*, 9981–9988. [CrossRef]
123. Kempegowda, R.M.; Malavalli, M.K.; Malimath, G.H.; Naik, L.; Manjappa, K.B. Synthesis and Photophysical Properties of Multi-Functional Bisimidazolyl Phenol Zinc (II) Complex: Application in OLED, Anti-Counterfeiting and Latent Finger Print Detection. *ChemistrySelect* **2021**, *6*, 3033–3039. [CrossRef]
124. Wu, T.; Xie, M.; Huang, J.; Yan, Y. Putting Ink into Polyion Micelles: Full-Color Anticounterfeiting with Water/Organic Solvent Dual Resistance. *ACS Appl. Mater. Interfaces* **2020**, *12*, 39578–39585. [CrossRef]
125. Sun, H.; Jiang, Y.; Nie, J.; Wei, J.-H.; Miao, B.X.; Zhao, Y.; Zhang, L.-F.; Ni, Z.-H. Multifunctional AIE-ESIPT dual mechanism tetraphenylethene-based Schiff base for inkless rewritable paper and a colorimetric/fluorescent dual-channel Zn^{2+} sensor. *Mater. Chem. Front.* **2021**, *5*, 347–354. [CrossRef]
126. Alam, P.; Leung, N.L.; Zhang, J.; Kwok, R.T.; Lam, J.W.; Tang, B.Z. AIE-based luminescence probes for metal ion detection. *Coord. Chem. Rev.* **2021**, *429*, 213693. [CrossRef]
127. Pasha, S.S.; Yadav, H.R.; Choudhury, A.R.; Laskar, I.R. Synthesis of an aggregation-induced emission (AIE) active salicylaldehyde based Schiff base: Study of mechanoluminescence and sensitive Zn(II) sensing. *J. Mater. Chem. C* **2017**, *5*, 9651–9658. [CrossRef]
128. Jagadesan, P.; Whittemore, T.; Beirl, T.; Turro, C.; McGrier, P.L. Excited-State Intramolecular Proton-Transfer Properties of Three Tris(N-Salicylideneaniline)-Based Chromophores with Extended Conjugation. *Chem. Eur. J.* **2017**, *23*, 917–925. [CrossRef] [PubMed]

129. Shyamal, M.; Mazumdar, P.; Maity, S.; Samanta, S.; Sahoo, G.P.; Misra, A. Highly Selective Turn-On Fluorogenic Chemosensor for Robust Quantification of Zn(II) Based on Aggregation Induced Emission Enhancement Feature. *ACS Sens.* **2016**, *1*, 739–747. [CrossRef]
130. Qin, J.-C.; Wang, B.-D.; Yang, Z.-Y.; Yu, K.-C. A ratiometric fluorescent chemosensor for Zn^{2+} in aqueous solution through an ESIPT coupled AIE process. *Sens. Actuators B Chem.* **2016**, *224*, 892–898. [CrossRef]
131. Gabr, M.T.; Pigge, F.C. A selective fluorescent sensor for Zn^{2+} based on aggregation-induced emission (AIE) activity and metal chelating ability of bis(2-pyridyl)-diphenylethylene. *Dalton Trans.* **2016**, *45*, 14039. [CrossRef]
132. Wang, D.; Li, S.-M.; Li, Y.-F.; Zheng, X.-J.; Jina, L.-P. Hydrogen bond-assisted aggregation-induced emission and application in the detection of the Zn(II) ion. *Dalton Trans.* **2016**, *45*, 8316. [CrossRef]
133. Li, X.; Zhao, R.; Tang, X.; Shi, Y.; Li, C.; Wang, Y. One-Pot Click Access to a Cyclodextrin Dimer-Based Novel Aggregation Induced Emission Sensor and Monomer-Based Chiral Stationary Phase. *Sensors* **2016**, *16*, 1985. [CrossRef]
134. Zhao, Q.; Gong, G.-F.; Yang, H.-L.; Zhang, Q.-P.; Yao, H.; Zhang, Y.-M.; Lin, Q.; Qu, W.-J.; Wei, T.-B. Pillar[5]arene-based supramolecular AIE hydrogel with white light emission for ultrasensitive detection and effective separation of multianalytes. *Polym. Chem.* **2020**, *11*, 5455–5462. [CrossRef]
135. Xu, Z.-H.; Wang, Y.; Wang, Y.; Li, J.Y.; Luo, W.-F.; Wu, W.-N.; Fan, Y.-C. AIE active salicylaldehyde-based hydrazone: A novel single-molecule multianalyte (Al^{3+} or Cu^{2+}) sensor in different solvents. *Spectrochim. Acta Part A Mol. Biomol. Spectrosc.* **2019**, *212*, 146–154. [CrossRef]
136. Wang, Y.; Mao, P.-D.; Wu, W.-N.; Mao, X.-J.; Fan, Y.-C.; Zhao, X.-L.; Xu, Z.-Q.; Xu, Z.-H. New pyrrole-based single-molecule multianalyte sensor for Cu^{2+}, Zn^{2+}, and Hg^{2+} and its AIE activity. *Sens. Actuators B Chem.* **2018**, *255*, 3085–3092. [CrossRef]
137. Maity, S.; Shyamal, M.; Maity, R.; Mudi, N.; Hazra, P.; Giri, P.K.; Samanta, S.S.; Pyne, S.; Misra, A. An antipyrine based fluorescent probe for distinct detection of Al^{3+} and Zn^{2+} and its AIEE behaviour. *Photochem. Photobiol. Sci.* **2020**, *19*, 681–694. [CrossRef]
138. Yan, L.; Qing, T.; Li, R.; Wang, Z.; Qi, Z. Synthesis and optical properties of aggregation-induced emission (AIE) molecules based on the ESIPT mechanism as pH- and Zn^{2+}-responsive fluorescent sensors. *RSC Adv.* **2016**, *6*, 63874–63879. [CrossRef]
139. Wu, M.; Yang, D.-D.; Zheng, H.-W.; Liang, Q.-F.; Li, J.-B.; Kang, Y.; Li, S.; Jiao, C.; Zheng, X.-J.; Jin, L.-P. A multi-binding site hydrazone-based chemosensor for Zn(II) and Cd(II): A new strategy for the detection of metal ions in aqueous media based on aggregation-induced emission. *Dalton Trans.* **2021**, *50*, 1507–1513. [CrossRef]
140. Samanta, S.; Manna, U.; Ray, T.; Das, G. An aggregation-induced emission (AIE) active probe for multiple targets: A fluorescent sensor for Zn^{2+} and Al^{3+} & a colorimetric sensor for Cu^{2+} and F^-. *Dalton Trans.* **2015**, *44*, 18902–18910. [CrossRef] [PubMed]
141. Wang, E.; Liu, S.; Lam, J.W.Y.; Tang, B.Z.; Wang, X.; Wang, F. Deciphering Structure–Functionality Relationship of Polycarbonate-Based Polyelectrolytes by AIE Technology. *Macromolecules* **2020**, *53*, 5839–5846. [CrossRef]
142. Lin, N.; Zhang, Q.; Xia, X.; Liang, M.; Zhang, S.; Zheng, L.; Cao, Q.; Ding, Z. A highly zinc-selective ratiometric fluorescent probe based on AIE luminogen functionalized coordination polymer nanoparticles. *RSC Adv.* **2017**, *7*, 21446–21451. [CrossRef]
143. Li, Y.; Hu, X.; Zhang, X.; Cao, H.; Huang, Y. Unconventional application of gold nanoclusters/Zn-MOF composite for fluorescence turn-on sensitive detection of zinc ion. *Anal. Chim. Acta* **2018**, *1024*, 145–152. [CrossRef]
144. Jiang, T.; Lu, N.; Hang, Y.; Yang, J.; Mei, J.; Wang, J.; Hua, J.; Tian, H. Dimethoxy triarylamine-derived terpyridine–zinc complex: A fluorescence light-up sensor for citrate detection based on aggregation-induced emission. *J. Mater. Chem. C* **2016**, *4*, 10040–10046. [CrossRef]
145. Wang, A.; Fan, R.; Dong, Y.; Song, Y.; Zhou, Y.; Zheng, J.; Du, X.; Xing, K.; Yang, Y. Novel Hydrogen-Bonding Cross-Linking Aggregation-Induced Emission: Water as a Fluorescent "Ribbon" Detected in a Wide Range. *ACS Appl. Mater. Interfaces* **2017**, *9*, 15744–15757. [CrossRef]
146. Li, B.; Zhang, Y.; Yan, B.; Xiao, D.; Zhou, X.; Dong, J.; Zhou, Q. A self-healing supramolecular hydrogel with temperature-responsive fluorescence based on an AIE luminogen. *RSC Adv.* **2020**, *10*, 7118–7124. [CrossRef]
147. Zhao, X.; Li, W.; Wu, T.; Liu, P.; Wang, W.; Xu, G.; Xu, S.; Luo, X. Zinc ion-triggered aggregation induced emission enhancement of dual ligand co-functionalized gold nanoclusters based novel fluorescent nanoswitch for multi-component detection. *Anal. Chim. Acta* **2019**, *1079*, 192–199. [CrossRef]
148. Lei, Z.; Zhou, J.; Liang, M.; Xiao, Y.; Liu, Z. Aggregation-Induced Emission of Au/Ag Alloy Nanoclusters for Fluorescence Detection of Inorganic Pyrophosphate and Pyrophosphatase Activity. *Front. Bioeng. Biotechnol.* **2021**, *8*, 1530. [CrossRef] [PubMed]
149. Jiang, Y.; Sun, L.; Du, J.; Liu, Y.; Shi, H.; Liang, Z.; Li, J. Multifunctional Zinc Metal–Organic Framework Based on Designed H4TCPP Ligand with Aggregation-Induced Emission Effect: CO_2 Adsorption, Luminescence, and Sensing Property. *Cryst. Growth Des.* **2017**, *17*, 2090–2096. [CrossRef]
150. Li, X.; Xie, S.; Hu, Y.; Xiang, J.; Wang, L.; Li, R.; Chen, M.; Wang, F.; Liu, Q.; Chen, X. AIEgen modulated per-functionalized flower-like IRMOF-3 frameworks with tunable light emission and excellent sensing properties. *Chem. Commun.* **2021**, *57*, 2392–2395. [CrossRef]
151. Yu, L.; Chen, H.; Yue, J.; Chen, X.; Sun, M.; Tan, H.; Asiri, A.M.; Alamry, K.A.; Wang, X.; Wang, S. Metal–Organic Framework Enhances Aggregation-Induced Fluorescence of Chlortetracycline and the Application for Detection. *Anal. Chem.* **2019**, *91*, 5913–5921. [CrossRef] [PubMed]
152. Fan, C.; Zhang, X.; Li, N.; Xu, C.; Wu, R.; Zhu, B.; Zhang, G.; Bi, S.; Fan, Y. Zn-MOFs based luminescent sensors for selective and highly sensitive detection of Fe^{3+} and tetracycline antibiotic. *J. Pharm. Biomed. Anal.* **2020**, *188*, 113444. [CrossRef]

153. Xie, S.; Liu, Q.; Zhu, F.; Chen, M.; Wang, L.; Xiong, Y.; Zhu, Y.; Zheng, Y.; Chen, X. AIE-active metal–organic frameworks: Facile preparation, tunable light emission, ultrasensitive sensing of copper(II) and visual fluorescence detection of glucose. *J. Mater. Chem. C* **2020**, *8*, 10408–10415. [CrossRef]
154. Ying, X.-D.; Chen, J.-X.; Tu, D.-Y.; Zhuang, Y.-C.; Wu, D.; Shen, L. Tetraphenylpyrazine-Based Luminescent Metal–Organic Framework for Chemical Sensing of Carcinoids Biomarkers. *ACS Appl. Mater. Interfaces* **2021**, *13*, 6421–6429. [CrossRef]
155. Wei, C.; Wei, X.; Hu, Z.; Yang, D.; Mei, S.; Zhang, G.; Su, D.; Zhang, W.; Guo, R. A fluorescent probe for Cd^{2+} detection based on the aggregation-induced emission enhancement of aqueous Zn–Ag–In–S quantum dots. *Anal. Methods* **2019**, *11*, 2559–2564. [CrossRef]
156. Zhao, W.; Ma, Y.; Ye, J.; Jin, J. A closed bipolar electrochemiluminescence sensing platform based on quantum dots: A practical solution for biochemical analysis and detection. *Sens. Actuators B Chem.* **2020**, *311*, 127930. [CrossRef]
157. Naskar, B.; Dhara, A.; Maiti, D.K.; Kukułka, M.; Mitoraj, M.P.; Srebro-Hooper, M.; Prodhan, C.; Chaudhuri, K.; Goswami, S. Aggregation-Induced Emission-Based Sensing Platform for Selective Detection of Zn^{2+}: Experimental and Theoretical Investigations. *ChemPhysChem* **2019**, *20*, 1630–1639. [CrossRef]
158. Yan, L.; Wen, X.; Fan, Z. A large-Stokes-shift fluorescent probe for Zn^{2+} based on AIE, and application in live cell imaging. *Anal. Bioanal. Chem.* **2020**, *412*, 1453–1463. [CrossRef]
159. Wang, J.; Lu, L.; Wang, C.; Wang, M.; Ju, J.; Zhu, J.; Sun, T. An AIE and PET fluorescent probe for effective Zn(ii) detection and imaging in living cells. *New J. Chem.* **2020**, *44*, 15426–15431. [CrossRef]
160. Xiong, J.; Li, Z.; Tan, J.-H.; Ji, S.; Sun, J.; Li, X.; Huo, Y. Two new quinoline-based regenerable fluorescent probes with AIE characteristics for selective recognition of Cu^{2+} in aqueous solution and test strips. *Analyst* **2018**, *143*, 4870–4886. [CrossRef]
161. Wang, D.; Li, S.-M.; Zheng, J.-Q.; Kong, D.-Y.; Zheng, X.-J.; Fang, D.-C.; Jin, L.-P. Coordination-Directed Stacking and Aggregation-Induced Emission Enhancement of the Zn(II) Schiff Base Complex. *Inorg. Chem.* **2017**, *56*, 984–990. [CrossRef]
162. Mehdi, H.; Gong, W.; Guo, H.; Watkinson, M.; Ma, H.; Wajahat, A.; Ning, G. Aggregation-Induced Emission (AIE) Fluorophore Exhibits a Highly Ratiometric Fluorescent Response to Zn^{2+} in vitro and in Human Liver Cancer Cells. *Chem. Eur. J.* **2017**, *23*, 13067–13075. [CrossRef]
163. Lin, L.; Hu, Y.; Zhang, L.; Huang, Y.; Zhao, S. Photoluminescence light-up detection of zinc ion and imaging in living cells based on the aggregation induced emission enhancement of glutathione-capped copper nanoclusters. *Biosens. Bioelectron.* **2017**, *94*, 523–529. [CrossRef]
164. Chao, D.; Ni, S. Nanomolar pyrophosphate detection and nucleus staining in living cells with simple terpyridine–Zn(II) complexes. *Sci. Rep.* **2016**, *6*, 26477. [CrossRef] [PubMed]
165. Feng, Z.; Li, D.; Zhang, M.; Shao, T.; Shen, Y.; Tian, X.; Zhang, Q.; Li, S.; Wu, J.; Tian, Y. Enhanced three-photon activity triggered by the AIE behaviour of a novel terpyridine-based Zn(ii) complex bearing a thiophene bridge. *Chem. Sci.* **2019**, *10*, 7228–7232. [CrossRef]
166. Leung, A.C.S.; Zhao, E.; Kwok, R.T.K.; Lam, J.W.Y.; Leung, C.W.T.; Deng, H.; Tang, B.Z. An AIE-based bioprobe for differentiating the early and late stages of apoptosis mediated by H_2O_2. *J. Mater. Chem. B* **2016**, *4*, 5510–5514. [CrossRef]
167. Jiang, S.; Qiu, J.; Chen, S.; Guo, H.; Yang, F. Double-detecting fluorescent sensor for ATP based on Cu^{2+} and Zn^{2+} response of hydrazono-bis-tetraphenylethylene. *Spectrochim. Acta Part A Mol. Biomol. Spectrosc.* **2020**, *227*, 117568. [CrossRef] [PubMed]
168. Tian, D.; Li, F.; Zhu, Z.; Zhang, L.; Zhu, J. An AIE-based metallo-supramolecular assembly enabling an indicator displacement assay inside living cells. *Chem. Commun.* **2018**, *54*, 8921–8924. [CrossRef] [PubMed]
169. Janeta, M.; Lis, T.; Szafert, S. Zinc Imine Polyhedral Oligomeric Silsesquioxane as a Quattro-Site Catalyst for the Synthesis of Cyclic Carbonates from Epoxides and Low-Pressure CO_2. *Chem. Eur. J.* **2020**, *26*, 13686–13697. [CrossRef] [PubMed]
170. Di Iulio, C.; Jones, M.; Mahon, M.F.; Apperley, D. Zinc(II) Silsesquioxane Complexes and Their Application for the Ring-Opening Polymerization ofrac-Lactide. *Inorg. Chem.* **2010**, *49*, 10232–10234. [CrossRef] [PubMed]
171. Zhang, X.; Liu, G.; Chen, H.; Fan, L.; Li, B. Structural diversities, magnetic, luminescence and photocatalytic properties of seven inorganic-organic hybrid supramolecular complexes based on 3,5-dimethyl-2,6-bis(3-(pyrid-2-yl)-1,2,4-triazolyl) pyridine. *Inorg. Chim. Acta* **2017**, *465*, 61–69. [CrossRef]
172. Gong, C.; Zeng, X.; Zhu, C.; Shu, J.; Xiao, P.; Xu, H.; Liu, L.; Zhang, J.; Zeng, Q.; Xie, J. A series of organic–inorganic hybrid materials consisting of flexible organic amine modified polyoxomolybdates: Synthesis, structures and properties. *RSC Adv.* **2016**, *6*, 106248–106259. [CrossRef]
173. Buvaylo, E.A.; Kokozay, V.N.; Linnik, R.P.; Vassilyeva, O.Y.; Skelton, B.W. Hybrid organic–inorganic chlorozincate and a molecular zinc complex involving the in situ formed imidazo[1,5-a]pyridinium cation: Serendipitous oxidative cyclization, structures and photophysical properties. *Dalton Trans.* **2015**, *44*, 13735–13744. [CrossRef]
174. Zhang, W.; Ding, X.; Cheng, H.; Yin, C.; Yan, J.; Mou, Z.; Wang, W.; Cui, D.; Fan, C.; Sun, N. Dual-Targeted Gold Nanoprism for Recognition of Early Apoptosis, Dual-Model Imaging and Precise Cancer Photothermal Therapy. *Theranostics* **2019**, *9*, 5610–5625. [CrossRef]
175. Wei, Y.; Wang, L.; Huang, J.; Zhao, J.; Yan, Y. Multifunctional Metallo-Organic Vesicles Displaying Aggregation-Induced Emission: Two-Photon Cell-Imaging, Drug Delivery, and Specific Detection of Zinc Ion. *ACS Appl. Nano Mater.* **2018**, *1*, 1819–1827. [CrossRef]

MDPI
St. Alban-Anlage 66
4052 Basel
Switzerland
Tel. +41 61 683 77 34
Fax +41 61 302 89 18
www.mdpi.com

Molecules Editorial Office
E-mail: molecules@mdpi.com
www.mdpi.com/journal/molecules

www.ingramcontent.com/pod-product-compliance
Lightning Source LLC
LaVergne TN
LVHW070849170726
843515LV00005B/604

9783036527239